VOLUME ONE HUNDRED AND NINE

ADVANCES IN VIRUS RESEARCH

Proteomics Approaches to Unravel Virus - Vertebrate Host Interactions

VOLUME ONE HUNDRED AND NINE

ADVANCES IN VIRUS RESEARCH

Proteomics Approaches to Unravel Virus - Vertebrate Host Interactions

Edited by

GISA GEROLD

Molecular and Clinical Infection Research
Department of Biochemistry & Research Center for Emerging Infections and Zoonoses (RIZ)
University of Veterinary Medicine Hannover
Hannover, Germany

ELSEVIER

ACADEMIC PRESS
An imprint of Elsevier

Academic Press is an imprint of Elsevier
50 Hampshire Street, 5th Floor, Cambridge, MA 02139, United States
525 B Street, Suite 1650, San Diego, CA 92101, United States
The Boulevard, Langford Lane, Kidlington, Oxford OX5 1GB, United Kingdom
125 London Wall, London, EC2Y 5AS, United Kingdom

First edition 2021

ISBN: 978-0-12-823042-8
ISSN: 0065-3527

For information on all Academic Press publications
visit our website at https://www.elsevier.com/books-and-journals

Publisher: Zoe Kruze
Acquisitions Editor: Ashlie M. Jackman
Developmental Editor: Shellie Bryant
Production Project Manager: James Selvam
Cover Designer: Alan Studholme

Typeset by SPi Global, India

Contents

Contributors

Luca D. Bertzbach
Institute of Virology, Freie Universität Berlin, Berlin, Germany

Belén Carriquí-Madroñal
Department of Biochemistry & Research Center for Emerging Infections and Zoonoses (RIZ), University of Veterinary Medicine Hannover; Institute for Experimental Virology, TWINCORE, Centre for Experimental and Clinical Infection Research, a Joint Venture Between the Medical School Hannover and the Helmholtz Centre for Infection Research, Hannover, Germany

Danielle L. Chappell
Lineberger Comprehensive Cancer Center; Department of Pharmacology, The University of North Carolina at Chapel Hill, Chapel Hill, NC, United States

Ileana M. Cristea
Department of Molecular Biology, Princeton University, Princeton, NJ, United States

Blossom Damania
Lineberger Comprehensive Cancer Center; Department of Pharmacology; Department of Microbiology and Immunology, The University of North Carolina at Chapel Hill, Chapel Hill, NC, United States

Gisa Gerold
Department of Biochemistry & Research Center for Emerging Infections and Zoonoses (RIZ), University of Veterinary Medicine Hannover; Institute for Experimental Virology, TWINCORE, Centre for Experimental and Clinical Infection Research, a Joint Venture Between the Medical School Hannover and the Helmholtz Centre for Infection Research, Hannover, Germany; Department of Clinical Microbiology, Virology; Wallenberg Centre for Molecular Medicine (WCMM), Umeå University, Umeå, Sweden

Todd M. Greco
Department of Molecular Biology, Princeton University, Princeton, NJ, United States

Axel Karger
Institute of Molecular Virology and Cell Biology, Friedrich-Loeffler-Institut, Greifswald-Insel Riems, Germany

Benedikt B. Kaufer
Institute of Virology, Freie Universität Berlin, Berlin, Germany

Marjolein Kikkert
Molecular Virology Laboratory, Department of Medical Microbiology, LUMC Center for Infectious Diseases (LU-CID), Leiden University Medical Center, Leiden, The Netherlands

Kevin Klann
Institute of Biochemistry II, Faculty of Medicine, Goethe University, Frankfurt am Main, Germany

Dawei Liu
Department of Molecular Biology, Princeton University, Princeton, NJ, United States

Xavier Martiáñez-Vendrell
Molecular Virology Laboratory, Department of Medical Microbiology, LUMC Center for Infectious Diseases (LU-CID), Leiden University Medical Center, Leiden, The Netherlands

Emanuela S. Milani
Department of Health Sciences and Technology (D-HEST), ETH Zurich, Institute of Translational Medicine (ITM), Zurich; Swiss Institute of Bioinformatics (SIB), Lausanne, Switzerland

Christian Münch
Institute of Biochemistry II, Faculty of Medicine, Goethe University; Frankfurt Cancer Institute; Cardio-Pulmonary Institute, Frankfurt am Main, Germany

Bokai Song
Department of Molecular Biology, Princeton University, Princeton, NJ, United States

Georg Tascher
Institute of Biochemistry II, Faculty of Medicine, Goethe University, Frankfurt am Main, Germany

Fabian Wendt
Department of Health Sciences and Technology (D-HEST), ETH Zurich, Institute of Translational Medicine (ITM), Zurich; Swiss Institute of Bioinformatics (SIB), Lausanne, Switzerland

Maria C. White
Lineberger Comprehensive Cancer Center, The University of North Carolina at Chapel Hill, Chapel Hill, NC, United States

Bernd Wollscheid
Department of Health Sciences and Technology (D-HEST), ETH Zurich, Institute of Translational Medicine (ITM), Zurich; Swiss Institute of Bioinformatics (SIB), Lausanne, Switzerland

Francisco José Zapatero-Belinchón
Department of Biochemistry & Research Center for Emerging Infections and Zoonoses (RIZ), University of Veterinary Medicine Hannover; Institute for Experimental Virology, TWINCORE, Centre for Experimental and Clinical Infection Research, a Joint Venture Between the Medical School Hannover and the Helmholtz Centre for Infection Research, Hannover, Germany; Department of Clinical Microbiology, Virology; Wallenberg Centre for Molecular Medicine (WCMM), Umeå University, Umeå, Sweden

Preface

After an era of genomics in the 20th century, we have entered an era of proteomics with far-reaching impact in virology, as the technology provides tools to study the interaction of viruses with their host cells on several levels. Mass spectrometry (MS)-based proteomics can unravel host factors for virus infection, elucidate mechanisms of virus propagation and spread, determine underlying causes of virus-induced pathology including tumorigenesis, and discover innate immune sensing mechanisms of the host, which have the potential to limit virus infection.

To understand the previous limitations and the now arising tremendous opportunities of proteomics studies, one needs to take into account that the cellular proteome is orders of magnitude more complex than the cellular genome and transcriptome. Each protein is expressed at a specific time, localizes to specific subcellular areas, may have several isoforms, can alter its function by interaction with different partners, and can be post-translationally modified. Hence, it was not until computing power and sensitivity as well as quantitative accuracy of MS technologies had increased extensively in the past two decades that unbiased shotgun proteomics approaches became feasible. With the advent of faster sample acquisition and data analysis, MS-based proteomics has matured into a more widely used and essential element in the toolbox of virologists.

While the German physicist Wilhelm Wien discovered the principle of separating biomolecules according to their mass to charge ratio already in 1899, it took almost a century until technological advances such as ionization methods for biological macromolecules started the modern era of high-resolution quantitative MS-based proteomics. The latter development was awarded the Nobel Prize in Chemistry in 2002, 6 years after the term proteome was coined. In particular, shotgun proteomics, i.e., the identification and quantification of proteins in a complex mixture has nowadays the potential to not only uncover mechanisms of virus infection but also discover biomarkers and potential therapeutic targets.

Complementary approaches including transcriptome analysis by RNA sequencing and high-resolution imaging are available today and allow studying complex molecular mechanisms from different angles to assemble a systems biology view of virus infection. The challenge of the new century will clearly not be data acquisition per se, but acquisition of biologically meaningful data, which is backed up by orthogonal approaches and interpreted correctly. While single-cell proteomics is not yet available broadly, proteomics is rapidly catching up with other -omics technologies and has reached a level of complexity that allows systems biology studies. This volumes attempts to give an overview of currently available proteomics methods and developments that will further our knowledge on the interaction of viruses with their host cells and organisms.

In Chapter 1, Klann et al. give an overview of global proteome changes during virus infection of host cells and broadly describe proteomic approaches to understand protein network plasticity during infection. While proteomics can elucidate biological mechanisms of infection, it can also unravel how the innate immune system senses virus infection and how viruses cause disease. To analyze such global changes in the proteome, mass spectrometry imaging (MSI) is a powerful technique allowing the visualization of spatial distribution of proteins in tissue sections from infected organisms. Chapter 2 by Bertzbach et al. discuss the applications of matrix-assisted laser desorption/ionization (MALD) MSI in understanding virus-host interactions, in cancer diagnosis and biomarker development. They further delineate how MSI can unravel antiviral drug distribution and kinetics in whole organisms. In Chapter 3, Zapatero-Belinchón et al. describe proximity labeling as novel method to perform spatiotemporal proteomics at subcellular resolution. The authors cover different proximity labeling methods, which allow studying dynamics of protein complexes in living cells and organisms. They provide an overview of applications in infection biology and discuss future applications of the technology. In Chapter 4, Wendt et al. specifically focus on the host cell surfaceome and how it contributes to signaling events during the infection process. The dynamic interactions at cell surfaces range from virus entry to immune evasion, virion release, and virus spread to naïve cells. The authors detail technologies to uncover the cellular surfaceome and more specifically protein interactions at the plasma membrane including virus-receptor interactions.

Protein complex formation but also protein localization and function can be guided by proteolytic cleavage. Host and viral proteases hence impact infection of host cells and it has become increasingly clear that both host and viral proteases cleave a large repertoire of proteins. Hence, Martiánez-Vendrell et al. explore in Chapter 5 proteomics methods to systematically identify substrates of host and virus proteases during infection. The authors give examples for the role of specific proteases during the course of virus infection. Another important determinant of protein localization, turnover, and function is its landscape of posttranslational modifications. Song et al. discuss in Chapter 6 techniques to map posttranslational modifications by MS and focus on the role of protein modifications during innate immune sensing of DNA viruses and retroviruses.

The authors finally give an overview of how virus-induced modifications of host DNA sensors aid in immune evasion. A specific group of DNA viruses associated with cancer is discussed in Chapter 7, where Chappell et al. highlight proteomics studies on tumor induction by the subfamily *Gammaherpesvirinae*. The authors introduce the use of proteomics to understand the biology and pathogenesis of herpesvirus infection. They moreover describe how protein analytics furthered the understanding of gammaherpesvirus-induced oncogenic processes as well as clinical applications in diagnostics and therapy of gammaherpesvirus-associated cancers.

Thus, proteomics in infection biology comes full circle (Fig. 1). It spans from spatial proteomics at tissue and subcellular level to methods of

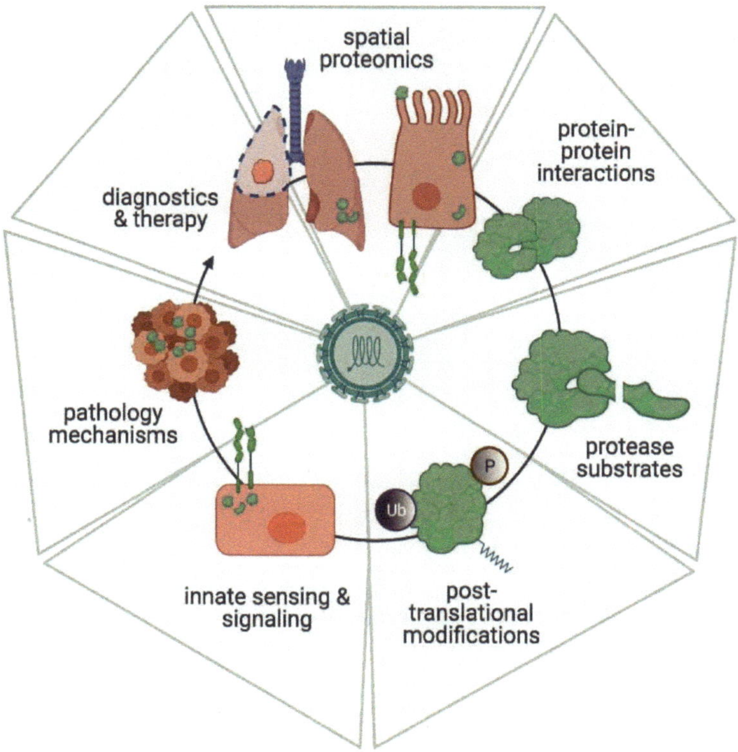

Fig. 1 Overview of proteomics approaches and their applications in studying virus-host interactions. From top in clockwise direction. Mass spectrometry-based proteomics can identify protein distribution in tissues, map protein complexes at subcellular resolution, and identify protein interactions, protease substrates, and posttranslational modifications. Thereby, the technology can elucidate molecular mechanisms of virus infection, innate immune sensing, and evasion during infection as well as pathology caused by viruses. Ultimately, proteomics can identify diagnostic biomarkers and pave the way toward therapy development. Future proteomics and multi-omics studies will likely contribute to an integrative systems biology view of virus-host interactions.

determining protein-protein interactions and detailed protein analytics of posttranslational modifications and proteolytic processing. These analyses ultimately lead to a broad understanding of virus propagation, virus-induced immunity, and pathology, thereby providing a systems biology view of virus infection. With the now available speed and accuracy of data acquisition, proteomics will ultimately pave the way toward diagnostic and therapeutic developments to detect and treat disease states. Biochemists and infection biologists can look forward to an exciting era with exceptional studies on protein networks operating during virus infection and an increasing number of bench to bedside applications of proteomics in the 21st century.

GISA GEROLD
Molecular and Clinical Infection Research
Department of Biochemistry & Research Center for Emerging Infections and Zoonoses (RIZ)
University of Veterinary Medicine Hannover
Hannover, Germany

CHAPTER ONE

Virus systems biology: Proteomics profiling of dynamic protein networks during infection

Kevin Klann[a], Georg Tascher[a], and Christian Münch[a,b,c,*]
[a]Institute of Biochemistry II, Faculty of Medicine, Goethe University, Frankfurt am Main, Germany
[b]Frankfurt Cancer Institute, Frankfurt am Main, Germany
[c]Cardio-Pulmonary Institute, Frankfurt am Main, Germany
*Corresponding author: e-mail address: ch.muench@em.uni-frankfurt.de

Contents

Abstract

The host cell proteome undergoes a variety of dynamic changes during viral infection, elicited by the virus itself or host cell defense mechanisms. Studying these changes on a global scale by integrating functional and physical interactions within protein networks during infection is an important tool to understand pathology. Indeed, proteomics studies dissecting protein signaling cascades and interaction networks upon infection showed how global information can significantly improve understanding of disease mechanisms of diverse viral infections. Here, we summarize and give examples of different experimental designs, proteomics approaches and bioinformatics analyses that allow profiling proteome changes and host–pathogen interactions to gain a molecular systems view of viral infection.

Advances in Virus Research, Volume 109
ISSN 0065-3527
https://doi.org/10.1016/bs.aivir.2020.12.001

1. Host cell proteome modulation during viral infection

The viral infection process is guided by an intricate extra- and intracellular network of virus–host interactions (Fig. 1). Already during extracellular interaction with cells, a virus may elicit multiple changes to the cellular proteome. Binding of cell surface receptors to viruses, such as pattern recognition receptors (PRRs) recognizing viral epitopes, triggers various signaling cascades (depending on the cell type) and thereby drives a range of immune responses (Monteiro and Lepenies, 2017). Viral entry into cells depends on various host cell proteins exposed on the surface that promote uptake (Cruz-Oliveira et al., 2015; Londrigan et al., 2012; Miller et al., 2012; Monteiro and Lepenies, 2017). Interaction of the virus with these attachment factors can lead to forward signaling events, triggering cytoskeleton remodeling and subsequent pinocytosis of the viral particles via prominent cellular signaling pathways (e.g., PI3K) (Wolfrum and Greber, 2013). Once the virus passed the plasma membrane, a number of additional host-virus interactions are necessary for the virus to reach its cellular destination and ultimately start its replicative cycle. Genome-wide studies on influenza A virus (IAV) infection revealed specific host genes associated with each stage of the life cycle (Watanabe et al., 2010). During the viral infection cycle, a plethora of changes to the proteome landscape are triggered, changing function and interaction of host proteins. These changes occur on multiple levels, requiring global approaches, such as proteomics, to gain a complete understanding of the host–virus interaction network during the different infection stages (Fig. 1). Revealing these mechanisms will uncover new potential avenues for medical intervention. As an example, hepatitis C virus (HCV) reroutes the host metabolism by changes in the proteome to support the viral replication cycle (Diamond et al., 2010). Targeting these metabolic routes provides promising therapy approaches (Ikeda and Kato, 2007).

Here, we will explore the use of proteomics for systems level studies of viral infection.

1.1 Dynamic protein modifications during viral infection

Cellular signaling events are among the first responses of a host cell to pathogen infection. However, optimized host cell signaling is also indispensable for the viral replication cycle and will be hijacked to promote virus maturation (Hale et al., 2006). While viruses strive to remodel cellular signaling pathways in their favor, the host cell will attempt to counteract the virus by

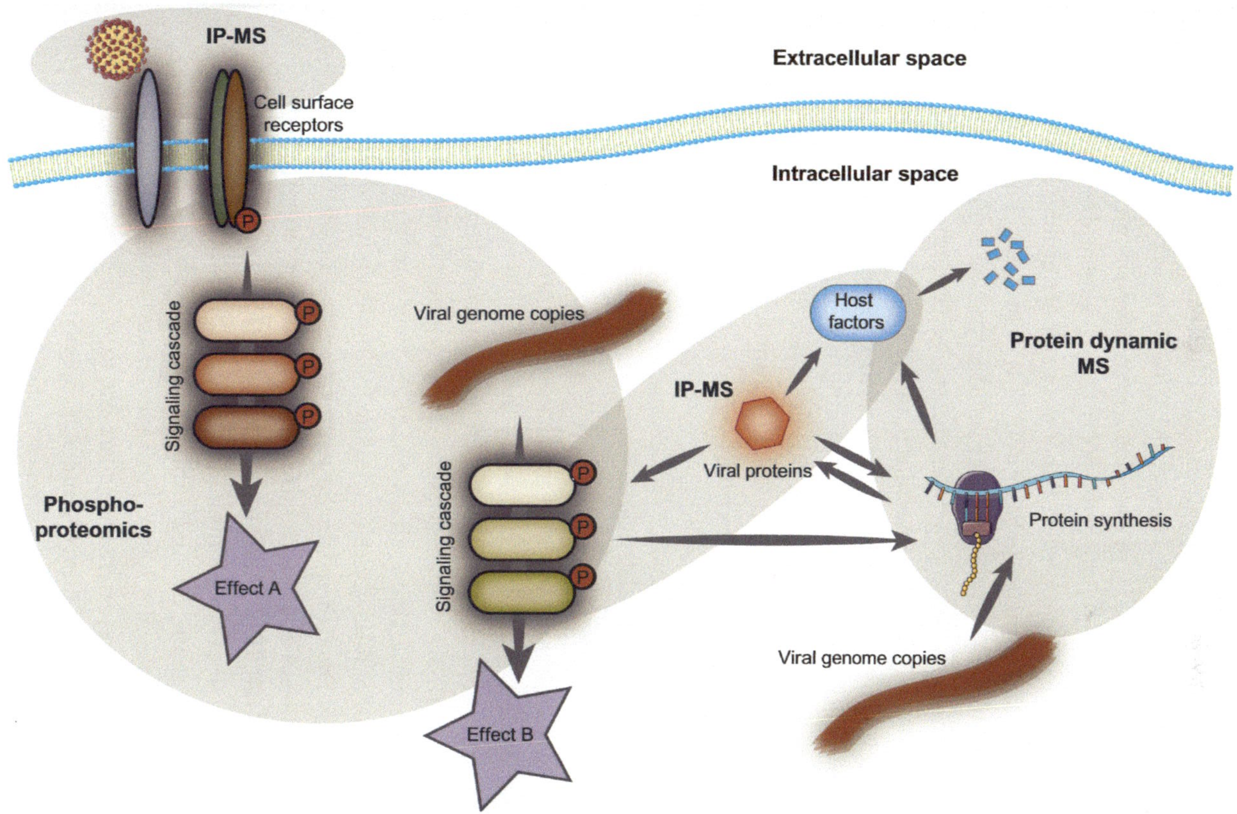

Fig. 1 Dynamic proteome changes during viral infection. Viral infection can cause various dynamic changes to the host proteome. During interaction of the virus with cell surface proteins, signaling cascades are triggered that drive specific host responses and change protein modification and expression. Inside cells, the viral particles, genome and proteins, can interact with or be recognized by host cell proteins and elicit additional signaling changes. Also viral proteins are (transcribed and) synthesized, changing protein synthesis behavior and causing several effects, such as changes in signaling, modification, protein synthesis and degradation, via interaction with host proteins. This figure highlights some examples of these changes, together with proteomics methods that can be applied to measure these changes in a quantitative manner. Quantification of these effects allows system-wide analysis of viral infection.

signaling antiviral responses and preventing the attempts of the virus to rearrange cellular behavior. This results in a molecular tug-of-war over host cell pathways. These modulated pathways can range from prominent signaling hubs, such as mammalian target of rapamycin (mTOR) (Kindrachuk et al., 2015) or mitogen-activated protein kinase (MAPK) (Kindrachuk et al., 2015; Wolfrum and Greber, 2013), to specific interference with metabolism (Popescu and Dubuisson, 2010; Ritter et al., 2010). Most cellular phenotypes triggered by pathogens are in one way or another the result of signaling events, largely driven by posttranslational modification (PTM) of proteins. The most prominent (and most studied) PTM controlling cellular signaling is phosphorylation that drives numerous kinase-dependent pathways (Oppermann et al., 2009; Ubersax and Ferrell Jr, 2007). In addition, ubiquitination and acetylation have been studied by proteomics in the context of viral infection and have been shown to play an important role during infection (Grossegesse et al., 2018; Hu et al., 2018; Liu et al., 2015; Stukalov et al., 2020). Every layer of PTM modifications revealed provides novel insight into disease biology and adds to the global understanding of infection.

1.2 Changes in proteome abundance

Dynamic changes not only happen on the layer of protein PTMs. A common cellular antiviral strategy is the global attenuation of protein synthesis to prevent viral replication (Balachandran et al., 2000; Dauber and Wolff, 2009; Mohr and Sonenberg, 2012). This effect is often mediated by the integrated stress response (ISR), which activates by phosphorylation of the eukaryotic initiation factor 2 alpha (eIF2α or EIF2S1) via four kinases that activate upon a range of cellular stresses, including folding stress in the endoplasmic reticulum (ER, recognized by protein kinase R-like ER kinase [PERK]) and occurrence of double-stranded RNA (dsRNA, recognized by protein kinase R [PKR]) (Pakos-Zebrucka et al., 2016; Taniuchi et al., 2016). Phosphorylation of eIF2α results in limited availability of the ternary translation initiation complex, causing a reduction in global protein production (Krishnamoorthy et al., 2001). However, many viruses evade or even take advantage of these defense mechanisms to ensure viral protein translation. For example, HCV mRNA interacts directly with host translation components through RNA structures in their 3′ untranslated region (UTR). The association with the host 40S ribosomal subunit stimulates several rounds of translation of the same mRNA to overcome a limited protein synthesis machinery (Bai et al., 2013). Additionally, HCV can restrict PKR activity directly by viral proteins that target the kinase activity and block the inhibitory effect on protein synthesis (Gale et al., 1998; Khaperskyy et al., 2014; Taylor et al., 1999). As another example, IAV nucleoprotein (NP) and the host-shutoff protein polymerase-acidic protein-X (PA-X) inhibit the formation of stress granules (which form with translationally inactive mRNAs and ribosomes) (Khaperskyy et al., 2014). Hence, the cellular translational landscape is shaped by both virus and host cell. This complicates understanding which cellular effects may be driven by pro- or antiviral processes and are thus predominantly triggered by the virus or host cell, respectively. Quantitative measurements of translational control unveil the immediate response and carry information about response pathways, which would not be observable by monitoring total protein levels within the same timeframe. The combination of these information aids to understand the nature and regulation of observed effects and identify key aspects of cellular rearrangements required for viral replication (Bojkova et al., 2020).

Some host cell changes are primarily visible on the level of protein synthesis while others are more pronounced monitoring rapid protein degradation processes in response to the infection. Viral proteins can directly target host factors for degradation. In those cases, the viral proteins interfere with the host degradation machinery, often the ubiquitin-proteasome pathway (by eliciting dynamic changes in protein ubiquitination) and target specific host proteins for degradation, for example, to inhibit immune response signaling (Didcock et al., 1999; Morrison et al., 2013; Nightingale et al., 2018). As discussed earlier, the host cell and virus are constantly struggling for the control over the cellular proteome. Degradation events are no exception, with viral effector proteins attempting to degrade specific host cell proteins and host cell machineries striving to degrade viral proteins (Di Pietro et al., 2013; Li et al., 2018).

1.3 Differential protein interaction

Changes to the dynamic host cell proteome can be quantified by the measurements of PTMs and protein abundances, including synthesis and degradation rates. However, protein networks sometimes change their function by differential protein–protein interaction (PPI), often guided by changes in PTMs, rather than by changes in their total abundance (Ideker and Krogan, 2012). Therefore, profiling viral protein interactions during infection can provide direct evidence of how the virus interferes with cellular function (Jäger et al., 2012). Strikingly, PPI networks may reveal possible antiviral therapeutic strategies (Gordon et al., 2020; Li et al., 2012) and shed new light on infection mechanisms (Jäger et al., 2012; Shah et al., 2018; Wang et al., 2017).

Overall, the host cell proteome is highly dynamic upon infection and changes are observable on multiple layers (Fig. 1). Profiling those layers by proteomics, either separately or simultaneously, provides with datasets to better understand pathogenesis of viral infections. Here, the analysis of proteins in their functional and physical context network can significantly improve data interpretation (Fig. 2).

2. Viral systems biology

2.1 A systemic view on the proteome

The classical view of cell biology from the 1940s emphasizing "one-gene-one-function" (Tatum and Beadle, 1942) still influences analysis workflows in all methodologies, including proteomics. This view implies a linear flow of information and function that shapes a single directional pathway. Many data analysis workflows concentrate on the identification of single differentially expressed proteins to explain observed phenotypes. While this might be perfectly suitable to identify biomarkers, the identification of single genes often cannot, or only partially, explain the complex phenotypes that are seen upon perturbation (e.g., infection). Only few proteins exhibit a large biological effect size (i.e., fold changes). A large fraction of the proteome changes in a very tight window to ensure correct function of the cell (Harper and Bennett, 2016). As an example, already a 1.4-fold overexpression of β-tubulin leads to the loss of cell viability and cytoskeleton structures (Katz et al., 1990; Weinstein and Solomon, 1990), while an excess of α-tubulin is well tolerated. Aneuploidy studies in yeast showed that at least 20% of the proteome do not scale with their copy numbers, suggesting tight regulation (Dephoure et al., 2014). During proteomics experiments, the dynamic range is additionally biased by inherent physical and chemical properties of the peptides (Bantscheff et al., 2007). Therefore, the assumption that all critical proteins exhibit large fold changes, with smaller effects (e.g., 30%–60%) representing noise-effects, draws a simplified picture of the cellular events and masks key effects.

To overcome this conceptual limitation, the study of proteins in their biological context proves a powerful tool (Bensimon et al., 2012). Many omics studies in past years changed the classical view on how cellular pathways act (Arkin and Schaffer, 2011), observing the cellular environment

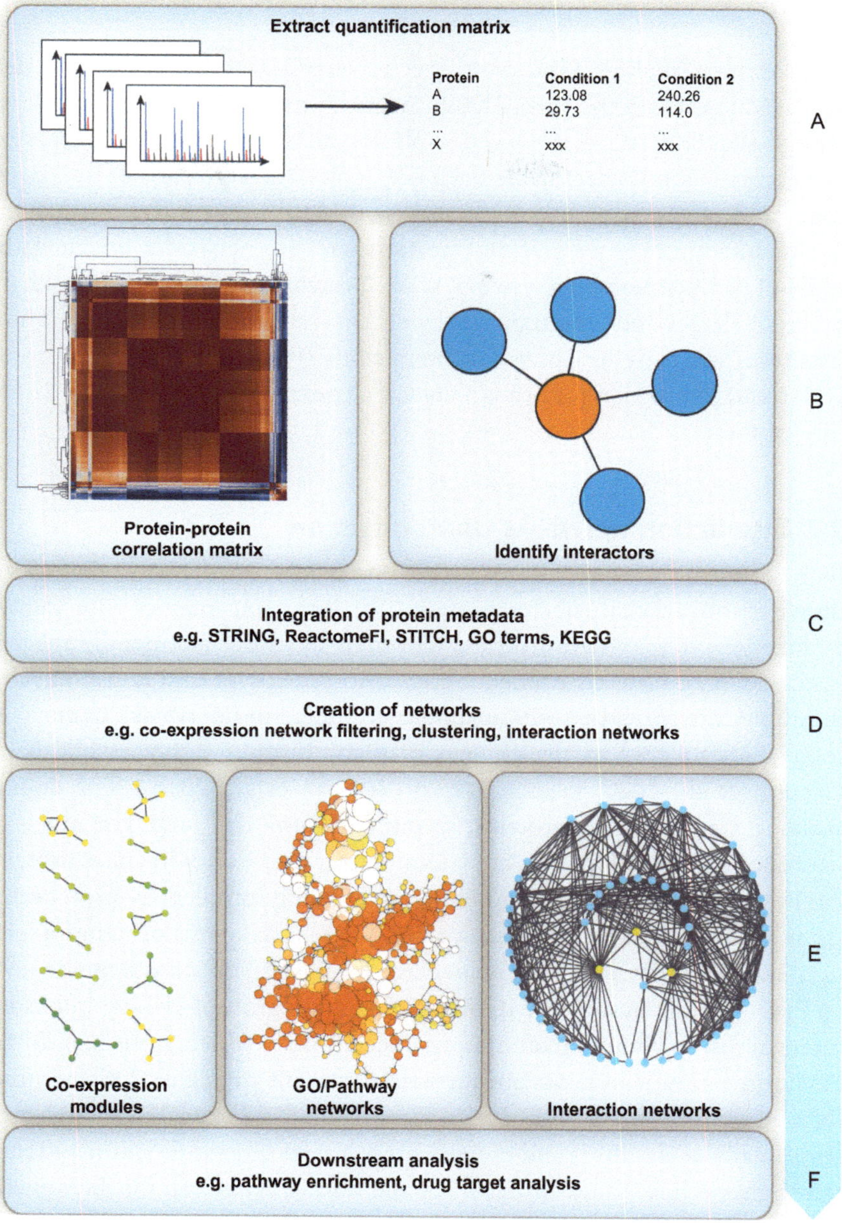

Fig. 2 Example network analysis of proteomics data. (A) After quantitative proteomics measurements of the samples, quantification values are extracted and processed. (B) In interaction studies, the interaction partners are identified, while coexpression workflows focus on correlation analysis. A matrix containing the metrics later used for network creation is build. (C) Now, additional protein meta-data, such as pathways can be added and (D) subsequently the networks are created. (E) Depending on the nature of the data, these networks may represent coexpression modules, networks of pathways and functions or interaction networks. (F) Using the metadata, the networks can be studied further (e.g., by pathway enrichment of highly connected gene clusters or diffusion modeling).

as a complex nonlinear and multidirectional pathway and interaction network (Milo et al., 2002; Vidal, 2009). Available network information is constantly evolving and is now accessible for all areas of research. Modern proteomic technologies enable deep (i.e., covering large parts of the proteome) and quantitative measurements of the cell upon infection that can quantitatively cover a significant portion of these networks, enabling the study of pathogenesis on a systemic scale. Networks in this context can have different shapes and appearances, depending on the analyzed facet. These structures might reflect physical interactions (PPIs), functional (pathway, GO-term) interaction (or other metadata), coexpression behavior or a mixture of these.

2.2 Interaction networks upon infection

During infection, interference with the host machineries largely relies on the direct actions of viral effector proteins. Viral proteins interact with various endogenous host proteins to carry out their function. To identify mechanisms defining how virus infection leads to phenotype and reveal possible drug interference targets, the mapping of interaction networks has proven to be a useful tool in the context of many diseases (Davis et al., 2015; Gordon et al., 2020; Jäger et al., 2012; Shah et al., 2018). The coding sequences of all viral proteins, identified from the viral genome, are expressed in a suitable cellular system and interactions with host proteins (and eventually other viral proteins) subsequently profiled by proteomics. As an example, comparative analysis of different flavivirus protein interaction networks revealed direct evidence of observed phenotypes on the level of PPIs (Shah et al., 2018). Besides the confirmation of already published interactions, a systemic map of new interactions revealed a mechanism for flavivirus immune evasion and possible treatment strategies. Other studies focused on mapping PPIs of HIV (Jäger et al., 2012) and herpesvirus proteins (Davis et al., 2015) and thereby predict viral protein function and shed new light on how these viruses shape the host cell function for their needs. Furthermore, recent analysis of SARS-CoV-2 viral protein interaction networks led to the identification of several compounds that inhibit viral replication (Gordon et al., 2020). Overall, identification of interactions or interaction partners by proteomics and subsequent targeting with small molecule compounds has proven a viable option to discover treatment strategies for viral infection.

2.3 Profiling signaling networks

While interpretation of PPI networks is usually relatively straight forward and well established, the situation proves more difficult when studying signaling networks. Currently, around 300,000 nonredundant phosphosites are known in vertebrates (Hornbeck et al., 2012) (~177,000 in humans [PhosphoNet]), depending on the database, while many more have been predicted, but not confirmed yet (Vlastaridis et al., 2017). However, the function of the majority of these sites remains unknown. Historically, signal transduction pathways have been studied with a strong focus on single signpost modification sites for which antibodies had been raised. However, the past years of high-throughput measurements of protein modification showed that proteins typically harbor multisite modifications (Whitmarsh and Davis, 2016). In fact, often multiple modification events are needed to induce a functional change, or they counteract each other. The resulting complexity is enormous, making data analysis complicated. Nevertheless, analyzing the network of protein modification changes can reveal previously unknown relations and host–pathogen interactions (Munday et al., 2010; Soderholm et al., 2016; Weber et al., 2019; Yángüez et al., 2018). Instead of focusing on identification of single modification sites, systematic network analysis can reveal global signal rewiring inside the host cells. Even when the exact function of the majority of quantified modifications is unknown, the accumulation of differentially modified proteins associated with a specific pathway is a powerful tool to discover important pathways upon infection (Klann et al., 2020a). Recent phosphoproteomics studies profiling cells infected with SARS-CoV-2 showed extensive rearrangements of the host signaling cascades (Bouhaddou et al., 2020; Klann et al., 2020a; Stukalov et al., 2020). By pinpointing differentially modified pathway networks, all studies discovered important signaling pathways during infection. Subsequent interference with those pathways significantly modulated SARS-CoV-2 infection and thereby opened new avenues for potential therapies. However, due to the more complicated workflow and data analysis, only few deep phosphoproteomics studies have been conducted in viral research so far.

2.4 Protein dynamic networks

Viral infection modulates the rates of synthesis and degradation of a variety of proteins within the host cell. Comprehensive and global proteomic analysis of these changes can lead to new insights into host–virus interaction mechanisms. Degradation proteomics of human cytomegalovirus (HCMV) showed that of all proteins down-regulated on total protein level between 32% and 54% were primarily regulated via accelerated degradation (Nightingale et al., 2018), rather than expression changes. Of those proteins, the majority was targeted to the proteasome, suggesting the ubiquitin proteasome system as the common route of degradation during HCMV infection. Further analysis of regulatory events revealed how HCMV targets the host antiviral protein HTLF for degradation thus revealing a new interaction potentially emendable for therapeutic intervention.

Another study, dissecting the dynamic response of cells after IAV infection, showed major rearrangements of cellular protein synthesis during the course of infection (Bogdanow et al., 2019). After 12 h of infection, viral proteins accounted for 20%–40% total amount of proteins synthesized in cells, while the synthesis of host proteins was significantly reduced. This study shows how viruses can efficiently overtake host cell machineries and hijack them in their favor. Additionally, our recent work using quantitative temporal translation profiling of SARS-CoV-2 revealed pharmacologically accessible pathways that are key for viral replication (Bojkova et al., 2020).

Profiling protein dynamics on a global scale by proteomics is a relatively new field of research. Thus, only few studies on viral infections have been conducted so far. However, the available studies show remarkably well how profiling protein dynamics (degradation (Nightingale et al., 2018), translation (Bogdanow et al., 2019; Bojkova et al., 2020)) can lead to new understanding of the mechanisms underlying viral infection and provide critical resources for future research and the development of novel therapies.

3. Proteomics methodologies to measure dynamic protein networks

Interference of a virus with the host cell and perturbation of cellular homeostasis occurs on several layers of the proteome during the course of an infection. Hence, the investigation of affected host proteins and the involved networks and pathways ideally should cover as much of these layers and as comprehensively as possible to obtain a complete picture of the infection process. To this end, liquid-chromatography coupled to mass spectrometry (LC-MS) based proteomics has come a long way during the last two decades and became the method of choice for interrogating proteomes in a sensitive, quantitative, unbiased and nowadays almost comprehensive fashion (Aebersold and Mann, 2016). It can and has been applied to a variety of experimental setups and scientific questions fueling bio(medical) research, including virology (Greco et al., 2014). "Top-Down"-proteomics workflows, which analyze intact proteins and even whole viral particles by MS, have gained increased attention in recent years; however, they are still limited by the number of proteins that can be quantified within a sample, restraining its potential to gain system-wide insight (Durbin et al., 2016; Snijder et al., 2013). Thus, we will focus on "bottom-up" or "shotgun"-proteomics workflows that allow deep and quantitative proteomic analyses to obtain cellular network information. Bottom-up proteomics entails digestion of proteins within a sample by endoproteases (predominantly trypsin) and analyses of the resulting peptides by LC-MS. Peptides are separated on the chromatographic column, followed by measurement within the mass spectrometer. The peptides are identified using the combination of the molecular mass of the intact peptide and the produced peptide fragments by comparing them to theoretical values derived from in silico digestion of sequence databases (Eng et al., 2011). This method allows sequencing and precise quantification of peptides, including their localization and abundance of their PTMs, which can then be inferred to the protein level using well-established bioinformatic approaches (Li and Radivojac, 2012).

Here, we review the current state of the art of proteomics workflows as an important part of the toolbox for modern virology.

3.1 Methods to measure dynamic proteome changes

3.1.1 Interaction proteomics

For better understanding of viral infection, describing the physical interactions between virus and host proteins as a first step in leveraging the host cell's machinery can be crucial. Coimmunoprecipitation experiments coupled to mass spectrometry analysis are perfectly suited to both establish interaction networks and describe their dynamics during infection. These experiments require specific antibodies targeting the viral protein(s) of interest (the "bait" protein) or tagging-approaches in order to immobilize the bait to an affinity-matrix. These approaches are often called affinity purification coupled to mass spectrometry (AP-MS), guided by the aim to purify the "bait" protein with their interactors ("prey" proteins) still attached after stringent washes and specific elution from the affinity resin followed by digestion and MS analysis (Gingras et al., 2007). Given the relatively high amount of unspecific binders to the affinity-resin used for immunoprecipitation, which are actually helpful for global abundance normalization procedures being a relatively constant background, it was recently proposed to rather consider it an affinity enrichment coupled to MS (AE-MS). Especially, when the digestion step is performed "on-bead", i.e., without an elution step and often less stringent washes, this can help in preserving low-affinity interactions and thus increasing the sensitivity in discovering interaction partners of the bait protein (Keilhauer et al., 2015). Even though tedious and time-consuming, large-scale AP-MS experiments to build proteome–wide protein interaction networks in a high throughput manner are now feasible and ongoing. For example, more than 57,000 interactions between 11,000 proteins have been mapped for human embryonic kidney 293T cells within the BioPlex project so far and an extension of these networks to even more proteins and mammalian cell lines is underway (Huttlin et al., 2015, 2017). Another important approach for studying rewiring of protein complexes is based on coelution of proteins within a given complex during separation approaches, which are based on their molecular weight under native conditions, such as native polyacrylamide gels or size-exclusion chromatography. Individual fractions are then analyzed by MS and the constituents of a complex can be defined by the elution profile of individual proteins along the fractions, where coeluting proteins are considered members of the same complex (Kirkwood et al., 2013). This approach emerged very recently and thus has not yet been used to study viral infection so far, but might add an interesting additional layer to the knowledge of host–virus interactions.

3.1.1 Posttranslational modification proteomics

AP-MS experiments are also helpful for characterizing changes in a protein's modification status. Enrichment of the protein of interest considerably reduces the complexity of the sample when compared to a whole cell extract. Thus, it facilitates the detection of the typically low stoichiometric PTMs with the modified form of a protein usually comprising only a few percent of all protein molecules present in the sample. This is also the reason why, for holistic PTM-profiling, enrichment strategies are still indispensable. Phosphorylation is probably the most comprehensively studied PTM because of its important role as a molecular switch in cellular signaling pathways, but also because enrichment strategies for phosphoproteins have been available for decades (Riley and Coon, 2016). The most common approaches use either immobilized metal affinity chromatography (IMAC) or metal oxide affinity chromatography (MOAC) that use metal-ions forming complexes with the peptide's phosphate group, allowing for efficient capture. Fe^{3+} and Ti^{4+}

are the most common agents used in IMAC, while TiO_2 is the commonly used MOAC-agent. Affinity purification can be performed on chromatography systems or on home-made/commercially available spin-columns to study the phosphoproteome. Although exhibiting high specificity above 90% and reaching a depth of several tens of thousands of phosphosites, modern phosphoproteomics studies using IMAC or MOAC tend to undersample one important group of phosphosites—phospho-tyrosines. For these, antibody-based enrichment strategies are available. A combination of IMAC/MOAC together with anti phospho-tyrosine antibodies allows to increase the depth of the phosphoproteome (Sharma et al., 2014). Also for other important PTMs, enrichment strategies are available, mainly using antibodies targeting the modification itself. These include acetylated peptides (Kim et al., 2006) and remnants after proteolytic digestion of ubiquitinated proteins (Akimov et al., 2018; Ordureau et al., 2015; Udeshi et al., 2013) or SUMOylated proteins (Hendriks et al., 2018), enabling the profiling ofthousands of these modification sites within an experiment. Another important aspect of PTM-profiling is the determination of modification stoichiometries, i.e., the fraction of modified protein relative to its total amount in the cell. This can be of importance for more detailed insights into PTM-based networks. Thus, total proteome profiling is usually performed in parallel to the determination of PTM-abundances to determine modification stoichiometry.

3.1.2 Whole-cell proteomics

LC-MS technology has reached a state where it is now possible to accomplish near-complete proteome coverage within relatively short timeframes. It has been shown that around 90% of all proteins in a yeast proteome can be covered within an hour, identifying around 4000 proteins from 4500 open reading frames (Hebert et al., 2014). However, the proteome of a typical mammalian cell can theoretically be comprised of more than 20,000 different gene products, excluding the additional complexity from isoforms generated by alternative splicing. This poses tremendous challenges to instrumentation as these have to be fast and sensitive enough to capture the several hundreds of thousands of peptides eluting from the column in the course of one measurement run. Moreover, proteomic samples exhibit a very high dynamic range with protein copy numbers spanning up to six orders of magnitude in cells and even more in body fluids such as serum (Harper and Bennett, 2016; Schwanhäusser et al., 2011; Schwenk et al., 2017). Hence, in order to enable deep proteome coverage in mammalian cells, which is essential for building meaningful protein networks, prefractionation of the samples is necessary. In general, the longer the instrument time, the higher the proteome coverage. Thus, either using longer gradients or analyzing more fractions is required. As a result, there is always a compromise to be made between instrument time and proteome coverage, with instrument time per experiment usually being the limiting factor. Independent of the proteomics strategy applied for quantification, a "standard" proteomics experiment of human cells usually covers between 8000 and 10,000 proteins, for which abundances can be determined in most samples.

One additional aspect of monitoring the proteome is the ability to analyze protein synthesis and degradation rates. To measure protein synthesis by MS, approaches exploiting noncanonical amino acids (e.g., azidohomoalanine) and click chemistry (Dieterich et al., 2006), heavy amino acids (Schwanhäusser et al., 2009), as well as combinations of both, are being used (Rothenberg et al., 2018). Dynamics of incorporation of heavy and light amino acids using SILAC is also useful for profiling global protein degradation (Schwanhäusser et al., 2011), even providing cellular spatial resolution showing that turnover can be localization-dependent within the cell (Boisvert et al., 2012). In addition, when combined with chemical labeling approaches, the throughput in such studies can be heavily increased (Klann et al., 2020b; Klann and Münch, 2020; Savitski et al., 2018; Welle et al., 2016). When more detailed knowledge of the degradation pathway is needed, approaches tailored to the analysis of digestion products were established, e.g., for the investigation of proteasomal degradation (Wolf-Levy et al., 2018). This provides an important additional level of information invisible to global abundance profiling as effects of virus infection on protein translation and degradation.

3.2 Quantification strategies

LC-MS-based abundance measurements are not quantitative per se because of analyte-specific differences in retention behavior, ionization properties and detector responses. Therefore, quantification by MS is performed relative to a standard. This standard has to exhibit identical chemical properties, such that the standard and the analyte behave identical during sample preparation, chromatography, ionization and detection in the MS. These standards can be the same peptide in different LC-MS runs used for relative quantification between samples, or introduced by differential labeling approaches using molecules containing heavy isotopes (^{13}C, ^{15}N), which are then analyzed in the same LC-MS run.

3.2.1 Label-free quantification

Simply comparing the MS signal of the same peptide in different runs is the basis of label-free quantification (LFQ). It had long been hampered by irreproducible sample preparation, drifts in retention-times during LC separations as well as the lack of normalization approaches to account for slight abundance differences between the runs, especially for fractionated samples. Meanwhile, robust quantification algorithms have been established and LFQ is nowadays an accurate and sensitive way of quantifying proteins (Cox et al., 2014; Millikin et al., 2018). An important drawback is that samples have to be measured separately and require extensive instrument time for deep proteome measurements, even with optimized workflows (Bekker-Jensen et al., 2017). Alternatively, quantification can be carried out taking advantage of stable heavy isotopes introduced into proteins by metabolic or chemical labeling approaches (Bantscheff et al., 2007).

3.2.2 Isotope-based quantification

The incorporation of heavy isotopes into the proteome under study can be achieved by exploiting metabolic pathways in living cells. The most important metabolic labeling approach in mammalian cells is stable isotope labeling with amino acids in cell culture (SILAC). There, essential amino acids in the cell culture medium (usually lysine and arginine), are replaced by analogs containing several naturally-occurring heavy carbon or nitrogen atoms (Ong et al., 2002). This results in a mass shift of the peptide signal during MS analysis, which can be used for very accurate ratio-determination after mixing the samples. Mixing occurs very early in the experimental workflow, thus minimizing experimental biases due to lysis and digestion steps. However, to ensure complete labeling of the proteome, cells have to be cultured for a sufficient number of passages, which initially restricted the method to cell lines. This has been overcome by the super-SILAC approach in which a reference proteome is generated by mixing different cell lines of a similar biological origin enabling SILAC-based analysis of primary cells and tissues (Geiger et al., 2010). Mixing samples reduces the required instrument time to a certain extent, but replicates still have to be measured independently and only up to three different samples can be combined.

Chemical labeling approaches using isobaric tags like isobaric tags for relative and absolute quantification (iTRAQ) or tandem mass tags (TMT) allow multiplexing of up to 16 different samples (Thompson et al., 2019). These molecules contain three different building blocks: a reporter region, balancer region, and an NHS-ester. The reporter and balancer region have the same nominal mass (i.e., they are isobaric). Thus, peptides originating from the different samples are indistinguishable when intact. However, they show differences in the distribution of heavy isotopes that make it possible to create up to 16 different tags (of the same mass) that can be quantified relatively to each other during analysis by MS. The NHS ester allows to label all lysines and peptide N-termini of all peptides in a solution with the desired version of the label. After labeling individual samples, these are pooled and analyzed by MS, where the different reporter ions are released upon fragmentation of a peptide. A great advantage of isobaric labeling-based proteomics approaches is the possibility of pooling samples together and thus requiring far less starting material per sample, allowing deep proteomics studies even if sample material is limited. Moreover, signal intensity can be "boosted" by adding samples to the pool with, e.g., a higher peptide amount or a specific composition enriched in the peptides of interest and thus surpassing the threshold for triggering fragmentation. This approach has already proven successful for, e.g., single cell proteomics (Budnik et al., 2018), (tyrosine-)phosphoproteomics (Chua et al., 2020; Yi et al., 2019) and translation proteomics (Bojkova et al., 2020; Klann et al., 2020b).

3.2.3 Translation and degradation quantification

To assess translation and degradation separately and on a whole proteome scale, the ability to differentiate between newly synthesized and preexisting proteins is crucial. This can be achieved by incubation of the cells with heavy SILAC medium for a usually short period of time (minutes to hours). In these pulsed SILAC experiments, newly translated proteins will incorporate the heavy amino acids and can now be distinguished from their older (i.e., light) "siblings," while the fraction of "light" protein to total protein, i.e., heavy and light, can be exploited to infer degradation rates (Schwanhäusser et al., 2009). When combining pulsed-SILAC experiments with TMT, it is also possible to measure acute changes in translation after very short periods of time by adding a boost-channel comprised of heavy peptides, allowing increased coverage of the otherwise very low abundant population (Klann and Münch, 2020; Klann et al., 2020b).

Taken together, it can be said that, as in other areas of biology, the virologist is "spoiled for choice" regarding the plethora of different quantification approaches available for the investigation of dynamic protein networks upon virus infection. Since the global quantification methods are nowadays almost on par, the choice should be made based on the approaches and especially experience available in the local proteomics facility or the collaborating mass spectrometry laboratory. Given enough expertise, basically all of the aforementioned quantification approaches may yield comparable and useful data to advance the field of virology.

3.3 Considerations before performing proteomics during infection

To quantitatively assess protein networks, some points need to be considered before performing proteomics experiments. For most applications, the timing of sample harvest after infection is critical. Most dynamic cellular processes upon infection are transient. Thus, the effects observed during virus entry will be most likely different than seen at later stages of the viral life cycle (Weekes et al., 2014). In addition, host cell stress responses triggered during infection, such as the ISR, are usually of a transient nature (Novoa et al., 2001). Thus, the timing of proteomics experiments is highly important. Usually, the work with live virus, particularly with highly pathogenic viruses, is restricted by the capacities of high safety level facilities. Additionally, deep proteomic experiments have a high demand on sample amounts, machine time and informatics after data acquisition. To efficiently use the existing resources, we recommend to first assess viral behavior in target cells (cell susceptibility, infection rates per MOI, viral replication time) and profiling of phenotypes (e.g., cytopathic effects or viral replication assays) for different time-points prior to carrying out proteomics. Based on these results the optimal timing of sample harvest for proteomics experiments can be evaluated.

Infection courses are heavily influenced by the multiplicity of infection (MOI) used for viral infection. Usage of low MOIs leads to infection of only a fraction of cells, which allows profiling several rounds of the viral life cycle. However, since the pace of the viral cycle is modulated by the host cellular state (e.g., cell cycle or circadian rhythm) (Bagga and Bouchard, 2014; Edgar et al., 2016; Eremenko et al., 1972; Painter et al., 2017), low MOIs often result in measurements of mixed effects across the cell population. Different population of cells display various stages of the infection cycle, averaging together in the final proteomics experiment, making interpretation difficult. Therefore, in such experimental approaches synchronization of cells and viral infection (i.e., adsorption) are crucial and dramatically increase data quality (Haim et al., 2005; Wyler et al., 2019). In contrast, a higher MOI ensures that nearly all cells are infected at the inoculation stage. Still, some cell types or subpopulations might be resistant to infection, as for example, seen for coronaviruses, where in airway epithelium only certain cell types are infected (Rockx et al., 2020). When using high MOIs, special consideration needs to be taken regarding the purity of the obtained virus used for inoculation. If virus is used directly from cell supernatants, debris and signaling molecules of the used reproduction cells might trigger cell non-autonomous stimulations. Concentration and purification of viral particles (e.g., ultracentrifugation) might overcome some of these limitations. Generally, we think there is no one-fits-all method how to infect cells for proteomic research. Depending on all of the mentioned factors, the conditions need to be adjusted (or at least taken into account during data interpretation) for each virus and cell type profiled. Finally, the ideal condition highly depends on the biological question: Are cells/tissue profiled for one viral cycle or multiple rounds of infection? Or is the cell population response to single infected cells examined? As already stated above, preliminary nonproteomics experiments describing viral infection kinetics and behavior are recommended to evaluate the best experimental approach.

Additional considerations are necessary regarding the statistical power of the resulting dataset. Large-scale proteomics experiments have several sources of variance that can lead to the identification of false discoveries (Maes et al., 2015; Piehowski et al., 2013). Thus, the sample size is a critical determinant of the resulting data's quality and usefulness. In general, it is advisable to use as many replicates as possible within the given capacities. Similar to other types of experiments, statistics of proteomic experiments greatly benefit from higher replicates. When profiling time-course experiments by proteomics, the applicability of a high replicate number is often limited due to the large number of samples required putting a high burden on input materials and mass spectrometry time. However, if the time-points are chosen in a way that renders most time-points partially dependent on each other, they can be used for each other's validation. In other words, if protein expression over time follows a certain pattern (e.g., constant increase/decrease, exponential, dose-dependent) or mathematical model, then the different time-points can be analyzed as dependent measurements from the same curve (Peddada et al., 2003). Taking this ordering information into account, for a given protein time-points measurements that follow a determined trend (with good coefficient of determination) will have less propensity to yield false discoveries and increase inference power (Aryee et al., 2009; Peddada et al., 2003). If the measurements for this protein would be solely influenced by statistical noise, then this would be in most cases independent of the time-point, making a good curve fit just by chance unlikely (but notably not impossible). Therefore, one could reduce sample size in time-course profiling experiments, without losing much statistical power. This is also reflected by the fact, that statistical power of inference rises with more time-points, when the number of replicates stays constant (Berk et al., 2011).

3.4 Introduction to network inference

Depending on the nature of the experimental design, several options exist to create and analyze proteins in their network environment. Networks can be created de novo from proteomics data using different metrics like coexpression (correlation) or coprecipitation in AP-MS experiments. Furthermore, networks can also be created or expanded using already existing databases, such as interaction data from STRING (von Mering et al., 2005) or protein function (Reactome) data (Jassal et al., 2020). Here, proteins can be analyzed together with their functional interactors and pathways. Additional layers may be added from the databases to visualize drug targets or already known associations with viral infections.

Proteomic datasets, especially when performed on multiple levels, often contain an easily overwhelming amount of data that needs to be sorted and interpreted. While single conditions, such as mock vs infected, are easy to process, experimental setups with multiple different conditions and/or time-points create a multidimensional complexity that is hard to grasp by solely interpreting bare tables (Larance and Lamond, 2015). Thus, structuring and visualizing the data becomes crucial for useful biological interpretation. There are many ways to perform analysis with multidimensional data and the number of sophisticated statistics packages for environments such as R and python is rising. The advantages and limitations of these approaches have been reviewed elsewhere (De Smet and Marchal, 2010; Larance and Lamond, 2015; Marbach et al., 2010). Summarized in a simplified manner, network inference searches for connections between quantified proteins across data points, for example, correlation across several conditions represented by the Pearson coefficient (notably, the metric can differ depending on biological question and algorithm). Proteins showing a correlated behavior across conditions are connected via an edge in the network, depending on the cut-off used (Fig. 3). Modules of highly connected genes may then be analyzed for pathway enrichment. During analysis of interaction networks, the metric used to create the network is usually the enrichment of a given protein in an IP compared to a control. If a certain threshold requirement is met, a protein will be considered as an interactor of the bait. Performance of

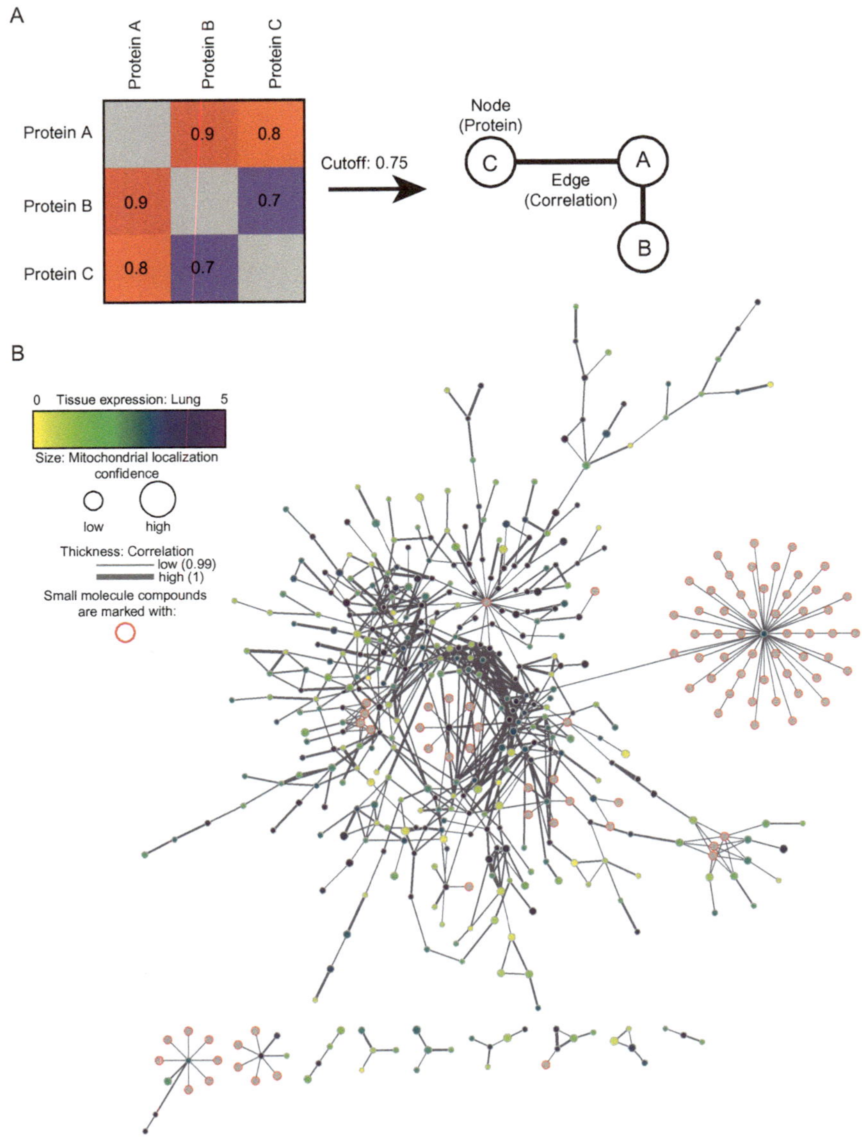

Fig. 3 From correlation matrices to multilayered network views. (A) After calculation of the correlation matrix, the calculated metrics are used to create a network. In this example, the nodes represent the different proteins, while the edges show correlation between the proteins. Usually these networks are created using a cut-off value for the connection, that helps to filter for real connections. (B) Example coexpression network (unpublished data) implementing additional metadata and drug–target interactions: Edge thickness represents correlation over the profiled conditions. Protein nodes are colored according to their expression in the lung. Additionally, node size represents the confidence score for mitochondrial localization. All available small molecules from the ChEMBL database targeting proteins in the coexpression network have been added to the network (*red circled nodes*).

AP-MS experiments on several viral (or host) proteins revealed novel interaction networks (Jäger et al., 2012; Shah et al., 2018).

During recent years, the accessibility of system biological tools has significantly improved. Many user-friendly software releases (e.g., Perseus (Tyanova et al., 2016) and Cytoscape (Shannon et al., 2003)) made it easier to use such tools without experience in writing commands or scripts. However, many sophisticated algorithms remain only available as command-line tools or R/Python packages, which requires at least basic usage of the latter.

4. Conclusions

Studying proteins in their functional environment (i.e., network) can be a successful way to analyze the effect of infections on host cells. Viral proteins often exhibit their functions by distinct interactions with multiple host cell proteins, leading to a disease phenotype. Hence, creating an interaction network during infection gives rise to new insights into molecular disease mechanisms and is able to reveal possible drug interference approaches (Davis et al., 2015; Gordon et al., 2020; Jäger et al., 2012; Shah et al., 2018; Stukalov et al., 2020). Infection also usually triggers various changes in host cell signaling networks. Dissecting changes in these networks leads to important knowledge of functional host–pathogen interactions (Bouhaddou et al., 2020; Klann et al., 2020a; Lapek et al., 2017; Stukalov et al., 2020). Signaling changes can also lead to various effects on protein synthesis and degradation that occur dynamically during the infection course. Measurement of these changes adds an additional layer of information toward understanding viral infections (Bogdanow et al., 2019; Bojkova et al., 2020; Nightingale et al., 2018). Overall, the comprehensive measurement of dynamic protein networks is important to understand viral infections. Consequently, expanding our proteomics toolset and application to quantify changes in the multidimensional and dynamic proteome upon perturbation is required and will eventually lead to an extensive increase in our knowledge of host–pathogen interactions. Many examples in the recent years have shown that network biology of viral infections can successfully discover new therapeutic avenues and shed new light on disease biology. For most viruses, such information remains missing. Filling this gap and carrying out analyses of dynamic proteome changes for a wide variety of viruses will allow a better understanding of their biology in infected host cells and the processes underlying disease.

Acknowledgments

We acknowledge support from the European Research Council under the European Union's Seventh Framework Programme (ERC StG 803565), the Deutsche Forschungsgemeinschaft (DFG, German Research Foundation) Emmy Noether Programme (MU 4216/1-1) and Project-ID 259130777 - SFB1177, the Johanna Quandt Young Academy at Goethe, Aventis Foundation Bridge Award and the Goethe Corona Fonds.

References

Aebersold, R., Mann, M., 2016. Mass-spectrometric exploration of proteome structure and function. Nature 537, 347–355.

Akimov, V., Barrio-Hernandez, I., Hansen, S.V.F., Hallenborg, P., Pedersen, A.K., Bekker-Jensen, D.B., Puglia, M., Christensen, S.D.K., Vanselow, J.T., Nielsen, M.M., et al., 2018. Ubisite approach for comprehensive mapping of lysine and n-terminal ubiquitination sites. Nat. Struct. Mol. Biol. 25, 631–640.

Arkin, A.P., Schaffer, D.V., 2011. Network news: innovations in 21st century systems biology. Cell 144, 844–849.

Aryee, M.J., Gutiérrez-Pabello, J.A., Kramnik, I., Maiti, T., Quackenbush, J., 2009. An improved empirical bayes approach to estimating differential gene expression in microarray time-course data: BETR (Bayesian Estimation of Temporal Regulation). BMC Bioinformatics 10, 409.

Bagga, S., Bouchard, M.J., 2014. Cell cycle regulation during viral infection. Methods Mol. Biol. 1170, 165–227.

Bai, Y., Zhou, K., Doudna, J.A., 2013. Hepatitis C virus 3′UTR regulates viral translation through direct interactions with the host translation machinery. Nucleic Acids Res. 41, 7861–7874.

Balachandran, S., Roberts, P.C., Brown, L.E., Truong, H., Pattnaik, A.K., Archer, D.R., Barber, G.N., 2000. Essential role for the dsRNA-dependent protein kinase PKR in innate immunity to viral infection. Immunity 13, 129–141.

Bantscheff, M., Schirle, M., Sweetman, G., Rick, J., Kuster, B., 2007. Quantitative mass spectrometry in proteomics: a critical review. Anal. Bioanal. Chem. 389, 1017–1031.

Bekker-Jensen, D.B., Kelstrup, C.D., Batth, T.S., Larsen, S.C., Haldrup, C., Bramsen, J.B., Sørensen, K.D., Høyer, S., Ørntoft, T.F., Andersen, C.L., et al., 2017. An optimized shotgun strategy for the rapid generation of comprehensive human proteomes. Cell Syst. 4, 587–599.e4.

Bensimon, A., Heck, A.J.R., Aebersold, R., 2012. Mass spectrometry-based proteomics and network biology. Annu. Rev. Biochem. 81, 379–405.

Berk, M., Ebbels, T., Montana, G., 2011. A statistical framework for biomarker discovery in metabolomic time course data. Bioinformatics 27, 1979–1985.

Bogdanow, B., Wang, X., Eichelbaum, K., Sadewasser, A., Husic, I., Paki, K., Budt, M., Hergeselle, M., Vetter, B., Hou, J., et al., 2019. The dynamic proteome of influenza A virus infection identifies M segment splicing as a host range determinant. Nat. Commun. 10, 5518.

Boisvert, F.M., Ahmad, Y., Gierliński, M., Charrière, F., Lamont, D., Scott, M., Barton, G., Lamond, A.I., 2012. A quantitative spatial proteomics analysis of proteome turnover in human cells. Mol. Cell. Proteomics 11, 1–15.

Bojkova, D., Klann, K., Koch, B., Widera, M., Krause, D., Ciesek, S., Cinatl, J., Münch, C., 2020. Proteomics of SARS-CoV-2-infected host cells reveals therapy targets. Nature 583 (7816), 469–472.

Bouhaddou, M., Memon, D., Meyer, B., White, K.M., Rezelj, V.V., Correa Marrero, M., Polacco, B.J., Melnyk, J.E., Ulferts, S., Kaake, R.M., et al., 2020. The global phosphorylation landscape of SARS-CoV-2 infection. Cell 182 (3), 685–712.

Budnik, B., Levy, E., Harmange, G., Slavov, N., 2018. SCoPE-MS: mass spectrometry of single mammalian cells quantifies proteome heterogeneity during cell differentiation. Genome Biol. 19, 1–12.

Chua, X.Y., Mensah, T., Aballo, T., Mackintosh, S.G., Edmondson, R.D., Salomon, A.R., 2020. Tandem mass tag approach utilizing pervanadate BOOST channels delivers deeper quantitative characterization of the tyrosine phosphoproteome. Mol. Cell. Proteomics 19 (4), 730–743.

Cox, J., Hein, M.Y., Luber, C.A., Paron, I., Nagaraj, N., Mann, M., 2014. Accurate proteome-wide label-free quantification by delayed normalization and maximal peptide ratio extraction, termed MaxLFQ. Mol. Cell. Proteomics 13, 2513–2526.

Cruz-Oliveira, C., Freire, J.M., Conceição, T.M., Higa, L.M., Castanho, M.A.R.B., Da Poian, A.T., 2015. Receptors and routes of dengue virus entry into the host cells. FEMS Microbiol. Rev. 39, 155–170.

Dauber, B., Wolff, T., 2009. Activation of the antiviral kinase PKR and viral countermeasures. Viruses 1, 523–544.

Davis, Z.H., Verschueren, E., Jang, G.M., Kleffman, K., Johnson, J.R., Park, J., Von Dollen, J., Maher, M.C., Johnson, T., Newton, W., et al., 2015. Global mapping of herpesvirus-host protein complexes reveals a transcription strategy for late genes. Mol. Cell 57, 349–360.

De Smet, R., Marchal, K., 2010. Advantages and limitations of current network inference methods. Nat. Rev. Microbiol. 8, 717–729.

Dephoure, N., Hwang, S., O'Sullivan, C., Dodgson, S.E., Gygi, S.P., Amon, A., Torres, E.M., 2014. Quantitative proteomic analysis reveals posttranslational responses to aneuploidy in yeast. Elife 3, e03023.

Di Pietro, A., Kajaste-Rudnitski, A., Oteiza, A., Nicora, L., Towers, G.J., Mechti, N., Vicenzi, E., 2013. TRIM22 inhibits influenza A virus infection by targeting the viral nucleoprotein for degradation. J. Virol. 87, 4523–4533.

Diamond, D.L., Syder, A.J., Jacobs, J.M., Sorensen, C.M., Walters, K.A., Proll, S.C., McDermott, J.E., Gritsenko, M.A., Zhang, Q., Zhao, R., et al., 2010. Temporal proteome and lipidome profiles reveal hepatitis C virus-associated reprogramming of hepatocellular metabolism and bioenergetics. PLoS Pathog. 6, e1000719.

Didcock, L., Young, D.F., Goodbourn, S., Randall, R.E., 1999. The V protein of simian virus 5 inhibits interferon signalling by targeting STAT1 for proteasome-mediated degradation. J. Virol. 73, 9928 LP–9933.

Dieterich, D.C., Link, A.J., Graumann, J., Tirrell, D.A., Schuman, E.M., 2006. Selective identification of newly synthesized proteins in mammalian cells using bioorthogonal noncanonical amino acid tagging (BONCAT). Proc. Natl. Acad. Sci. U. S. A. 103 (25), 9482–9487.

Durbin, K.R., Fornelli, L., Fellers, R.T., Doubleday, P.F., Narita, M., Kelleher, N.L., 2016. Quantitation and identification of thousands of human proteoforms below 30 kDa. J. Proteome Res. 15, 976–982.

Edgar, R.S., Stangherlin, A., Nagy, A.D., Nicoll, M.P., Efstathiou, S., O'Neill, J.S., Reddy, A.B., 2016. Cell autonomous regulation of herpes and influenza virus infection by the circadian clock. Proc. Natl. Acad. Sci. U. S. A. 201601895.

Eng, J.K., Searle, B.C., Clauser, K.R., Tabb, D.L., 2011. A face in the crowd: recognizing peptides through database search. Mol. Cell. Proteomics 10, 1–24.

Eremenko, T., Benedetto, A., Volpe, P., 1972. Virus infection as a function of the host cell life cycle: replication of poliovirus RNA. J. Gen. Virol. 16, 61–68.

Gale, M., Blakely, C.M., Kwieciszewski, B., Tan, S.-L., Dossett, M., Tang, N.M., Korth, M.J., Polyak, S.J., Gretch, D.R., Katze, M.G., 1998. Control of PKR protein kinase by Hepatitis C virus nonstructural 5A protein: molecular mechanisms of kinase regulation. Mol. Cell. Biol. 18, 5208 LP–5218.
Geiger, T., Cox, J., Ostasiewicz, P., Wisniewski, J.R., Mann, M., 2010. Super-SILAC mix for quantitative proteomics of human tumor tissue. Nat. Methods 7 (5), 383–385.
Gingras, A.C., Gstaiger, M., Raught, B., Aebersold, R., 2007. Analysis of protein complexes using mass spectrometry. Nat. Rev. Mol. Cell Biol. 8 (8), 645–654.
Gordon, D.E., Jang, G.M., Bouhaddou, M., Xu, J., Obernier, K., White, K.M., O'Meara, M.J., Rezelj, V.V., Guo, J.Z., Swaney, D.L., et al., 2020. A SARS-CoV-2 protein interaction map reveals targets for drug repurposing. Nature 583 (7816), 459–468.
Greco, T.M., Diner, B.A., Cristea, I.M., 2014. The impact of mass spectrometry–based proteomics on fundamental discoveries in virology. Annu. Rev. Virol. 1, 581–604.
Grossegesse, M., Doellinger, J., Fritsch, A., Laue, M., Piesker, J., Schaade, L., Nitsche, A., 2018. Global ubiquitination analysis reveals extensive modification and proteasomal degradation of cowpox virus proteins, but preservation of viral cores. Sci. Rep. 8, 1807.
Haim, H., Steiner, I., Panet, A., 2005. Synchronized infection of cell cultures by magnetically controlled virus. J. Virol. 79, 622–625.
Hale, B.G., Jackson, D., Chen, Y.-H., Lamb, R.A., Randall, R.E., 2006. Influenza A virus NS1 protein binds p85β and activates phosphatidylinositol-3-kinase signaling. Proc. Natl. Acad. Sci. U. S. A. 103, 14194 LP–14199.
Harper, J.W., Bennett, E.J., 2016. Proteome complexity and the forces that drive proteome imbalance. Nature 537 (7620), 328–338.
Hebert, A.S., Richards, A.L., Bailey, D.J., Ulbrich, A., Coughlin, E.E., Westphall, M.S., Coon, J.J., 2014. The one hour yeast proteome. Mol. Cell. Proteomics 13, 339–347.
Hendriks, I.A., Lyon, D., Su, D., Skotte, N.H., Daniel, J.A., Jensen, L.J., Nielsen, M.L., 2018. Site-specific characterization of endogenous SUMOylation across species and organs. Nat. Commun. 9 (1), 2456.
Hornbeck, P.V., Kornhauser, J.M., Tkachev, S., Zhang, B., Skrzypek, E., Murray, B., Latham, V., Sullivan, M., 2012. PhosphoSitePlus: a comprehensive resource for investigating the structure and function of experimentally determined post-translational modifications in man and mouse. Nucleic Acids Res. 40, D261–D270.
Hu, D., Xue, S., Zhao, C., Wei, M., Yan, H., Quan, Y., Yu, W., 2018. Comprehensive profiling of lysine acetylome in baculovirus infected silkworm (Bombyx mori) cells. Proteomics 18, 1700133.
Huttlin, E.L., Ting, L., Bruckner, R.J., Gebreab, F., Gygi, M.P., Szpyt, J., Tam, S., Zarraga, G., Colby, G., Baltier, K., et al., 2015. The BioPlex network: a systematic exploration of the human interactome. Cell.
Huttlin, E.L., Bruckner, R.J., Paulo, J.A., Cannon, J.R., Ting, L., Baltier, K., Colby, G., Gebreab, F., Gygi, M.P., Parzen, H., et al., 2017. Architecture of the human interactome defines protein communities and disease networks. Nature 545 (7655), 505–509.
Ideker, T., Krogan, N.J., 2012. Differential network biology. Mol. Syst. Biol. 8, 565.
Ikeda, M., Kato, N., 2007. Modulation of host metabolism as a target of new antivirals. Adv. Drug Deliv. Rev. 59, 1277–1289.
Jäger, S., Cimermancic, P., Gulbahce, N., Johnson, J.R., McGovern, K.E., Clarke, S.C., Shales, M., Mercenne, G., Pache, L., Li, K., et al., 2012. Global landscape of HIV-human protein complexes. Nature 481, 365–370.
Jassal, B., Matthews, L., Viteri, G., Gong, C., Lorente, P., Fabregat, A., Sidiropoulos, K., Cook, J., Gillespie, M., Haw, R., et al., 2020. The reactome pathway knowledgebase. Nucleic Acids Res. 48, D498–D503.
Katz, W., Weinstein, B., Solomon, F., 1990. Regulation of tubulin levels and microtubule assembly in Saccharomyces cerevisiae: consequences of altered tubulin gene copy number. Mol. Cell. Biol. 10, 5286 LP–5294.

Keilhauer, E.C., Hein, M.Y., Mann, M., 2015. Accurate protein complex retrieval by affinity enrichment mass spectrometry (AE-MS) rather than affinity purification mass spectrometry (AP-MS). Mol. Cell. Proteomics 14, 120–135.

Khaperskyy, D.A., Emara, M.M., Johnston, B.P., Anderson, P., Hatchette, T.F., McCormick, C., 2014. Influenza A virus host shutoff disables antiviral stress-induced translation arrest. PLoS Pathog. 10, e1004217.

Kim, S.C., Sprung, R., Chen, Y., Xu, Y., Ball, H., Pei, J., Cheng, T., Kho, Y., Xiao, H., Xiao, L., et al., 2006. Substrate and functional diversity of lysine acetylation revealed by a proteomics survey. Mol. Cell 23, 607–618.

Kindrachuk, J., Ork, B., Hart, B.J., Mazur, S., Holbrook, M.R., Frieman, M.B., Traynor, D., Johnson, R.F., Dyall, J., Kuhn, J.H., et al., 2015. Antiviral potential of ERK/MAPK and PI3K/AKT/mTOR signaling modulation for middle east respiratory syndrome coronavirus infection as identified by temporal kinome analysis. Antimicrob. Agents Chemother. 59, 1088–1099.

Kirkwood, K.J., Ahmad, Y., Larance, M., Lamond, A.I., 2013. Characterization of native protein complexes and protein isoform variation using size fractionation-based quantitative proteomics. Mol. Cell. Proteomics 12 (12), 3851–3873.

Klann, K., Münch, C., 2020. Instrument logic increases identifications during mutliplexed translatome measurements. Anal. Chem. 92, 8041–8045.

Klann, K., Bojkova, D., Tascher, G., Ciesek, S., Münch, C., Cinatl, J., 2020a. Growth factor receptor signaling inhibition prevents SARS-CoV-2 replication. Mol. Cell. 2020.05.14.095661.

Klann, K., Tascher, G., Münch, C., 2020b. Functional translatome proteomics reveal converging and dose-dependent regulation by mTORC1 and eIF2α. Mol. Cell 77, 913–925.e4.

Krishnamoorthy, T., Pavitt, G.D., Zhang, F., Dever, T.E., Hinnebusch, A.G., 2001. Tight binding of the phosphorylated α subunit of initiation factor 2 (eIF2α) to the regulatory subunits of guanine nucleotide exchange factor eIF2B is required for inhibition of translation initiation. Mol. Cell. Biol. 21, 5018 LP–5030.

Lapek, J.D., Lewinski, M.K., Wozniak, J.M., Guatelli, J., Gonzalez, D.J., 2017. Quantitative temporal viromics of an inducible HIV-1 model yields insight to global host targets and phospho-dynamics associated with protein Vpr. Mol. Cell. Proteomics 16, 1447–1461.

Larance, M., Lamond, A.I., 2015. Multidimensional proteomics for cell biology. Nat. Rev. Mol. Cell Biol. 16, 269–280.

Li, Y.F., Radivojac, P., 2012. Computational approaches to protein inference in shotgun proteomics. BMC Bioinformatics 13 (suppl. 1), S4.

Li, S., Yu, X., Guo, Y., Kong, L., 2012. Interaction networks of hepatitis C virus NS4B: implications for antiviral therapy. Cell. Microbiol. 14, 994–1002.

Li, F., Zhang, C., Li, Y., Wu, G., Hou, X., Zhou, X., Wang, A., 2018. Beclin1 restricts RNA virus infection in plants through suppression and degradation of the viral polymerase. Nat. Commun. 9, 1268.

Liu, X., Liu, S., Bode, L., Liu, C., Zhang, L., Wang, X., Li, D., Lei, Y., Peng, X., Cheng, Z., et al., 2015. Persistent human Borna disease virus infection modifies the acetylome of human oligodendroglia cells towards higher energy and transporter levels. Virology 485, 58–78.

Londrigan, S.L., Tate, M.D., Brooks, A.G., Reading, P.C., 2012. Cell-surface receptors on macrophages and dendritic cells for attachment and entry of influenza virus. J. Leukoc. Biol. 92, 97–106.

Maes, E., Valkenborg, D., Baggerman, G., Willems, H., Landuyt, B., Schoofs, L., Mertens, I., 2015. Determination of variation parameters as a crucial step in designing TMT-based clinical proteomics experiments. PLoS One 10, e0120115.

Marbach, D., Prill, R.J., Schaffter, T., Mattiussi, C., Floreano, D., Stolovitzky, G., 2010. Revealing strengths and weaknesses of methods for gene network inference. Proc. Natl. Acad. Sci. U. S. A. 107, 6286 LP–6291.

Miller, E.H., Obernosterer, G., Raaben, M., Herbert, A.S., Deffieu, M.S., Krishnan, A., Ndungo, E., Sandesara, R.G., Carette, J.E., Kuehne, A.I., et al., 2012. Ebola virus entry requires the host-programmed recognition of an intracellular receptor. EMBO J. 31, 1947–1960.

Millikin, R.J., Solntsev, S.K., Shortreed, M.R., Smith, L.M., 2018. Ultrafast peptide label-free quantification with FlashLFQ. J. Proteome Res. 17, 386–391.

Milo, R., Shen-Orr, S., Itzkovitz, S., Kashtan, N., Chklovskii, D., Alon, U., 2002. Network motifs: simple building blocks of complex networks. Science (80-) 298, 824–827.

Mohr, I., Sonenberg, N., 2012. Host translation at the nexus of infection and immunity. Cell Host Microbe 12, 470–483.

Monteiro, J.T., Lepenies, B., 2017. Myeloid C-type lectin receptors in viral recognition and antiviral immunity. Viruses 9 (3), 59.

Morrison, J., Laurent-Rolle, M., Maestre, A.M., Rajsbaum, R., Pisanelli, G., Simon, V., Mulder, L.C.F., Fernandez-Sesma, A., García-Sastre, A., 2013. Dengue virus Co-opts UBR4 to degrade STAT2 and antagonize type I interferon signaling. PLoS Pathog. 9, e1003265.

Munday, D.C., Emmott, E., Surtees, R., Lardeau, C.-H., Wu, W., Duprex, W.P., Dove, B.K., Barr, J.N., Hiscox, J.A., 2010. Quantitative proteomic analysis of A549 cells infected with human respiratory syncytial virus. Mol. Cell Proteomics 9, 2438 LP–2459.

Nightingale, K., Lin, K.M., Ravenhill, B.J., Davies, C., Nobre, L., Fielding, C.A., Ruckova, E., Fletcher-Etherington, A., Soday, L., Nichols, H., et al., 2018. High-definition analysis of host protein stability during human cytomegalovirus infection reveals antiviral factors and viral evasion mechanisms. Cell Host Microbe 24, 447–460.e11.

Novoa, I., Zeng, H., Harding, H.P., Ron, D., 2001. Feedback inhibition of the unfolded protein response by GADD34-mediated dephosphorylation of eIF2α. J. Cell Biol. 153, 1011–1021.

Ong, S.E., Blagoev, B., Kratchmarova, I., Kristensen, D.B., Steen, H., Pandey, A., Mann, M., 2002. Stable isotope labeling by amino acids in cell culture, SILAC, as a simple and accurate approach to expression proteomics. Mol. Cell. Proteomics 1 (5), 376–386.

Oppermann, F.S., Gnad, F., Olsen, J.V., Hornberger, R., Greff, Z., Kéri, G., Mann, M., Daub, H., 2009. Large-scale proteomics analysis of the human kinome. Mol. Cell Proteomics 8, 1751 LP–1764.

Ordureau, A., Münch, C., Harper, J.W., 2015. Quantifying ubiquitin signaling. Mol. Cell 58 (4), 660–676.

Painter, M.M., Zaikos, T.D., Collins, K.L., 2017. Quiescence promotes latent HIV infection and resistance to reactivation from latency with histone deacetylase inhibitors. J. Virol. 91. e01080-17.

Pakos-Zebrucka, K., Koryga, I., Mnich, K., Ljujic, M., Samali, A., Gorman, A.M., 2016. The integrated stress response. EMBO Rep. 17, 1374 LP–1395.

Peddada, S.D., Lobenhofer, E.K., Li, L., Afshari, C.A., Weinberg, C.R., Umbach, D.M., 2003. Gene selection and clustering for time-course and dose–response microarray experiments using order-restricted inference. Bioinformatics 19, 834–841.

Piehowski, P.D., Petyuk, V.A., Orton, D.J., Xie, F., Moore, R.J., Ramirez-Restrepo, M., Engel, A., Lieberman, A.P., Albin, R.L., Camp, D.G., et al., 2013. Sources of technical variability in quantitative LC-MS proteomics: human brain tissue sample analysis. J. Proteome Res. 12, 2128–2137.

Popescu, C.-I., Dubuisson, J., 2010. Role of lipid metabolism in hepatitis C virus assembly and entry. Biol. Cell 102, 63–74.

Riley, N.M., Coon, J.J., 2016. Phosphoproteomics in the age of rapid and deep proteome profiling. Anal. Chem. 88, 74–94.

Ritter, J.B., Wahl, A.S., Freund, S., Genzel, Y., Reichl, U., 2010. Metabolic effects of influenza virus infection in cultured animal cells: Intra- and extracellular metabolite profiling. BMC Syst. Biol. 4, 61.

Rockx, B., Kuiken, T., Herfst, S., Bestebroer, T., Lamers, M.M., Oude Munnink, B.B., de Meulder, D., van Amerongen, G., van den Brand, J., Okba, N.M.A., et al., 2020. Comparative pathogenesis of COVID-19, MERS, and SARS in a nonhuman primate model. Science (80-) 368, 1012 LP–1015.

Rothenberg, D.A., Taliaferro, J.M., Huber, S.M., Begley, T.J., Dedon, P.C., White, F.M., 2018. A proteomics approach to profiling the temporal translational response to stress and growth. IScience 9, 367–381.

Savitski, M.M., Zinn, N., Faelth-Savitski, M., Poeckel, D., Gade, S., Becher, I., Muelbaier, M., Wagner, A.J., Strohmer, K., Werner, T., et al., 2018. Multiplexed proteome dynamics profiling reveals mechanisms controlling protein homeostasis. Cell 173, 260–274.e25.

Schwanhäusser, B., Gossen, M., Dittmar, G., Selbach, M., 2009. Global analysis of cellular protein translation by pulsed SILAC. Proteomics 9 (1), 205–209.

Schwanhäusser, B., Busse, D., Li, N., Dittmar, G., Schuchhardt, J., Wolf, J., Chen, W., Selbach, M., 2011. Global quantification of mammalian gene expression control. Nature 473, 337–342.

Schwenk, J.M., Omenn, G.S., Sun, Z., Campbell, D.S., Baker, M.S., Overall, C.M., Aebersold, R., Moritz, R.L., Deutsch, E.W., 2017. The human plasma proteome draft of 2017: building on the human plasma peptideatlas from mass spectrometry and complementary assays. J. Proteome Res. 16, 4299–4310.

Shah, P.S., Link, N., Jang, G.M., Sharp, P.P., Zhu, T., Swaney, D.L., Johnson, J.R., Von Dollen, J., Ramage, H.R., Satkamp, L., et al., 2018. Comparative flavivirus-host protein interaction mapping reveals mechanisms of dengue and Zika virus pathogenesis. Cell 175, 1931–1945.e18.

Shannon, P., Markiel, A., Ozier, O., Baliga, N.S., Wang, J.T., Ramage, D., Amin, N., Schwikowski, B., Ideker, T., 2003. Cytoscape: a software environment for integrated models of biomolecular interaction networks. Genome Res. 13, 2498–2504.

Sharma, K., D'Souza, R.C.J., Tyanova, S., Schaab, C., Wiśniewski, J.R., Cox, J., Mann, M., 2014. Ultradeep human phosphoproteome reveals a distinct regulatory nature of tyr and Ser/Thr-based signaling. Cell Rep. 8, 1583–1594.

Snijder, J., Rose, R.J., Veesler, D., Johnson, J.E., Heck, A.J.R., 2013. Studying 18 MDa virus assemblies with native mass spectrometry. Angew. Chem. Int. Ed. 52 (14), 4020–4023.

Soderholm, S., Kainov, D.E., Ohman, T., Denisova, O.V., Schepens, B., Kulesskiy, E., Imanishi, S.Y., Corthals, G., Hintsanen, P., Aittokallio, T., et al., 2016. Phosphoproteomics to characterize host response during influenza a virus infection of human macrophages. Mol. Cell. Proteomics 15, 3203–3219.

Stukalov, A., Girault, V., Grass, V., Bergant, V., Karayel, O., Urban, C., Haas, D.A., Huang, Y., Oubraham, L., Wang, A., et al., 2020. Multi-level proteomics reveals host-perturbation strategies of SARS-CoV-2 and SARS-CoV. BioRxiv. 2020.06.17.156455.

Taniuchi, S., Miyake, M., Tsugawa, K., Oyadomari, M., Oyadomari, S., 2016. Integrated stress response of vertebrates is regulated by four eIF2α kinases. Sci. Rep. 6, 32886.

Tatum, E.L., Beadle, G.W., 1942. Genetic control of biochemical reactions in neurospora: an "Aminobenzoicless" mutant. Proc. Natl. Acad. Sci. U. S. A. 28, 234–243.

Taylor, D.R., Shi, S.T., Romano, P.R., Barber, G.N., Lai, M.M.C., 1999. Inhibition of the interferon-inducible protein kinase PKR by HCV E2 protein. Science (80-) 285, 107–110.

Thompson, A., Wölmer, N., Koncarevic, S., Selzer, S., Böhm, G., Legner, H., Schmid, P., Kienle, S., Penning, P., Höhle, C., et al., 2019. TMTpro: design, synthesis, and initial evaluation of a proline-based isobaric 16-Plex tandem mass tag reagent set. Anal. Chem. 91, 15941–15950.

Tyanova, S., Temu, T., Sinitcyn, P., Carlson, A., Hein, M.Y., Geiger, T., Mann, M., Cox, J., 2016. The Perseus computational platform for comprehensive analysis of (prote)omics data. Nat. Methods 13, 731–740.

Ubersax, J.A., Ferrell Jr., J.E., 2007. Mechanisms of specificity in protein phosphorylation. Nat. Rev. Mol. Cell Biol. 8, 530–541.

Udeshi, N.D., Svinkina, T., Mertins, P., Kuhn, E., Mani, D.R., Qiao, J.W., Carr, S.A., 2013. Refined preparation and use of anti-diglycine remnant (k-ε-gg) antibody enables routine quantification of 10,000s of ubiquitination sites in single proteomics experiments. Mol. Cell. Proteomics 12, 825–831.

Vidal, M., 2009. A unifying view of 21st century systems biology. FEBS Lett. 583, 3891–3894.

Vlastaridis, P., Kyriakidou, P., Chaliotis, A., Van de Peer, Y., Oliver, S.G., Amoutzias, G.D., 2017. Estimating the total number of phosphoproteins and phosphorylation sites in eukaryotic proteomes. Gigascience 6, 1–11.

von Mering, C., Jensen, L.J., Snel, B., Hooper, S.D., Krupp, M., Foglierini, M., Jouffre, N., Huynen, M.A., Bork, P., 2005. STRING: known and predicted protein-protein associations, integrated and transferred across organisms. Nucleic Acids Res. 33, D433–D437.

Wang, L., Fu, B., Li, W., Patil, G., Liu, L., Dorf, M.E., Li, S., 2017. Comparative influenza protein interactomes identify the role of plakophilin 2 in virus restriction. Nat. Commun. 8, 13876.

Watanabe, T., Watanabe, S., Kawaoka, Y., 2010. Cellular networks involved in the Influenza virus life cycle. Cell Host Microbe 7, 427–439.

Weber, A., Dam, S., Saul, V.V., Kuznetsova, I., Müller, C., Fritz-Wolf, K., Becker, K., Linne, U., Gu, H., Stokes, M.P., et al., 2019. Phosphoproteome analysis of cells infected with adapted and nonadapted Influenza A virus reveals novel pro- and antiviral signaling networks. J. Virol. 93. e00528-19.

Weekes, M.P., Tomasec, P., Huttlin, E.L., Fielding, C.A., Nusinow, D., Stanton, R.J., Wang, E.C.Y., Aicheler, R., Murrell, I., Wilkinson, G.W.G., et al., 2014. Quantitative temporal viromics: an approach to investigate host-pathogen interaction. Cell 157, 1460–1472.

Weinstein, B., Solomon, F., 1990. Phenotypic consequences of tubulin overproduction in Saccharomyces cerevisiae: differences between alpha-tubulin and beta-tubulin. Mol. Cell Biol. 10, 5295 LP–5304.

Welle, K.A., Zhang, T., Hryhorenko, J.R., Shen, S., Qu, J., Ghaemmaghami, S., 2016. Time-resolved analysis of proteome dynamics by tandem mass tags and stable isotope labeling in cell culture (TMT-SILAC) hyperplexing. Mol. Cell. Proteomics 15, 3551–3563.

Whitmarsh, A.J., Davis, R.J., 2016. Multisite phosphorylation by MAPK. Science (80-) 354, 179 LP–180.

Wolf-Levy, H., Javitt, A., Eisenberg-Lerner, A., Kacen, A., Ulman, A., Sheban, D., Dassa, B., Fishbain-Yoskovitz, V., Carmona-Rivera, C., Kramer, M.P., et al., 2018. Revealing the cellular degradome by mass spectrometry analysis of proteasome-cleaved peptides. Nat. Biotechnol. 36, 1110–1122.

Wolfrum, N., Greber, U.F., 2013. Adenovirus signalling in entry. Cell. Microbiol. 15, 53–62.

Wyler, E., Franke, V., Menegatti, J., Kocks, C., Boltengagen, A., Praktiknjo, S., Walch-Rückheim, B., Bosse, J., Rajewsky, N., Grässer, F., et al., 2019. Single-cell RNA-sequencing of herpes simplex virus 1-infected cells connects NRF2 activation to an antiviral program. Nat. Commun. 10, 4878.

Yángüez, E., Hunziker, A., Dobay, M.P., Yildiz, S., Schading, S., Elshina, E., Karakus, U., Gehrig, P., Grossmann, J., Dijkman, R., et al., 2018. Phosphoproteomic-based kinase profiling early in influenza virus infection identifies GRK2 as antiviral drug target. Nat. Commun. 9, 3679.

Yi, L., Tsai, C.F., Dirice, E., Swensen, A.C., Chen, J., Shi, T., Gritsenko, M.A., Chu, R.K., Piehowski, P.D., Smith, R.D., et al., 2019. Boosting to amplify signal with isobaric labeling (BASIL) strategy for comprehensive quantitative phosphoproteomic characterization of small populations of cells. Anal. Chem. 91 (9), 5794–5801.

CHAPTER TWO

Applications of mass spectrometry imaging in virus research

Luca D. Bertzbach[a], **Benedikt B. Kaufer**[a,*], **and Axel Karger**[b,*]
[a]Institute of Virology, Freie Universität Berlin, Berlin, Germany
[b]Institute of Molecular Virology and Cell Biology, Friedrich-Loeffler-Institut, Greifswald-Insel Riems, Germany
*Corresponding authors: e-mail address: b.kaufer@fu-berlin.de; axel.karger@fli.de

Contents

Abstract

Mass spectrometry imaging (MSI) is a label-free molecular imaging technique allowing an untargeted detection of a broad range of biomolecules and xenobiotics. MSI enables imaging of the spatial distribution of proteins, peptides, lipids and metabolites from a wide range of samples. To date, this technique is commonly applied to tissue sections in cancer diagnostics and biomarker development, but also molecular histology in general. Advances in the methodology and bioinformatics improved the resolution of MS images below the single cell level and increased the flexibility of the workflow. However, MSI-based research in virology is just starting to gain momentum and its full potential has not been exploited yet. In this review, we discuss the main applications of

Advances in Virus Research, Volume 109
ISSN 0065-3527
https://doi.org/10.1016/bs.aivir.2020.10.002

MSI in virology. We review important aspects of matrix-assisted laser desorption/ionization (MALDI) MSI, the most widely used MSI technique in virology. In addition, we summarize relevant literature on MSI studies that aim to unravel virus-host interactions and virus pathogenesis, to elucidate antiviral drug kinetics and to improve current viral disease diagnostics. Collectively, these studies strongly improve our general understanding of virus-induced changes in the proteome, metabolome and metabolite distribution in host tissues of humans, animals and plants upon infection. Furthermore, latest MSI research provided important insights into the drug distribution and distribution kinetics, especially in antiretroviral research. Finally, MSI-based investigations of oncogenic viruses greatly increased our knowledge on tumor mass signatures and facilitated the identification of cancer biomarkers.

Abbreviations

CoxV	coxsackievirus
DESI	desorption electrospray ionization
EBV	Epstein Barr virus
FFPE	formalin-fixation and paraffin embedding
HBV	hepatitis B virus
HCC	hepatocellular carcinoma
HCV	hepatitis C virus
HIV	human immunodeficiency virus
HPV	human papillomavirus
IR-MALDESI	infrared matrix-assisted laser desorption electrospray ionization imaging
LAESI	laser ablation electrospray ionization
LA-ICP	laser ablation inductively coupled plasma
LCM	laser capture microdissection
LDI	laser desorption ionization
LTR	retroviral long terminal repeat
MALDI	matrix-assisted laser desorption/ionization
MALDI-FTICR	MALDI-Fourier transform ion cyclotron resonance
MDV	Marek's disease virus
MMTV	mouse mammary tumor virus
MSI	mass spectrometry imaging
PyMT	polyoma middle T oncoprotein
SAMDI	self-assembled monolayer desorption/ionization
SIMS	secondary ion mass spectrometry
TOF	time of flight
ZIKV	Zika virus

1. Introduction

Mass spectrometry (MS)-based proteomics facilitated fundamental discoveries in virus research and had an immense impact on life sciences. Identification of proteins by MS provided important insights into the cellular protein-protein interaction networks and their functions. The characterization of viral interactomes elucidated the mechanisms that allow viruses to usurp cellular pathways. This knowledge may pave the way for the development of urgently needed targeted intervention strategies against life-threatening pathogens like Ebola virus (Batra et al., 2018). The systematic evaluation of protein-protein interactions across human viruses using bioinformatics is only beginning to unfold (Lasso et al., 2019). Moreover, the MS-based characterization of host responses to viral infections using systems virology (Law et al., 2013) shed light on viral diseases and their pathogenesis. For example, a network analysis of temporally resolved proteomes elucidated the reprogramming of host cell protein expression in the course of Influenza virus infections in great detail (Ranadheera et al., 2018). Proteome analyses of virus particles revealed interactions of viral and host proteins during morphogenesis (Alejo et al., 2018; Greco et al., 2014; Hutchinson et al., 2014; Michael et al., 2006a,b; Mirzakhanyan and Gershon, 2019). Diagnostic applications of MS have also been implemented (Wee et al., 2019) including genotyping (Michael et al., 2009; Zhu et al., 2018) or proteotyping assays as potential tools for epidemiologic studies and disease surveillance (Fernandes and Downard, 2014).

All these eminent achievements are based on the capability of MS-based proteomics to identify high numbers of proteins in a given sample with high specificity. A prerequisite of this approach is the availability of genome sequence data of the organisms of interest. In this respect, mass spectrometry imaging (MSI, also referred to as imaging mass spectrometry; IMS) is different. An illustration of a common MSI workflow is given in Fig. 1. First, MSI is more than a proteomic approach, as it detects multiple classes of biomolecules including proteins, peptides, nucleic acids, carbohydrates, xenobiotics, lipids, and other metabolites. A recent review entitled "Mass spectrometry imaging of biomolecular information" (Spengler, 2015), highlighted the potential of MSI and brings it to the point. Secondly, MSI analyses can answer many questions completely independent of sequence data. Images can be constructed based on the intensity distribution of single masses but also on more complicated mass signatures, which are calculated from the spectral data by statistical procedures. These spectra-based classifications of tissue areas using statistical models (Deininger et al., 2010) allow the use of MSI for diagnostic purposes, even if the identity of specific single or multiple biomarker molecules has not been revealed by analyses of the tissue regions of interest following the MSI process (panel "Marker Identification" in Fig. 1). The number of masses included into statistical classification models can vary in a broad range. In a study of

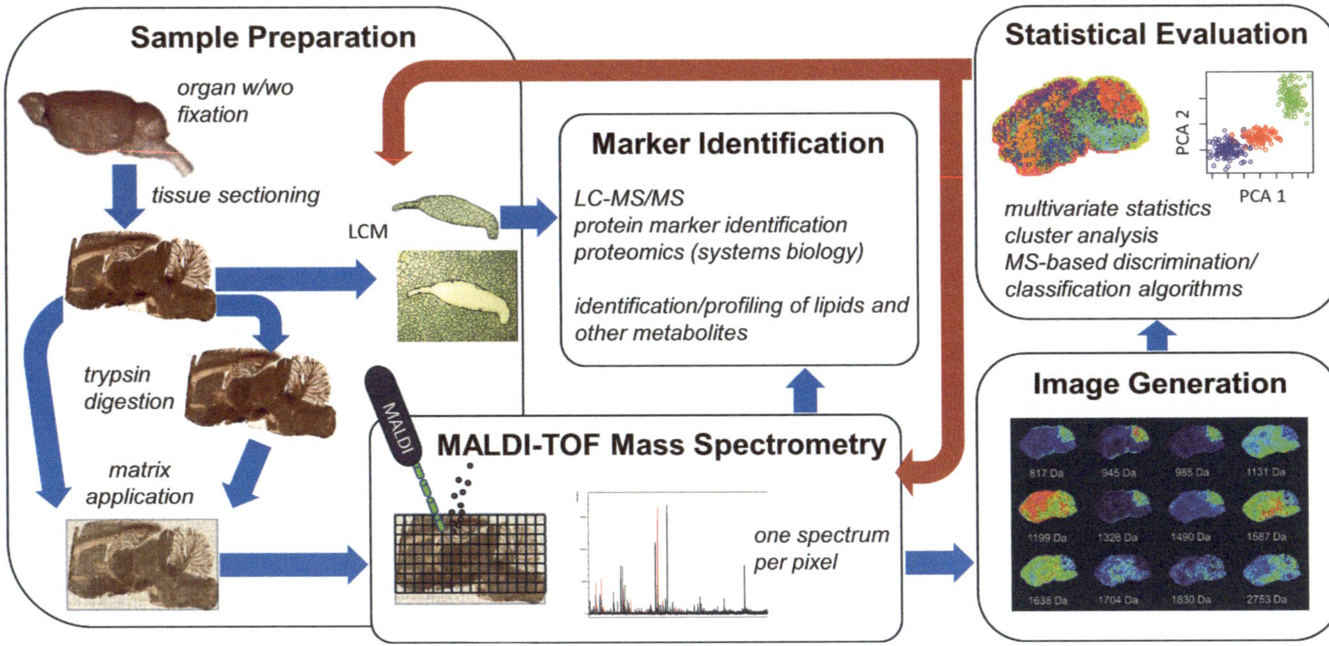

Fig. 1 The basic MSI workflow. This schematic depicts a common MALDI MSI workflow (blue arrows). The description of the different MSI variants are included in the main text. Starting material are tissue sections from an animal (or human) organ, which may or may not be formalin-fixed. Native material can be optionally digested with a protease to allow the measurement of peptides. In the next steps, an appropriate matrix layer is deposited and the spectra are generated pixel by pixel. Image generation can either be performed on the basis of the intensity distributions of individual masses or using statistical approaches like principal component analysis (PCA) to discriminate certain regions of interest. The information gained by the statistical evaluation can be fed back to the histopathology laboratory to prepare tissue regions (preferentially by Laser capture microdissection, LCM) which are characterized by the presence of certain marker masses or mass signatures. Finally, marker biomolecules can be identified by LC-MS/MS analysis of LCM samples or by MALDI-TOF MS directly from the regions of interest of a tissue section. Information feedback loops are highlighted as red arrows.

hepatocellular carcinoma (Marquardt et al., 2015) four masses were used for model generation while 46 peptide masses, partially of known origin, were used to generate a model for the discrimination of pancreatic ductal adenocarcinoma (Casadonte et al., 2019). In contrast to conventional *in situ* imaging tools used for histopathology, MSI is an untargeted approach that can be implemented if targeted tools like specific antibodies or hybridization probes are not available. In addition, MSI does not require labeling, as the ionization and MS measurement of biomolecules is straightforward. Finally, MSI is highly multiplexed and the so-gained information commonly exceeds that of conventional histologic staining procedures. These untargeted "open view" features, the unbiased multiplexed characterization and classification of certain areas within tissue sections, and the correlation of biomolecular characteristics with histopathologic structures make MSI an excellent tool for biomarker discovery, especially in cancer research (Arentz et al., 2017). Identification of specific biomarkers (proteins or other biomolecules) is a goal in many MSI studies and can be achieved if sequence data is available. Once the molecules behind the discriminating masses have been identified, a less complex technique than MSI can be used for routine diagnostics. Indeed, MSI has been successfully applied to identify tumor markers for a number of cancers and can be used for tumor diagnosis (Casadonte et al., 2017; Holzlechner et al., 2020; Kriegsmann et al., 2015; Quanico et al., 2017b). In addition to diagnostics, studies on prostate cancer have shown that MSI markers can also differentiate benign and malignant forms and indicate how aggressive the cancer is and how likely metastasis will occur (Cazares et al., 2009; Kurreck et al., 2018; Steurer et al., 2013), highlighting the potential of MSI as an innovative histopathological tool. For diagnostic applications large numbers of samples can be screened by MSI using tissue arrays (Casadonte et al., 2017). MSI-based methods recently caught attention in virus research, but have not reached their full potential yet. Our literature review summarizes the main applications of MSI in virology. It introduces the most widely used MSI technique in virology, matrix-assisted laser desorption/ionization (MALDI) MSI and discusses how MSI can be used to investigate virus-host interactions, antiviral drug kinetics, and establish novel diagnostic tools for viral diseases.

2. MALDI imaging

2.1 State of the art

The MSI workflow has been comprehensively reviewed (Aichler and Walch, 2015; Caprioli, 2019; Spengler, 2015). Therefore, only a short description of the most frequently used MSI technique, MALDI MSI, will be provided to outline the principles and some of the pitfalls of this MSI workflow. Typically, tissue sections are prepared from cryo-conserved organs and overlaid with a MALDI matrix using commercially available spray devices, inkjet printing (Baluya et al., 2007), airbrush (Gemperline et al., 2014), sublimation (Fernandez et al., 2019; Hankin et al., 2007), matrix sheet transfer (Nakashima et al., 2019), immersion (Schafermann et al., 2020) or other matrix deposition techniques. After drying, the samples are inserted into a MALDI mass spectrometer, most frequently equipped with a time-of-flight (TOF) analyzer, and spectra are registered from a grid pattern within the area of interest (Fig. 1). Commercially available instruments allow a spatial resolution (lateral distance between individual mass measurements) of 10 μm, while sophisticated experimental instrumentation allows resolutions down to 1 μm (Caprioli, 2019; Zavalin et al., 2015). The spectra are subsequently processed by appropriate softwares to produce molecular images. The simplest way to generate images is to select a certain single mass, representing a chemical compound like a protein, peptide, metabolite, lipid, glycan or drug, and to color-code the tissue based on the respective intensity distribution and the position on the analyzed tissue section (Fig. 2) (Norris and Caprioli, 2013).

More elaborate image generation is based on statistical evaluation of the spectra. In this case, colors represent mass signatures, which describe tissue areas with common mass-spectrometric features that have been revealed by cluster analysis, principal component analysis or other statistical procedures (Deininger et al., 2010). This basic workflow allows (and requires) many adjustments and variations, depending on the specific organs or tissues analyzed, the (bio)-molecules of interest and the intended use of the generated data. The flexibility of the workflow is tremendous and only few of the possible variations will be mentioned in the following. Instead of tissue sections, cultured cells (Lima et al., 2018) or cytospin preparations (Schwamborn et al., 2011) can also be used. The first step of sample preparation strongly depends on the raw material and the targeted molecules. Imaging of lipids requires washing and fixation steps omitting organic solvents that would remove them from the sample. For formalin-fixed paraffin embedded (FFPE) material, dedicated procedures have been devised to reverse fixation and facilitate the release of peptides or proteins from the tissue (Judd et al., 2019). To generate peptide-based images, tissue sections are *in situ* digested before application of the matrix, usually by spray-application of trypsin and incubation at 37 °C (Hermann et al., 2020; Zubair et al., 2016). N-linked

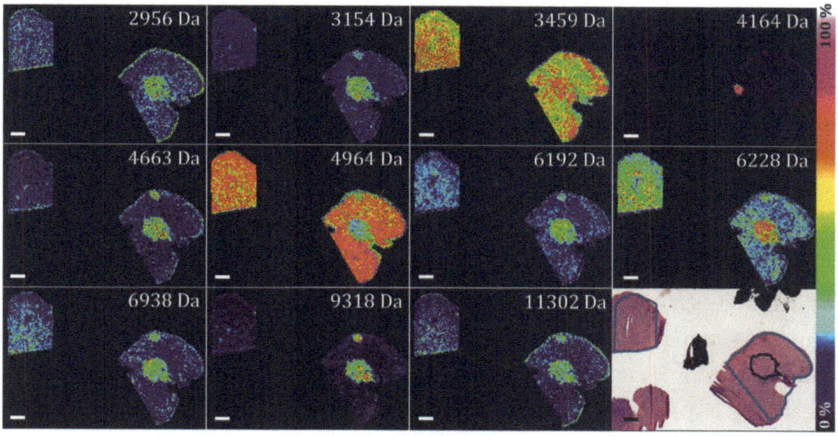

Fig. 2 Mass spectrometric imaging (MSI) analysis of Marek's disease virus (MDV)-induced lymphomas. Matrix-assisted laser desorption/ionization MSI-based protein analysis of an MDV-induced liver tumor with respective represented masses in each panel. The MSI protein scan was performed on cryosections with a mass-to-charge ratio (*m/z*) range of 2–12 kilodaltons (kDa). While 3459 Da and 4964 Da represent markers for liver tissue, especially 4663 Da and 9318 Da are highly specific masses for MDV-induced tumors (Pauker et al., 2019). The bottom right panel shows a hematoxylin and eosin-stained section of the same sample with MS-imaged regions defined by the blue borders and the tumor region by the black borders. Mass intensities are rainbow color coded from 0% (black) to 100% (white) and the colorbar on the right side indicates the relative intensity. Scale bars correspond to 1 mm.

glycans can be visualized after release from the protein by peptide-N-glycosidase F (PNGase F) treatment (Drake et al., 2018a; Powers et al., 2013). Notably, different MALDI matrices can be chosen to favor the ionization of the targeted analytes.

While the versatility of the MSI techniques mentioned above may lead to the impression that MSI is a perfect complementing and easy-to-use tool for every pathology laboratory, there are a number of limitations and pitfalls that should be mentioned. Some of them are linked to the MALDI ionization process and the design of MALDI-TOF mass spectrometers, others come with the acquisition of spectra from tissue samples. Generally, analytes have to be ionized and transferred to the gas phase to be amenable to MS. While very efficient matrices for MALDI are available, the number of peaks in a single spectrum is still limited by the ionization capacity of the MALDI process under the experimental conditions. For this reason, most biomolecules present in a tissue are still not visible in MALDI MSI. The selection of ions amenable for MALDI MSI is highly biased and for a given sample there is only limited control over the ionization process, for example by the choice of a certain matrix. Furthermore, the MALDI process requires a close interaction of the analytes with the matrix molecules which are deposited on the surface of tissue sections of a few micrometers thickness. To achieve high sensitivity, the analytes have to be extracted into the matrix layer which may be a limiting step for larger molecules like intact proteins. If peptide spectra are acquired, the choice of appropriate digestion conditions may be difficult. While the reproducible deposition of matrices to tissue sections has much improved recently by the availability of commercial devices, it has been shown that the tissue type itself can influence the morphology of the matrix crystals (Wiegelmann et al., 2016), introducing a potential cause for artifacts. It should also be mentioned that it may be hard to find a combination of experimental parameters which performs well for the given organ tissue, the analytes to be measured, and the intended purpose of the study. For example, there is a trade-off between sensitivity and lateral resolution which depends on the size of the laser focus and the amount of oversampling that is allowed (Maimo-Barcelo et al., 2019; Wiegelmann et al., 2016). Finally, as most MALDI matrices are small organic compounds themselves, there may be interferences with the measurement of small analyte molecules.

Different algorithms can be applied to generate MS images and to classify tissue areas based on their biomolecular features that are reflected in the spectra. For supervised clustering strategies, "training" areas of known composition are defined and statistical models are calculated for the classification of unknown tissue areas. Alternatively, unsupervised clustering strategies can be applied as well, and machine learning algorithms have been implemented (Zhang and Liu, 2018), as reviewed recently (Verbeeck et al., 2019).

Importantly, no sequence data of the sampled species is required to produce informative MS images. MSI is an untargeted and multiplexed analysis and MS images are constructed from the spatial distribution of single masses or from mass patterns, irrespective of their chemical nature. However, it is possible to identify proteins directly from the tissue surface by application of tandem mass spectrometry (MS/MS) to peptide peaks (Huber et al., 2018; Ryan et al., 2019) or after preparation of tissue areas with distinctive MSI features by laser capture microdissection (LCM) and subsequent conventional proteome analysis (Dewez et al., 2019; Pauker et al., 2019). Similarly, if small molecules like lipids are targeted, a register of molecular weights of the metabolites can aid to identify the molecules behind the masses, but small molecule identification is not strictly required to generate an image. On the other hand, MSI can be combined with the specificity of antibody-based techniques for targeted analysis using the "Tag-Mass" approach. Here, antibodies used to probe the tissues are conjugated with photocleavable mass tags, which produce specific mass signals upon activation with the laser during the MALDI process (Lemaire et al., 2007; Stauber et al., 2010; Thiery-Lavenant et al., 2013). In a similar setup, it was possible to image the distribution of specific transcripts after *in situ* hybridization using RNA probes with a photocleavable mass-tag (Lemaire et al., 2007). Beyond that, group-specific chemical modifications can be successfully used to preferentially target certain molecules for MSI. For example, modification of testosterone and 5α-dihydrotestosterone by the carbonyl reactive Girard's T reagent allowed the detection of these otherwise poorly ionizable compounds in mouse tissue (Cobice et al., 2016). Finally, benzophenone has been introduced as a reactive MALDI matrix, not only to ionize the sample but also to use its specific reaction with phospholipids to facilitate their identification (Waldchen et al., 2019). For some lipids, drugs, and secondary plant metabolites, MSI techniques based on fragment spectra have been described which also introduce a higher level of specificity into MSI applications (Garate et al., 2020; Li et al., 2014; Prentice et al., 2017, 2019; Takeo et al., 2019).

While MALDI MSI is the most frequently used technique, there are other MS technologies allowing MSI of tissue sections (Watrous and Dorrestein, 2011). They all have their individual special features with respect to spatial resolution, the molecules amenable to analysis, the option for MS/MS analysis and other parameters (Table 1). Just to mention a few, laser desorption ionization (LDI) is related to MALDI, but has the advantage that it does not require the application of a matrix. Secondary ion MS (SIMS) imaging allows much higher spatial resolutions in comparison to MALDI MSI (Gilmore et al., 2019). Laser ablation inductively coupled plasma (LA-ICP) MSI allows imaging of trace elements and isotopes (Becker et al., 2014). Desorption electrospray ionization (DESI) MSI facilitates scanning of a wide range of surfaces and is one of the most appropriate MSI technique for clinical or field applications (Eberlin et al., 2011; Parrot et al., 2018). Self-assembled monolayer desorption/ionization (SAMDI) MSI requires little to no sample preparation (Gurard-Levin et al., 2011). Nanoscale secondary ion mass spectrometry (nanoSIMS) allows acquisition with remarkably high resolutions and high sensitivity (He et al., 2017; Pasulka et al., 2018). Recent reviews provide detailed descriptions of these MSI techniques (Buchberger et al., 2018; Swales et al., 2019). Combinations

Table 1 Common mass spectrometry imaging techniques—strengths and limitations.

MSI technique	Acronym	Strengths	Limitations
Desorption electrospray ionization MSI	DESI MSI	No sample preparation, appropriate for clinical or field application	Low spatial resolution, detection of peptides and proteins
Infrared matrix-assisted laser desorption electrospray ionization imaging	IR-MALDESI	High salt content tolerance, ice crystals matrix reduces matrix artifacts	Frozen samples
Laser ablation electrospray ionization MS	LAESI IMS	No sample preparation	Low spatial resolution
Laser ablation inductively coupled plasma MSI	LA-ICP MSI	Determination of trace elements and isotopes	Imaging of only one ion at the time
Laser desorption ionization MSI	LDI MSI	No matrix required, high sample tolerance	No quantification
MALDI-Fourier transform ion cyclotron resonance MSI	MALDI-FTICR MSI	High mass accuracy	Matrix application, low processing of samples
Matrix-assisted laser desorption/ionization mass spectrometry imaging	MALDI MSI	Unlimited target size	Matrix application, matrix ion interference at low mass ranges
(nano)Secondary ion mass spectrometry imaging	(nano)SIMS	High spatial resolution	Limited to small molecules, small sampling area

of different MSI techniques in single samples are possible and lead to more complete and detailed description and localization of target biomolecules (Desbenoit et al., 2018).

2.2 MSI in virus research

While MS-based proteomics has always been a popular tool for the analysis of viral infections, publications presenting MSI data in the context of virus infections are rather scarce. Since the introduction of MSI as a tool for

"Molecular imaging of biological samples" by Caprioli and colleagues (Caprioli et al., 1997) it took several years until the first manuscript was published in the context of a viral infections (Han et al., 2011; Schwamborn et al., 2010). These studies focused on Hodgkin lymphoma and hepatocellular carcinomas (HCC), and tumor viruses remain in the focus of MSI studies of virus infections (Schwamborn and Caprioli, 2010). This may be due to the predominant use of MSI for tumor tissue imaging, which is a very successful approach if no tumor-specific markers are available yet. In contrast, it is a common strategy in virology to develop antibodies against viral proteins, if not already available, in the early stage of a study. This allows the specific detection of viral proteins in cells or tissues by immunohistochemistry and immunofluorescence microscopy (Pfankuche et al., 2018; Porta Siegel et al., 2018; Wang et al., 2012). These straightforward methods provide high-resolution images of virus infected cells and virus-induced pathological changes in tissues. Therefore, the application of MSI that provides lower resolution images, is more complicated and requires expensive MS equipment may not seem like the method of choice to study virus infections. However, MSI has been very successfully applied to study a number of viruses as summarized in the following sections. We will first focus on DNA, RNA and then reverse-transcribing viruses.

3. DNA viruses

3.1 Papillomaviridae

Members of the Papillomaviridae family infect numerous mammalian species, but also other vertebrates (Lopez-Bueno et al., 2016). While the majority of those infections are asymptomatic or cause warts or benign tumor lesions, infections with the oncogenic human papillomavirus (HPV) are an etiological factor for anogenital and oropharyngeal cancers (IARC Working Group on the Evaluation of Carcinogenic Risks to Humans, 2007; Roden and Stern, 2018). HPV infection can lead to a common sexually transmitted disease primarily in early adulthood and current HPV vaccines are successfully used to prevent cancer (de Sanjose et al., 2018). To aid in precise diagnosis and subsequent therapy decisions, a variety of screening tools including cytology, histology, PCR and antibody-based chemiluminescence detection are available (Burd, 2003). In a recent study, MSI facilitated the automated diagnosis and stratification of high-risk HPV-derived cervical carcinomas into different cervical Papanicolaou (Pap) classes using cytospin preparations of Pap smears (Schwamborn et al., 2011). This MS-based differentiation performed with high sensitivity and specificity using statistical models established with spectra of samples with defined Pap classes. Thus, MSI provides a feasible approach to screen and classify cervical cytology samples and simultaneously diagnose HPV infections routinely and in an automated manner. This elegant research is another example for the benefit provided by the combination of traditional morphology analyses and cytological evaluation with MALDI MSI-derived molecular signatures.

3.2 Herpesviridae

More and more cancers are found to originate from viral infections (Chang et al., 2017; Moore and Chang, 2010). Relevant oncogenic human herpesviruses include Epstein–Barr virus (EBV, HHV4) and Kaposi's sarcoma-associated herpesvirus (KSHV, HHV8) (Damania, 2007; Martin and Gutkind, 2008). EBV is associated with Hodgkin lymphoma and ovarian cancers (Flavell and Murray, 2000; Ingerslev et al., 2019), while KSHV causes Kaposi's sarcoma mainly in HIV-infected individuals (Mesri et al., 2010). In one of the very first MSI studies of oncogenic viruses, MALDI MSI was successfully applied to diagnose classical Hodgkin lymphoma and lymphadenitis (Schwamborn et al., 2010). In combination with classical MS analyses, this approach facilitated the identification of druggable targets and potential novel biomarkers in Hodgkin lymphoma patients (Schwamborn et al., 2010). A study on biomarker identification in ovarian cancers using MALDI imaging also indicated the involvement of EBV in disease etiology, as they affected similar cellular pathways (El Ayed et al., 2010). MSI-based investigations into KSHV pathogenesis, oncogenesis and diagnostics are missing to date.

Moreover, several authors discuss human herpesvirus 6 infections as a possible cause of myocarditis (Kuhl et al., 2005, 2015; Leveque et al., 2011; Olejniczak et al., 2020; Van Linthout and Tschope, 2018). In a recent review, van Linthout and Tschope outlined MSI-based analyses of biopsies from heart inflammations that are mainly caused by virus infections as efficient tools to tailor individualized therapies and thereby improve well-being of patients with viral-driven cardiac inflammation (Van Linthout and Tschope, 2018). They further suggested that MSI analyses might lead to the identification of new therapeutic targets and of novel myocarditis markers with diagnostic potential. However, this research is still ongoing.

In contrast to the situation in human medicine, the numbers of proteomic studies focusing on veterinary viruses is small (Ceciliani et al., 2014; Ghodasara et al., 2017; Kunec, 2013) and MSI studies are very scarce.

This is surprising as oncogenic herpesviruses as the Marek's disease virus (MDV) can play important roles for animal health (Truyen and Lochelt, 2006). MDV is a highly oncogenic virus in chickens, causes deadly T cell lymphoma and is a continuous threat to poultry farming worldwide (Bertzbach et al., 2020). In a recent proof of concept study, Pauker and colleagues used MALDI MSI to visualize MDV-induced tumors. They obtained characteristic mass profiles and precisely differentiated tumors from the surrounding tissue (Fig. 2). Additionally, they observed intratumoral microheterogeneity and differences in the mass signatures between tumor and normal tissue. Here, the margins of MDV-induced tumors form distinct clusters in contrast to microheterogeneous central tumor regions. While this phenomenon has been described in the context of human cancers, an unbiased MALDI MSI analysis provided important insights into the intratumor heterogeneity in a veterinary context. Moreover, the combination with noncontact laser capture microdissection followed by liquid chromatography-MALDI TOF/TOF MS identified highly discriminative mass markers for MDV-induced tumors (Pauker et al., 2019). This study identified and confirmed differential expression of proteins like interferon gamma-inducible protein 30 and a 70-kDa heat shock protein family member, however functional analyses are needed to understand their roles in MDV-induced T cell lymphoma formation.

3.3 Phycodnaviridae

Phycodnaviridae are very large double-stranded DNA viruses that infect eukaryotic algae and are linked to the disintegration of algal blooms worldwide (Wilson et al., 2009). To map virus-induced changes of the metabolic landscape in algae, Schleyer and colleagues combined MALDI MSI with conventional plaque assays to "in plaque"-MSI (Schleyer et al., 2019). With a computational co-localization approach, this study revealed enhanced synthesis of previously undescribed lipid species such as (C15:0)-based lipids upon infection (Schleyer et al., 2019). This method could also be applied to other systems and pioneered MSI applications in plant virus research.

A recent study combined nanoSIMS-based MSI with fluorescence-based click chemistry (biorthogonal non-canonical amino acid tagging; BONCAT) to detect and quantify newly synthesized ecologically relevant marine viruses in algae and bacteria (Pasulka et al., 2018). Remarkably, they could resolve carbon and nitrogen enrichment on a single viral particle level, thereby expanding the knowledge on host-derived carbon and nitrogen sources in marine (phyto-) plankton (Pasulka et al., 2018).

4. RNA viruses

4.1 Flaviviridae

The family *Flaviviridae* has four genera harboring several viruses including dengue, Zika virus (ZIKV) and West Nile virus (WNV) that gained great attention lately (Daep et al., 2014; Gerold et al., 2017). Flaviviruses cause disease in a wide range of mammals including humans and the majority are arthropod-borne, with hepatitis C virus (HCV) as a notable exception (Best, 2016; Daep et al., 2014). Until now, ZIKV and HCV infections were investigated using MSI. Research on HCV and HCV-related HCC goes back to the 1980s and current treatments successfully cure HCV-induced disease (Bartenschlager et al., 2016). In contrast, ZIKV is a relatively new threat and a public health concern for pregnant women due to a high risk of microcephaly of the unborn (Wen et al., 2017).

Over the last few years, research on ZIKV expanded drastically as a response to the emergence of the virus in the Asia-Pacific region and the western hemisphere (Lazear and Diamond, 2016). To unravel ZIKV-induced pathogenesis, both animal models and *in vitro* methods are used and supply fundamental data. Animal models of ZIKV infections include non-human primates, but primarily mice and other rodents (Caine et al., 2018; Dong and Liang, 2018). MSI studies of experimentally induced ZIKV encephalitis in mice uncovered infection-related dysregulation of metabolic pathways, mainly affecting the distribution of certain lipids (Lowen et al., 2018). To obtain such "in situ snapshots," Lowen and colleagues performed MALDI-Fourier transform ion cyclotron resonance (FTICR) MSI to visualize the distribution and abundances of particular long-chain fatty acids (acylcarnitines). Co-localization of these lipids with ZIKV RNA in certain brain areas of infected mice hint toward their value as lipid biomarkers (Lowen et al., 2018). Moreover, these changes may also be a key to the use ZIKV for oncolytic virotherapy, which is a promising approach to achieve tumor cell clearance (Zheng et al., 2019). ZIKV-infected cancer cells were shown to contain endogenous digoxin, a cardiac glycoside with anticancer activity (Platz et al., 2011). Measurements of digoxin levels in ZIKV-infected cancer cells by MALDI MSI revealed the accumulation in metabolic hot spots (Lima et al., 2018). Based on their findings and recent advantages in oncolytic virus therapy (Zhu et al., 2017) it is worth exploring the application of genetically engineered ZIKV in oncolytic therapy.

Finally, a sophisticated LDI-MSI-based diagnostic test allowed the detection of ZIKV using metal nanoparticles that bind to a panel of RNA viruses (Chu et al., 2019). This method could potentially be employed as a high-throughput screening tool for virus infections.

MALDI MSI was also applied to investigate chronic HCV infections that can result in severe liver damage. In a study using frozen HCC-positive liver tissue samples, MALDI MSI analysis allowed the distinction of liver tumors of different etiology from liver cirrhosis. It also identified a set of HCC specific markers including the highly overexpressed monomeric ubiquitin (Le Faouder et al., 2011). In addition, another study demonstrated the feasibility to assess the lipid distribution in HCV-induced HCC. In this case, much higher levels of phosphatidylcholines could be detected in HCV and non-HCV liver tumors compared to healthy liver tissue by MALDI MSI, suggesting phosphatidylcholine synthesis and metabolism as potential pathways that could be targeted to inhibit tumor progression (Morita et al., 2013).

4.2 Togaviridae

Most togavirus infections are mosquito-borne and prominent members of this family are chikungunya virus and several equine encephalitis viruses (Contigiani and Diaz, 2017). A proof of concept study using Venezuelan equine encephalitis virus confirmed that heat fixation inactivated these pathogens and was compatible with downstream MALDI MSI (Cazares et al., 2015). This indicates that MSI is a safe tool to analyze tissues from infected humans or animals under lower biosafety conditions after inactivation.

4.3 Picornaviridae

Members of the *Picornaviridae* family include human entero- and rhinoviruses, coxsackievirus, hepatitis A virus and some important veterinary pathogens such as foot-and-mouth disease virus (Jubelt and Lipton, 2014). They are non-enveloped positive-strand RNA viruses with a relatively small genome of about 7500 nucleotides in length (Bergelson and Coyne, 2013). MSI applications in *Picornaviridae* research mainly focus on virus-induced myocarditis studies in humans and respective animal models. A recent study in mice with acute coxsackievirus-induced myocarditis revealed distinct molecular profiles in infected and control mice, suggesting a disease-induced remodeling of the heart metabolism (Remels et al., 2018; Van Linthout and Tschope, 2018). MALDI MSI could visualize the spatial distribution and abundance of metabolites such as high-energy phosphate and nicotinamide adenine dinucleotide (NAD) in tissue sections and could distinguish inflammatory hot spots from non-leukocyte infiltrated areas in their samples.

Furthermore, a diagnostic test based on LDI-MSI was developed that allows specific detection of the human enterovirus 71 in the presence of other viruses (Chu et al., 2019).

4.4 Alphaflexiviridae

In a recent study on the maize Foxtail mosaic virus, MALDI MSI combined with Raman imaging revealed differences in the accumulation of plant carotenoids. In their proof of concept study, the authors used virus-induced gene silencing, a widely applied functional genomics method in plant biology (Lu et al., 2003), to show that this biochemical heterogeneity occurred with a mosaic spatial pattern that is indicative of virus spread within the plant (Burkhow et al., 2018).

5. Reverse transcribing viruses

5.1 Retroviridae

Research in retroviruses and antiretroviral drugs is a large field that is steadily expanding (Broder, 2010). Investigations into antiviral medication addressing pharmacokinetics and pharmacodynamics led to a range of approved and successful therapies. MSI based approaches provided important information on drug distributions and distribution kinetics. For instance, infrared matrix-assisted laser desorption electrospray ionization imaging (IR-MALDESI) facilitated a precise analysis of antiretroviral penetration in HIV positive lymphoid tissues (Barry et al., 2014), and monitoring of antiretroviral drugs in general (Dekker et al., 2009; Thompson et al., 2015, 2019). MALDI MSI provided regional spatial distribution of antiretroviral medication in brain tissue sections of rats (Mdanda et al., 2019; Ntshangase et al., 2019a,b) and in colorectal tissue samples of HIV infected patients (Seneviratne et al., 2018). In a recent study, Kevadiya and colleagues employed LA-ICP-MSI to reveal pharmacokinetics and distribution of antiretrovirals (Kevadiya et al., 2020). They used ^{177}lutetium labeled rod-shaped bismuth sulfide nanoparticles ("nanorods") as nano-theranostic probes to measure distribution of the non-nucleoside reverse transcriptase inhibitor rilpivirine (Kevadiya et al., 2020). Collectively, these studies highlight the importance and practicability of MSI-based analyses in antiretroviral drug research. In addition, MALDI MSI was used to investigate liver damages as adverse reaction of antiretroviral drug medication. Antiretroviral treatment with the HIV polymerase inhibitor nevirapine was shown to exacerbate liver fibrosis in an animal model (Brown et al., 2016).

Retroviral long terminal repeat (LTR) can initiate transcripts that increase the expression of transgenes in a tissue-specific manner (Thompson et al., 2016). In mouse models of breast cancer, the LTR of mouse mammary tumor virus is used to drive the (proto)-oncogenes polyoma middle T antigen (MMTV-PyMT) or the erb-b2 receptor tyrosine kinase 2 (MMTV-HER2) in transgenic mice (Fantozzi and Christofori, 2006; Guy et al., 1992; Muller et al., 1988). The MMTV promotor ensures mammary gland-specific expression of the (proto)-oncogenes, which results in tumor formation, progression and metastasis (Fantozzi and Christofori, 2006). Gao and colleagues recently analyzed metabolite utilization of primary breast cancer cells derived from MMTV-PyMT transgenic mice and a human breast cancer cell line using MALDI-FTICR MSI. They revealed that interactions of the CCL5 chemokine with the CCR5 receptor modulated anabolic processes in tumor cells that potentially drive oncogenesis (Gao et al., 2017). Furthermore, MMTV/HER2 transgenic mice were also used to detect molecular fingerprints that are predictive of therapeutic antitumor response (Reyzer et al., 2004). MALDI MSI revealed co-localization of anticancer drugs that correlated with proteomic markers of cellular responses in MMTV/HER2-induced mammary tumors (Reyzer et al., 2004).

5.2 Hepadnaviridae

Hepatitis B virus (HBV) belongs to the *Hepadnaviridae* that harbor a small, partially double-stranded DNA genome. The HBV polymerase is unique among viral polymerases: it possesses reverse transcriptase activity, RNAse activity and DNA polymerase activity to replicate the virus genome (Ganem and Schneider, 2001). HBV infections are closely associated to hepatitis, liver cirrhosis and HCC in chronically infected individuals (Karayiannis, 2017). Aberrant lipid metabolism is known to mediate liver inflammation (Moustafa et al., 2012) and viral hepatitis is associated with alterations of hepatic lipid composition (Li et al., 2015; Syed et al., 2010). In contrast to findings from human patients (Morita et al., 2013), Park and colleagues reported a dysregulation of phosphatidylcholine biosynthesis with decreased levels of several phosphatidylcholines in liver samples of HBV-infected mice (Park et al., 2014). This research provides valuable insights into HBV-mediated liver disease. Furthermore, MALDI MSI analyses of HBV-related liver tumors at different stages of tumor progression allowed classification of various tumor regions. This served as a histopathological analysis tool to elucidate the tumor microenvironments and the border regions between tumor and normal tissue (Han et al., 2011).

Finally, Le Faouder and colleagues used MALDI MSI to assess HBV-induced HCC and established biomarkers (Le Faouder et al., 2011). Notably, some of the *m/z* peaks could be confirmed in a more recent study aiming at developing advanced diagnostics for HCC (Marquardt et al., 2015).

6. Conclusions

MSI technologies provide molecular information on a large variety of analytes including their spatial distribution. Therefore, this label-free imaging approach is a promising option for the analyses of tissues and cells in biomedical research. Overall, MSI is mostly used in cancer research and consequentially for oncogenic viruses. Different MSI-based techniques, alone or in combination with various other proteomic and imaging approaches, were successfully applied to study virus infections, virus-induced tumors, and antiviral compounds. However, these were largely used in proof of-principle studies so far. The wide range of options for MSI applications in virology are just starting to gain traction and have immense potential (Table 2).

To date, antibody-based approaches, click chemistry and genetically modified viruses encoding fluorescent proteins dominate imaging techniques used in virus research. This is mainly due to their ease of use and their quick implementations. Moreover, they still allow visualization with highest resolutions (Müller et al., 2019; Witte et al., 2018). Importantly, in contrast to these standard techniques, MSI is an "open view" tool that is neither restricted to one or more defined analytes nor limited by the availability of antibodies, fluorescent chromophores or nucleic acid probes. Furthermore, MSI techniques are highly versatile and specific at the same time, as numerous masses can be simultaneously detected and discriminated. The possibility to use FFPE or heat inactivated samples for MSI is valuable

Table 2 Applications of mass spectrometry imaging in virus research.

Virus	Tumor	Main objective	Technique	Reference
CoxV	n.a.	Regulation of metabolites	MALDI MSI (cryo)	Remels et al. (2018)
E. huxleyi virus	n.a.	Metabolic cross-talk between host and virus	MALDI MSI and flow-probe *in situ* microextraction MSI	Schleyer et al. (2019)
E. huxleyi virus i.a.	n.a.	Carbon and hydrogen metabolism	nanoSIMS	Pasulka et al. (2018)
EBV	Hodgkin lymphoma	Diagnostic tool, biomarker identification	MALDI MSI (cryo)	Schwamborn et al. (2010)
FoMV	n.a.	Assessment of gene silencing consequences	MALDI MSI (cryo)	Burkhow et al. (2018)
HBV	HCC	Lipid analysis	MALDI MSI (cryo)	Park et al. (2014)
HBV	HCC	Diagnostic tool	MALDI MSI (cryo)	Han et al. (2011)
HBV	HCC	Diagnostic tool, biomarker identification	MALDI MSI (cryo)	Le Faouder et al. (2011)
HCV	HCC	Diagnostic tool, biomarker identification	MALDI MSI (cryo)	Le Faouder et al. (2011)
HCV	HCC	Lipid analysis	MALDI MSI (cryo)	Morita et al. (2013)
HIV	n.a.	Imaging of pharmacological compounds in tissue sections	IR-MALDESI MSI (cryo)	Barry et al. (2014)
HIV	n.a.	*In vitro* imaging of pharmacological compounds	MALDI MSI (cytospin)	Dekker et al. (2009)

Continued

Table 2 Applications of mass spectrometry imaging in virus research.—cont'd

Virus	Tumor	Main objective	Technique	Reference
HIV	n.a.	Imaging of pharmacological compounds in tissue sections	MALDI MSI (cryo)	Mdanda et al. (2019); Ntshangase et al. (2019a); Ntshangase et al. (2019b); Thompson et al. (2015); Thompson et al. (2019)
HIV	n.a.	Imaging of pharmacological compounds in tissue sections	MALDI MSI (OCT)	Seneviratne et al. (2018)
HIV	n.a.	Imaging of pharmacological compounds in tissue sections	LA-ICP MS (cryo)	Kevadiya et al. (2020)
HIV	n.a.	Assessment of unwanted side effects of antiretroviral treatment	MALDI MSI (cryo)	Brown et al. (2016)
HPV	Cervical carcinoma	Diagnostic tool	MALDI MSI (cytospin)	Schwamborn et al. (2011)
MDV	T cell lymphoma	Diagnostic tool, biomarker identification	MALDI MSI (FFPE)	Pauker et al. (2019)
MMTV	Breast cancer	Consequences of chemokine—receptor interactions	MALDI-FTICR MSI (cryo)	Gao et al. (2017)
MMTV	Breast cancer	Tumor response to molecular therapeutics	MALDI MSI (cryo)	Reyzer et al. (2004)
Parvovirus B19	n.a.	*In situ* tissue typing of heart biopsies	MALDI MSI (FFPE)	Van Linthout et al. (2018)

Table 2 Applications of mass spectrometry imaging in virus research.—cont'd

Virus	Tumor	Main objective	Technique	Reference
VEEV	n.a.	Inactivation of viruses and bacteria by MSI-compatible heat fixation	MALDI MSI (heat fixation)	Cazares et al. (2015)
ZIKV	n.a.	Lipid and small-molecule metabolite analysis	MALDI-FTICR MSI (FFPE)	Lowen et al. (2018)
ZIKV	n.a.	Induction of endogenous digoxin synthesis in glioblastoma cells. ZIKV for oncolytic treatment?	MALDI MSI (cultured cells)	Lima et al. (2018)

CoxV, Coxsackievirus; cryo, snap-frozen and cryomicrotome cut; EBV, Epstein Barr virus; FFPE, formalin-fixed paraffin-embedded; FoMV, foxtail mosaic virus; HBV, hepatitis B virus; HCC, hepatocellular carcinoma; HCV, hepatitis C virus; HIV, human immunodeficiency virus; HPV, human papillomavirus; MDV, Marek's disease virus; MMTV, mouse mammary tumor virus; OCT, optimal cutting temperature compound; VEEV, Venezuelan equine encephalitis virus; ZIKV, Zika virus.

with respect to biosafety. The relatively low spatial resolution of MSI images can be improved by recently established alternative MSI techniques like (nano)SIMS, or in combination with light or fluorescence microscopy and other microscopy techniques (Ahlf et al., 2014; Burkhow et al., 2018; Fuchs et al., 2018; Lin et al., 2020; Patterson et al., 2018). Antibody-based analyses can be combined with MSI (recently introduced as imaging mass cytometry (IMC)) to gain deeper insights into relevant processes in (patho)-physiology (Giesen et al., 2014; Wang et al., 2013). Another striking advantage of MSI is that it can be coupled with LCM and classical MS-based proteome analyses downstream (Pauker et al., 2019). In addition, applications of single cell MSI (Duenas et al., 2017; Gilmore et al., 2019; Yang et al., 2020), integrating the visualization of host cell reactions *in* and *ex vivo*, and the development of MSI techniques for 3D imaging of tissues have been developed (Porta Siegel et al., 2018; Seeley and Caprioli, 2012). Moreover, MSI-based lipid analyses are very informative and should be expanded to other viral infections. The studies on flavivirus infections highlight that this approach can provide valuable information for novel antiviral treatment options (Martin-Acebes et al., 2019;

Zhang et al., 2017). Future studies could also include glycan analyses (viral antigens) and other post-translational modifications during viral infections (Angel et al., 2018; Drake et al., 2018b; Mnatsakanyan et al., 2018).

Constantly developing avant-garde methodology will allow further in-depth MSI analyses, like chemometric analysis methods (Jaumot and Bedia, 2020; Mas et al., 2020; Mohammadi and Parastar, 2018), integration of multiomics workflows (Kappler and Lehmann, 2019; Quanico et al., 2017a) and assessments of heterogeneous tissue microenvironment that can even be expanded to include the extracellular matrix (Angel et al., 2019).

Taken together, improvements of sample preparation protocols and rapid advances in instrument technology and bioinformatics will provide ever more diverse MSI applications and MSI-based workflows designed for rapid and reliable biomarker discovery. Considering the extension of MSI to diagnostic and clinical applications, the application of a variety of MS techniques for MSI, and the combination of MSI with other imaging techniques, MSI-based research projects become increasingly multidisciplinary. This literature review highlights the great value of interdisciplinary studies combining clinical and wet lab virology with proteomics, image analysis and bioinformatics. This will not only provide valuable diagnostic tools but also further unravel virus-host interactions in humans, animals, plants and bacteria.

Acknowledgments

We cordially thank J.-P. Teifke and A. Hohmann (FLI, Insel Riems) for pathology and histology expertise. This work was funded by the Deutsche Forschungsgemeinschaft grants KA 3492/3-1 and KA 2900/3-1 awarded to B.B. Kaufer and A. Karger respectively, and the Volkswagen Foundation Lichtenberg grant A112662 awarded to B.B. Kaufer.

References

Ahlf, D.R., Masyuko, R.N., Hummon, A.B., Bohn, P.W., 2014. Correlated mass spectrometry imaging and confocal Raman microscopy for studies of three-dimensional cell culture sections. Analyst 139, 4578–4585.

Aichler, M., Walch, A., 2015. MALDI imaging mass spectrometry: current frontiers and perspectives in pathology research and practice. Lab. Invest. 95, 422–431.

Alejo, A., Matamoros, T., Guerra, M., Andres, G., 2018. A proteomic atlas of the African swine fever virus particle. J. Virol. 92. https://doi.org/10.1128/JVI.01293-18. e01293-18.

Angel, P.M., Mehta, A., Norris-Caneda, K., Drake, R.R., 2018. MALDI imaging mass spectrometry of N-glycans and tryptic peptides from the same formalin-fixed, paraffin-embedded tissue section. Methods Mol. Biol. 1788, 225–241.

Angel, P.M., Schwamborn, K., Comte-Walters, S., Clift, C.L., Ball, L.E., Mehta, A.S., Drake, R.R., 2019. Extracellular matrix imaging of breast tissue pathologies by MALDI-imaging mass spectrometry. Proteomics Clin. Appl. 13, e1700152.

Arentz, G., Mittal, P., Zhang, C., Ho, Y.Y., Briggs, M., Winderbaum, L., Hoffmann, M.K., Hoffmann, P., 2017. Applications of mass spectrometry imaging to cancer. Adv. Cancer Res. 134, 27–66.

Baluya, D.L., Garrett, T.J., Yost, R.A., 2007. Automated MALDI matrix deposition method with inkjet printing for imaging mass spectrometry. Anal. Chem. 79, 6862–6867.

Barry, J.A., Robichaud, G., Bokhart, M.T., Thompson, C., Sykes, C., Kashuba, A.D., Muddiman, D.C., 2014. Mapping antiretroviral drugs in tissue by IR-MALDESI MSI coupled to the Q exactive and comparison with LC-MS/MS SRM assay. J. Am. Soc. Mass Spectrom. 25, 2038–2047.

Bartenschlager, R.F., Rice, C.M., Sofia, M.J., 2016. Hepatitis C virus-from discovery to cure: the 2016 Lasker-DeBakey clinical medical research award. JAMA 316, 1254–1255.

Batra, J., Hultquist, J.F., Liu, D., Shtanko, O., Von Dollen, J., Satkamp, L., Jang, G.M., Luthra, P., Schwarz, T.M., Small, G.I., Arnett, E., Anantpadma, M., Reyes, A., Leung, D.W., Kaake, R., Haas, P., Schmidt, C.B., Schlesinger, L.S., LaCount, D.J., Davey, R.A., Amarasinghe, G.K., Basler, C.F., Krogan, N.J., 2018. Protein interaction mapping identifies RBBP6 as a negative regulator of Ebola virus replication. Cell 175, 1917–1930 e1913.

Becker, J.S., Matusch, A., Wu, B., 2014. Bioimaging mass spectrometry of trace elements—recent advance and applications of LA-ICP-MS: a review. Anal. Chim. Acta 835, 1–18.

Bergelson, J.M., Coyne, C.B., 2013. Picornavirus entry. Adv. Exp. Med. Biol. 790, 24–41.

Bertzbach, L.D., Conradie, A.M., You, Y., Kaufer, B.B., 2020. Latest insights into Marek's disease virus pathogenesis and tumorigenesis. Cancer 12, 647.

Best, S.M., 2016. Flaviviruses. Curr. Biol. 26, R1258–R1260.

Broder, S., 2010. The development of antiretroviral therapy and its impact on the HIV-1/AIDS pandemic. Antiviral Res. 85, 1–18.

Brown, H.R., Castellino, S., Groseclose, M.R., Elangbam, C.S., Mellon-Kusibab, K., Yoon, L.W., Gates, L.D., Krull, D.L., Cariello, N.F., Arrington-Brown, L., Tillman, T., Fowler, S., Shah, V., Bailey, D., Miller, R.T., 2016. Drug-induced liver fibrosis: testing Nevirapine in a viral-like liver setting using histopathology, MALDI IMS, and gene expression. Toxicol. Pathol. 44, 112–131.

Buchberger, A.R., DeLaney, K., Johnson, J., Li, L., 2018. Mass spectrometry imaging: a review of emerging advancements and future insights. Anal. Chem. 90, 240–265.

Burd, E.M., 2003. Human papillomavirus and cervical cancer. Clin. Microbiol. Rev. 16, 1–17.

Burkhow, S.J., Stephens, N.M., Mei, Y., Duenas, M.E., Freppon, D.J., Ding, G., Smith, S.C., Lee, Y.J., Nikolau, B.J., Whitham, S.A., Smith, E.A., 2018. Characterizing virus-induced gene silencing at the cellular level with in situ multimodal imaging. Plant Methods 14, 37.

Caine, E.A., Jagger, B.W., Diamond, M.S., 2018. Animal models of Zika virus infection during pregnancy. Viruses 10, 598.

Caprioli, R.M., 2019. Imaging mass spectrometry: a perspective. J. Biomol. Tech. 30, 7–11.

Caprioli, R.M., Farmer, T.B., Gile, J., 1997. Molecular imaging of biological samples: localization of peptides and proteins using MALDI-TOF MS. Anal. Chem. 69, 4751–4760.

Casadonte, R., Kriegsmann, M., Perren, A., Baretton, G., Deininger, S.O., Kriegsmann, K., Welsch, T., Pilarsky, C., Kriegsmann, J., 2019. Development of a class prediction model to discriminate pancreatic ductal adenocarcinoma from pancreatic neuroendocrine tumor by MALDI mass spectrometry imaging. Proteomics Clin. Appl. 13, e1800046.

Casadonte, R., Longuespee, R., Kriegsmann, J., Kriegsmann, M., 2017. MALDI IMS and cancer tissue microarrays. Adv. Cancer Res. 134, 173–200.

Cazares, L.H., Troyer, D., Mendrinos, S., Lance, R.A., Nyalwidhe, J.O., Beydoun, H.A., Clements, M.A., Drake, R.R., Semmes, O.J., 2009. Imaging mass spectrometry of a specific fragment of mitogen-activated protein kinase/extracellular signal-regulated kinase kinase kinase 2 discriminates cancer from uninvolved prostate tissue. Clin. Cancer Res. 15, 5541–5551.
Cazares, L.H., Van Tongeren, S.A., Costantino, J., Kenny, T., Garza, N.L., Donnelly, G., Lane, D., Panchal, R.G., Bavari, S., 2015. Heat fixation inactivates viral and bacterial pathogens and is compatible with downstream MALDI mass spectrometry tissue imaging. BMC Microbiol. 15, 101.
Ceciliani, F., Eckersall, D., Burchmore, R., Lecchi, C., 2014. Proteomics in veterinary medicine: applications and trends in disease pathogenesis and diagnostics. Vet. Pathol. 51, 351–362.
Chang, Y., Moore, P.S., Weiss, R.A., 2017. Human oncogenic viruses: nature and discovery. Philos. Trans. R. Soc. Lond. B Biol. Sci. 372. 20160264. https://doi.org/10.1098/rstb.2016.0264.
Chu, H.W., Lai, C.S., Ko, J.Y., Harroun, S.G., Chuang, C.I., Wang, R.Y.L., Unnikrishnan, B., Huang, C.C., 2019. Nanoparticle-based LDI-MS immunoassay for the multiple diagnosis of viral infections. ACS Sens. 4, 1543–1551.
Cobice, D.F., Livingstone, D.E., Mackay, C.L., Goodwin, R.J., Smith, L.B., Walker, B.R., Andrew, R., 2016. Spatial localization and quantitation of androgens in mouse testis by mass spectrometry imaging. Anal. Chem. 88, 10362–10367.
Contigiani, M.S., Diaz, L.A., 2017. Togaviridae. In: Marcondes, C. (Ed.), Arthropod Borne Diseases. Springer, Cham, Switzerland, pp. 115–135.
Daep, C.A., Munoz-Jordan, J.L., Eugenin, E.A., 2014. Flaviviruses, an expanding threat in public health: focus on dengue, West Nile, and Japanese encephalitis virus. J. Neurovirol. 20, 539–560.
Damania, B., 2007. DNA tumor viruses and human cancer. Trends Microbiol. 15, 38–44.
de Sanjose, S., Brotons, M., Pavon, M.A., 2018. The natural history of human papillomavirus infection. Best Pract. Res. Clin. Obstet. Gynaecol. 47, 2–13.
Deininger, S.O., Becker, M., Suckau, D., 2010. Tutorial: multivariate statistical treatment of imaging data for clinical biomarker discovery. Methods Mol. Biol. 656, 385–403.
Dekker, L.J., van Kampen, J.J., Reedijk, M.L., Burgers, P.C., Gruters, R.A., Osterhaus, A.D., Luider, T.M., 2009. A mass spectrometry based imaging method developed for the intracellular detection of HIV protease inhibitors. Rapid Commun. Mass Spectrom. 23, 1183–1188.
Desbenoit, N., Walch, A., Spengler, B., Brunelle, A., Rompp, A., 2018. Correlative mass spectrometry imaging, applying time-of-flight secondary ion mass spectrometry and atmospheric pressure matrix-assisted laser desorption/ionization to a single tissue section. Rapid Commun. Mass Spectrom. 32, 159–166.
Dewez, F., Martin-Lorenzo, M., Herfs, M., Baiwir, D., Mazzucchelli, G., De Pauw, E., Heeren, R.M.A., Balluff, B., 2019. Precise co-registration of mass spectrometry imaging, histology, and laser microdissection-based omics. Anal. Bioanal. Chem. 411, 5647–5653.
Dong, S., Liang, Q., 2018. Recent advances in animal models of Zika virus infection. Virol. Sin. 33, 125–130.
Drake, R., Angel, P., Mehta, A., 2018a. Mass spectrometry imaging of N-glycans in cancer tissues. Abstr. Pap. Am. Chem. Soc. 256. https://doi.org/10.1016/bs.acr.2016.11.009.
Drake, R.R., Powers, T.W., Norris-Caneda, K., Mehta, A.S., Angel, P.M., 2018b. In situ imaging of N-glycans by MALDI imaging mass spectrometry of fresh or formalin-fixed paraffin-embedded tissue. Curr. Protoc. Protein Sci. 94, e68.
Duenas, M.E., Essner, J.J., Lee, Y.J., 2017. 3D MALDI mass spectrometry imaging of a single cell: spatial mapping of lipids in the embryonic development of zebrafish. Sci. Rep. 7, 14946.

Eberlin, L.S., Ferreira, C.R., Dill, A.L., Ifa, D.R., Cooks, R.G., 2011. Desorption electrospray ionization mass spectrometry for lipid characterization and biological tissue imaging. Biochim. Biophys. Acta 1811, 946–960.

El Ayed, M., Bonnel, D., Longuespee, R., Castelier, C., Franck, J., Vergara, D., Desmons, A., Tasiemski, A., Kenani, A., Vinatier, D., Day, R., Fournier, I., Salzet, M., 2010. MALDI imaging mass spectrometry in ovarian cancer for tracking, identifying, and validating biomarkers. Med. Sci. Monit. 16, BR233–245.

Fantozzi, A., Christofori, G., 2006. Mouse models of breast cancer metastasis. Breast Cancer Res. 8, 212.

Fernandes, N.D., Downard, K.M., 2014. Incorporation of a proteotyping approach using mass spectrometry for surveillance of influenza virus in cell-cultured strains. J. Clin. Microbiol. 52, 725–735.

Fernandez, R., Garate, J., Martin-Saiz, L., Galetich, I., Fernandez, J.A., 2019. Matrix sublimation device for MALDI mass spectrometry imaging. Anal. Chem. 91, 803–807.

Flavell, K.J., Murray, P.G., 2000. Hodgkin's disease and the Epstein-Barr virus. Mol. Pathol. 53, 262–269.

Fuchs, K., Kiss, A., Bize, P.E., Duran, R., Denys, A., Hopfgartner, G., Borchard, G., Jordan, O., 2018. Mapping of drug distribution in the rabbit liver tumor model by complementary fluorescence and mass spectrometry imaging. J. Control. Release 269, 128–135.

Ganem, D., Schneider, R.J., 2001. Hepadnaviridae: the viruses and their replication. In: Knipe, D.M., Howley, P.M. (Eds.), Fields Virology. Lippincott, Williams & Wilkins, Philadelphia, pp. 2923–2969.

Gao, D., Cazares, L.H., Fish, E.N., 2017. CCL5-CCR5 interactions modulate metabolic events during tumor onset to promote tumorigenesis. BMC Cancer 17, 834.

Garate, J., Lage, S., Martin-Saiz, L., Perez-Valle, A., Ochoa, B., Boyano, M.D., Fernandez, R., Fernandez, J.A., 2020. Influence of lipid fragmentation in the data analysis of imaging mass spectrometry experiments. J. Am. Soc. Mass Spectrom. 31, 517–526.

Gemperline, E., Rawson, S., Li, L., 2014. Optimization and comparison of multiple MALDI matrix application methods for small molecule mass spectrometric imaging. Anal. Chem. 86, 10030–10035.

Gerold, G., Bruening, J., Weigel, B., Pietschmann, T., 2017. Protein interactions during the Flavivirus and Hepacivirus life cycle. Mol. Cell. Proteomics 16, S75–S91.

Ghodasara, P., Sadowski, P., Satake, N., Kopp, S., Mills, P.C., 2017. Clinical veterinary proteomics: techniques and approaches to decipher the animal plasma proteome. Vet. J. 230, 6–12.

Giesen, C., Wang, H.A., Schapiro, D., Zivanovic, N., Jacobs, A., Hattendorf, B., Schuffler, P.J., Grolimund, D., Buhmann, J.M., Brandt, S., Varga, Z., Wild, P.J., Gunther, D., Bodenmiller, B., 2014. Highly multiplexed imaging of tumor tissues with subcellular resolution by mass cytometry. Nat. Methods 11, 417–422.

Gilmore, I.S., Heiles, S., Pieterse, C.L., 2019. Metabolic imaging at the single-cell scale: recent advances in mass spectrometry imaging. Annu. Rev. Anal. Chem. (Palo Alto, Calif) 12, 201–224.

Greco, T.M., Diner, B.A., Cristea, I.M., 2014. The impact of mass spectrometry-based proteomics on fundamental discoveries in virology. Annu. Rev. Virol. 1, 581–604.

Gurard-Levin, Z.A., Scholle, M.D., Eisenberg, A.H., Mrksich, M., 2011. High-throughput screening of small molecule libraries using SAMDI mass spectrometry. ACS Comb. Sci. 13, 347–350.

Guy, C.T., Cardiff, R.D., Muller, W.J., 1992. Induction of mammary tumors by expression of polyomavirus middle T oncogene: a transgenic mouse model for metastatic disease. Mol. Cell Biol. 12, 954–961.

Han, E.C., Lee, Y.S., Liao, W.S., Liu, Y.C., Liao, H.Y., Jeng, L.B., 2011. Direct tissue analysis by MALDI-TOF mass spectrometry in human hepatocellular carcinoma. Clin. Chim. Acta 412, 230–239.

Hankin, J.A., Barkley, R.M., Murphy, R.C., 2007. Sublimation as a method of matrix application for mass spectrometric imaging. J. Am. Soc. Mass Spectrom. 18, 1646–1652.

He, C., Fong, L.G., Young, S.G., Jiang, H., 2017. NanoSIMS imaging: an approach for visualizing and quantifying lipids in cells and tissues. J. Invest. Med. 65, 669–672.

Hermann, J., Noels, H., Theelen, W., Lellig, M., Orth-Alampour, S., Boor, P., Jankowski, V., Jankowski, J., 2020. Sample preparation of formalin-fixed paraffin-embedded tissue sections for MALDI-mass spectrometry imaging. Anal. Bioanal. Chem. 412, 1263–1275.

Holzlechner, M., Eugenin, E., Prideaux, B., 2020. Mass spectrometry imaging to detect lipid biomarkers and disease signatures in cancer. Cancer Rep. 2. e1229. https://doi.org/10.1002/cnr2.1229.

Huber, K., Khamehgir-Silz, P., Schramm, T., Gorshkov, V., Spengler, B., Rompp, A., 2018. Approaching cellular resolution and reliable identification in mass spectrometry imaging of tryptic peptides. Anal. Bioanal. Chem. 410, 5825–5837.

Hutchinson, E.C., Charles, P.D., Hester, S.S., Thomas, B., Trudgian, D., Martinez-Alonso, M., Fodor, E., 2014. Conserved and host-specific features of influenza virion architecture. Nat. Commun. 5, 4816.

IARC Working Group on the Evaluation of Carcinogenic Risks to Humans, 2007. IARC Monographs on the Evaluation of Carcinogenic Risks to Humans, Human Papillomaviruses. vol. 90 International Agency for Research on Cancer, Lyon, France.

Ingerslev, K., Hogdall, E., Skovrider-Ruminski, W., Schnack, T.H., Lidang, M., Hogdall, C., Blaakaer, J., 2019. The prevalence of EBV and CMV DNA in epithelial ovarian cancer. Infect. Agent Cancer 14, 7.

Jaumot, J., Bedia, C., 2020. Mass spectrometry imaging: chemometric data analysis. In: Brown, S., Tauler, R., Walczak, B. (Eds.), Reference Module in Chemistry, Molecular Sciences and Chemical Engineering. Elsevier, pp. 381–394.

Jubelt, B., Lipton, H.L., 2014. Enterovirus/picornavirus infections. In: Tselis, A.C., Booss, J. (Eds.), Handbook of Clinical Neurology. Elsevier, pp. 379–416.

Judd, A.M., Gutierrez, D.B., Moore, J.L., Patterson, N.H., Yang, J., Romer, C.E., Norris, J.L., Caprioli, R.M., 2019. A recommended and verified procedure for in situ tryptic digestion of formalin-fixed paraffin-embedded tissues for analysis by matrix-assisted laser desorption/ionization imaging mass spectrometry. J. Mass Spectrom. 54, 716–727.

Kappler, L., Lehmann, R., 2019. Mass-spectrometric multi-omics linked to function—state-of-the-art investigations of mitochondria in systems medicine. TrAC Trends Anal. Chem. 119, 115635.

Karayiannis, P., 2017. Hepatitis B virus: virology, molecular biology, life cycle and intrahepatic spread. Hepatol. Int. 11, 500–508.

Kevadiya, B.D., Ottemann, B., Mukadam, I.Z., Castellanos, L., Sikora, K., Hilaire, J.R., Machhi, J., Herskovitz, J., Soni, D., Hasan, M., Zhang, W., Anandakumar, S., Garrison, J., McMillan, J., Edagwa, B., Mosley, R.L., Vachet, R.W., Gendelman, H.E., 2020. Rod-shape theranostic nanoparticles facilitate antiretroviral drug biodistribution and activity in human immunodeficiency virus susceptible cells and tissues. Theranostics 10, 630–656.

Kriegsmann, J., Kriegsmann, M., Casadonte, R., 2015. MALDI TOF imaging mass spectrometry in clinical pathology: a valuable tool for cancer diagnostics (review). Int. J. Oncol. 46, 893–906.

Kuhl, U., Lassner, D., Wallaschek, N., Gross, U.M., Krueger, G.R.F., Seeberg, B., Kaufer, B.B., Escher, F., Poller, W., Schultheiss, H.P., 2015. Chromosomally integrated human herpesvirus 6 in heart failure: prevalence and treatment. Eur. J. Heart Fail. 17, 9–19.

Kuhl, U., Pauschinger, M., Seeberg, B., Lassner, D., Noutsias, M., Poller, W., Schultheiss, H.P., 2005. Viral persistence in the myocardium is associated with progressive cardiac dysfunction. Circulation 112, 1965–1970.

Kunec, D., 2013. Proteomics applied to avian herpesviruses. Avian Dis. 57, 351–359.

Kurreck, A., Vandergrift, L.A., Fuss, T.L., Habbel, P., Agar, N.Y.R., Cheng, L.L., 2018. Prostate cancer diagnosis and characterization with mass spectrometry imaging. Prostate Cancer Prostatic Dis. 21, 297–305.

Lasso, G., Mayer, S.V., Winkelmann, E.R., Chu, T., Elliot, O., Patino-Galindo, J.A., Park, K., Rabadan, R., Honig, B., Shapira, S.D., 2019. A structure-informed atlas of human-virus interactions. Cell 178, 1526–1541 e1516.

Law, G.L., Korth, M.J., Benecke, A.G., Katze, M.G., 2013. Systems virology: host-directed approaches to viral pathogenesis and drug targeting. Nat. Rev. Microbiol. 11, 455–466.

Lazear, H.M., Diamond, M.S., 2016. Zika virus: new clinical syndromes and its emergence in the Western hemisphere. J. Virol. 90, 4864–4875.

Le Faouder, J., Laouirem, S., Chapelle, M., Albuquerque, M., Belghiti, J., Degos, F., Paradis, V., Camadro, J.M., Bedossa, P., 2011. Imaging mass spectrometry provides fingerprints for distinguishing hepatocellular carcinoma from cirrhosis. J. Proteome Res. 10, 3755–3765.

Lemaire, R., Stauber, J., Wisztorski, M., Van Camp, C., Desmons, A., Deschamps, M., Proess, G., Rudlof, I., Woods, A.S., Day, R., Salzet, M., Fournier, I., 2007. Tag-mass: specific molecular imaging of transcriptome and proteome by mass spectrometry based on photocleavable tag. J. Proteome Res. 6, 2057–2067.

Leveque, N., Boulagnon, C., Brasselet, C., Lesaffre, F., Boutolleau, D., Metz, D., Fornes, P., Andreoletti, L., 2011. A fatal case of human herpesvirus 6 chronic myocarditis in an immunocompetent adult. J. Clin. Virol. 52, 142–145.

Li, B., Bhandari, D.R., Janfelt, C., Rompp, A., Spengler, B., 2014. Natural products in Glycyrrhiza glabra (licorice) rhizome imaged at the cellular level by atmospheric pressure matrix-assisted laser desorption/ionization tandem mass spectrometry imaging. Plant J. 80, 161–171.

Li, H., Zhu, W., Zhang, L., Lei, H., Wu, X., Guo, L., Chen, X., Wang, Y., Tang, H., 2015. The metabolic responses to hepatitis B virus infection shed new light on pathogenesis and targets for treatment. Sci. Rep. 5, 8421.

Lima, E.O., Guerreiro, T.M., Melo, C., de Oliveira, D.N., Machado, D., Lancelloti, M., Catharino, R.R., 2018. MALDI imaging detects endogenous digoxin in glioblastoma cells infected by Zika virus-would it be the oncolytic key? J. Mass Spectrom. 53, 257–263.

Lin, L.E., Chen, C.L., Huang, Y.C., Chung, H.H., Lin, C.W., Chen, K.C., Peng, Y.J., Ding, S.T., Wang, M.Y., Shen, T.L., Hsu, C.C., 2020. Precision biomarker discovery powered by microscopy image fusion-assisted high spatial resolution ambient ionization mass spectrometry imaging. Anal. Chim. Acta 1100, 75–87.

Lopez-Bueno, A., Mavian, C., Labella, A.M., Castro, D., Borrego, J.J., Alcami, A., Alejo, A., 2016. Concurrence of iridovirus, polyomavirus, and a unique member of a new group of fish papillomaviruses in lymphocystis disease-affected gilthead sea bream. J. Virol. 90, 8768–8779.

Lowen, R.G., Bocan, T.M., Kane, C.D., Cazares, L.H., Kota, K.P., Ladner, J.T., Nasar, F., Pitt, L., Smith, D.R., Soloveva, V., Sun, M.G., Zeng, X., Bavari, S., 2018. Countering Zika virus: the USAMRIID response. Adv. Exp. Med. Biol. 1062, 303–318.

Lu, R., Martin-Hernandez, A.M., Peart, J.R., Malcuit, I., Baulcombe, D.C., 2003. Virus-induced gene silencing in plants. Methods 30, 296–303.

Maimo-Barcelo, A., Garate, J., Bestard-Escalas, J., Fernandez, R., Berthold, L., Lopez, D.H., Fernandez, J.A., Barcelo-Coblijn, G., 2019. Confirmation of sub-cellular resolution using oversampling imaging mass spectrometry. Anal. Bioanal. Chem. 411, 7935–7941.

Marquardt, C., Tolstik, T., Bielecki, C., Kaufmann, R., Crecelius, A.C., Schubert, U.S., Settmacher, U., Stallmach, A., Dirsch, O., 2015. MALDI imaging-based classification of hepatocellular carcinoma and non-malignant lesions in fibrotic liver tissue. Z. Gastroenterol. 53, 33–39.

Martin-Acebes, M.A., Jimenez de Oya, N., Saiz, J.C., 2019. Lipid metabolism as a source of druggable targets for antiviral discovery against Zika and other Flaviviruses. Pharmaceuticals (Basel) 12, 97.

Martin, D., Gutkind, J.S., 2008. Human tumor-associated viruses and new insights into the molecular mechanisms of cancer. Oncogene 27 (Suppl. 2), S31–S42.

Mas, S., Torro, A., Fernandez, L., Bec, N., Gongora, C., Larroque, C., Martineau, P., de Juan, A., Marco, S., 2020. MALDI imaging mass spectrometry and chemometric tools to discriminate highly similar colorectal cancer tissues. Talanta 208, 120455.

Mdanda, S., Ntshangase, S., Singh, S.D., Naicker, T., Kruger, H.G., Baijnath, S., Govender, T., 2019. Investigating time dependent brain distribution of nevirapine via mass spectrometric imaging. J. Mol. Histol. 50, 593–599.

Mesri, E.A., Cesarman, E., Boshoff, C., 2010. Kaposi's sarcoma and its associated herpesvirus. Nat. Rev. Cancer 10, 707–719.

Michael, K., Bottcher, S., Klupp, B.G., Karger, A., Mettenleiter, T.C., 2006a. Pseudorabies virus particles lacking tegument proteins pUL11 or pUL16 incorporate less full-length pUL36 than wild-type virus, but specifically accumulate a pUL36 N-terminal fragment. J. Gen. Virol. 87, 3503–3507.

Michael, K., Harder, T.C., Mettenleiter, T.C., Karger, A., 2009. Diagnosis and strain differentiation of avian influenza viruses by restriction fragment mass analysis. J. Virol. Methods 158, 63–69.

Michael, K., Klupp, B.G., Mettenleiter, T.C., Karger, A., 2006b. Composition of pseudorabies virus particles lacking tegument protein US3, UL47, or UL49 or envelope glycoprotein E. J. Virol. 80, 1332–1339.

Mirzakhanyan, Y., Gershon, P., 2019. The vaccinia virion: filling the gap between atomic and ultrastructure. PLoS Pathog. 15, e1007508.

Mnatsakanyan, R., Shema, G., Basik, M., Batist, G., Borchers, C.H., Sickmann, A., Zahedi, R.P., 2018. Detecting post-translational modification signatures as potential biomarkers in clinical mass spectrometry. Expert Rev. Proteomics 15, 515–535.

Mohammadi, S., Parastar, H., 2018. Quantitative analysis of multiple high-resolution mass spectrometry images using chemometric methods: quantitation of chlordecone in mouse liver. Analyst 143, 2416–2425.

Moore, P.S., Chang, Y., 2010. Why do viruses cause cancer? Highlights of the first century of human tumour virology. Nat. Rev. Cancer 10, 878–889.

Morita, Y., Sakaguchi, T., Ikegami, K., Goto-Inoue, N., Hayasaka, T., Hang, V.T., Tanaka, H., Harada, T., Shibasaki, Y., Suzuki, A., Fukumoto, K., Inaba, K., Murakami, M., Setou, M., Konno, H., 2013. Lysophosphatidylcholine acyltransferase 1 altered phospholipid composition and regulated hepatoma progression. J. Hepatol. 59, 292–299.

Moustafa, T., Fickert, P., Magnes, C., Guelly, C., Thueringer, A., Frank, S., Kratky, D., Sattler, W., Reicher, H., Sinner, F., Gumhold, J., Silbert, D., Fauler, G., Hofler, G., Lass, A., Zechner, R., Trauner, M., 2012. Alterations in lipid metabolism mediate inflammation, fibrosis, and proliferation in a mouse model of chronic cholestatic liver injury. Gastroenterology 142, 140–151 e112.

Müller, T.G., Sakin, V., Müller, B., 2019. A spotlight on viruses-application of click chemistry to visualize virus-cell interactions. Molecules 24, 481.

Muller, W.J., Sinn, E., Pattengale, P.K., Wallace, R., Leder, P., 1988. Single-step induction of mammary adenocarcinoma in transgenic mice bearing the activated c-neu oncogene. Cell 54, 105–115.

Nakashima, Y., Eto, F., Ishihara, K., Yamazaki, F., Sato, S., Sakurai, T., Kahyo, T., Setou, M., 2019. Development of sheet enhanced technique (Set) method for MALDI-IMS. Rapid Commun. Mass Spectrom. 34, e8703.

Norris, J.L., Caprioli, R.M., 2013. Analysis of tissue specimens by matrix-assisted laser desorption/ionization imaging mass spectrometry in biological and clinical research. Chem. Rev. 113, 2309–2342.

Ntshangase, S., Mdanda, S., Naicker, T., Kruger, H.G., Govender, T., Baijnath, S., 2019a. Rilpivirine as a potential candidate for the treatment of HIV-associated neurocognitive disorders (HAND). J. Mol. Histol. 50, 295–303.

Ntshangase, S., Mdanda, S., Singh, S.D., Naicker, T., Kruger, H.G., Baijnath, S., Govender, T., 2019b. Mass spectrometry imaging demonstrates the regional brain distribution patterns of three first-line antiretroviral drugs. ACS Omega 4, 21169–21177.

Olejniczak, M., Schwartz, M., Webber, E., Shaffer, A., Perry, T.E., 2020. Viral myocarditis-incidence, diagnosis and management. J. Cardiothorac. Vasc. Anesth. 34, 1591–1601.

Park, E.S., Lee, J.H., Hong, J.H., Park, Y.K., Lee, J.W., Lee, W.J., Lee, J.W., Kim, K.P., Kim, K.H., 2014. Phosphatidylcholine alteration identified using MALDI imaging MS in HBV-infected mouse livers and virus-mediated regeneration defects. PLoS One 9, e103955.

Parrot, D., Papazian, S., Foil, D., Tasdemir, D., 2018. Imaging the unimaginable: desorption electrospray ionization—imaging mass spectrometry (DESI-IMS) in natural product research. Planta Med. 84, 584–593.

Pasulka, A.L., Thamatrakoln, K., Kopf, S.H., Guan, Y., Poulos, B., Moradian, A., Sweredoski, M.J., Hess, S., Sullivan, M.B., Bidle, K.D., Orphan, V.J., 2018. Interrogating marine virus-host interactions and elemental transfer with BONCAT and nanoSIMS-based methods. Environ. Microbiol. 20, 671–692.

Patterson, N.H., Tuck, M., Lewis, A., Kaushansky, A., Norris, J.L., Van de Plas, R., Caprioli, R.M., 2018. Next generation histology-directed imaging mass spectrometry driven by autofluorescence microscopy. Anal. Chem. 90, 12404–12413.

Pauker, V.I., Bertzbach, L.D., Hohmann, A., Kheimar, A., Teifke, J.P., Mettenleiter, T.C., Karger, A., Kaufer, B.B., 2019. Imaging mass spectrometry and proteome analysis of Marek's disease virus-induced tumors. mSphere 4. e00569-18. https://doi.org/10.1128/mSphere.00569-18.

Pfankuche, V.M., Hahn, K., Bodewes, R., Hansmann, F., Habierski, A., Haverkamp, A.K., Pfaender, S., Walter, S., Baechlein, C., Postel, A., Steinmann, E., Becher, P., Osterhaus, A., Baumgartner, W., Puff, C., 2018. Comparison of different in situ hybridization techniques for the detection of various RNA and DNA viruses. Viruses 10, 384.

Platz, E.A., Yegnasubramanian, S., Liu, J.O., Chong, C.R., Shim, J.S., Kenfield, S.A., Stampfer, M.J., Willett, W.C., Giovannucci, E., Nelson, W.G., 2011. A novel two-stage, transdisciplinary study identifies digoxin as a possible drug for prostate cancer treatment. Cancer Discov. 1, 68–77.

Porta Siegel, T., Hamm, G., Bunch, J., Cappell, J., Fletcher, J.S., Schwamborn, K., 2018. Mass spectrometry imaging and integration with other imaging modalities for greater molecular understanding of biological tissues. Mol. Imaging Biol. 20, 888–901.

Powers, T., Jones, E., Denlinger, C., Troyer, D., Mehta, A., Drake, R., 2013. MALDI imaging mass spectrometry of N-glycans in frozen and FFPE tissues. Glycobiology 23, 1360.

Prentice, B.M., Chumbley, C.W., Caprioli, R.M., 2017. Absolute quantification of rifampicin by MALDI imaging mass spectrometry using multiple TOF/TOF events in a single laser shot. J. Am. Soc. Mass Spectrom. 28, 136–144.

Prentice, B.M., McMillen, J.C., Caprioli, R.M., 2019. Multiple TOF/TOF events in a single laser shot for multiplexed lipid identifications in MALDI imaging mass spectrometry. Int. J. Mass Spectrom. 437, 30–37.

Quanico, J., Franck, J., Wisztorski, M., Salzet, M., Fournier, I., 2017a. Integrated mass spectrometry imaging and omics workflows on the same tissue section using grid-aided, parafilm-assisted microdissection. Biochim. Biophys. Acta Gen. Subj. 1861, 1702–1714.

Quanico, J., Franck, J., Wisztorski, M., Salzet, M., Fournier, I., 2017b. Progress and potential of imaging mass spectrometry applied to biomarker discovery. Methods Mol. Biol. 1598, 21–43.

Ranadheera, C., Coombs, K.M., Kobasa, D., 2018. Comprehending a killer: the Akt/mTOR signaling pathways are temporally high-jacked by the highly pathogenic 1918 influenza virus. EBioMedicine 32, 142–163.

Remels, A.H.V., Derks, W.J.A., Cillero-Pastor, B., Verhees, K.J.P., Kelders, M.C., Heggermont, W., Carai, P., Summer, G., Ellis, S.R., de Theije, C.C., Heeren, R.M.A., Heymans, S., Papageorgiou, A.P., van Bilsen, M., 2018. NF-kappaB-mediated metabolic remodelling in the inflamed heart in acute viral myocarditis. Biochim. Biophys. Acta Mol. Basis Dis. 1864, 2579–2589.

Reyzer, M.L., Caldwell, R.L., Dugger, T.C., Forbes, J.T., Ritter, C.A., Guix, M., Arteaga, C.L., Caprioli, R.M., 2004. Early changes in protein expression detected by mass spectrometry predict tumor response to molecular therapeutics. Cancer Res. 64, 9093–9100.

Roden, R.B.S., Stern, P.L., 2018. Opportunities and challenges for human papillomavirus vaccination in cancer. Nat. Rev. Cancer 18, 240–254.

Ryan, D.J., Spraggins, J.M., Caprioli, R.M., 2019. Protein identification strategies in MALDI imaging mass spectrometry: a brief review. Curr. Opin. Chem. Biol. 48, 64–72.

Schafermann, J., Kliewer, G., Losch, J., Bednarz, H., Giampa, M., Niehaus, K., 2020. Immersion by rotation-based application of the matrix for fast and reproducible sample preparations and robust results in mass spectrometry imaging. J. Mass Spectrom. 55, e4488.

Schleyer, G., Shahaf, N., Ziv, C., Dong, Y., Meoded, R.A., Helfrich, E.J.N., Schatz, D., Rosenwasser, S., Rogachev, I., Aharoni, A., Piel, J., Vardi, A., 2019. In plaque-mass spectrometry imaging of a bloom-forming alga during viral infection reveals a metabolic shift towards odd-chain fatty acid lipids. Nat. Microbiol. 4, 527–538.

Schwamborn, K., Caprioli, R.M., 2010. MALDI imaging mass spectrometry–painting molecular pictures. Mol. Oncol. 4, 529–538.

Schwamborn, K., Krieg, R.C., Jirak, P., Ott, G., Knuchel, R., Rosenwald, A., Wellmann, A., 2010. Application of MALDI imaging for the diagnosis of classical Hodgkin lymphoma. J. Cancer Res. Clin. Oncol. 136, 1651–1655.

Schwamborn, K., Krieg, R.C., Uhlig, S., Ikenberg, H., Wellmann, A., 2011. MALDI imaging as a specific diagnostic tool for routine cervical cytology specimens. Int. J. Mol. Med. 27, 417–421.

Seeley, E.H., Caprioli, R.M., 2012. 3D imaging by mass spectrometry: a new frontier. Anal. Chem. 84, 2105–2110.

Seneviratne, H.K., Hendrix, C.W., Fuchs, E.J., Bumpus, N.N., 2018. MALDI mass spectrometry imaging reveals heterogeneous distribution of tenofovir and tenofovir diphosphate in colorectal tissue of subjects receiving a tenofovir-containing enema. J. Pharmacol. Exp. Ther. 367, 40–48.

Spengler, B., 2015. Mass spectrometry imaging of biomolecular information. Anal. Chem. 87, 64–82.

Stauber, J., Ayed, M.E., Wisztorski, M., Salzet, M., Fournier, I., 2010. Specific MALDI-MSI: tag-mass. Methods Mol. Biol. 656, 339–361.

Steurer, S., Borkowski, C., Odinga, S., Buchholz, M., Koop, C., Huland, H., Becker, M., Witt, M., Trede, D., Omidi, M., Kraus, O., Bahar, A.S., Seddiqi, A.S., Singer, J.M., Kwiatkowski, M., Trusch, M., Simon, R., Wurlitzer, M., Minner, S., Schlomm, T.,

Sauter, G., Schluter, H., 2013. MALDI mass spectrometric imaging based identification of clinically relevant signals in prostate cancer using large-scale tissue microarrays. Int. J. Cancer 133, 920–928.
Swales, J.G., Hamm, G., Clench, M.R., Goodwin, R.J.A., 2019. Mass spectrometry imaging and its application in pharmaceutical research and development: a concise review. Int. J. Mass Spectrom. 437, 99–112.
Syed, G.H., Amako, Y., Siddiqui, A., 2010. Hepatitis C virus hijacks host lipid metabolism. Trends Endocrinol. Metab. 21, 33–40.
Takeo, E., Sugiura, Y., Uemura, T., Nishimoto, K., Yasuda, M., Sugiyama, E., Ohtsuki, S., Higashi, T., Nishikawa, T., Suematsu, M., Fukusaki, E., Shimma, S., 2019. Tandem mass spectrometry imaging reveals distinct accumulation patterns of steroid structural isomers in human adrenal glands. Anal. Chem. 91, 8918–8925.
Thiery-Lavenant, G., Zavalin, A.I., Caprioli, R.M., 2013. Targeted multiplex imaging mass spectrometry in transmission geometry for subcellular spatial resolution. J. Am. Soc. Mass Spectrom. 24, 609–614.
Thompson, C.G., Bokhart, M.T., Sykes, C., Adamson, L., Fedoriw, Y., Luciw, P.A., Muddiman, D.C., Kashuba, A.D., Rosen, E.P., 2015. Mass spectrometry imaging reveals heterogeneous efavirenz distribution within putative HIV reservoirs. Antimicrob. Agents Chemother. 59, 2944–2948.
Thompson, C.G., Rosen, E.P., Prince, H.M.A., White, N., Sykes, C., de la Cruz, G., Mathews, M., Deleage, C., Estes, J.D., Charlins, P., Mulder, L.R., Kovarova, M., Adamson, L., Arora, S., Dellon, E.S., Peery, A.F., Shaheen, N.J., Gay, C., Muddiman, D.C., Akkina, R., Garcia, J.V., Luciw, P., Kashuba, A.D.M., 2019. Heterogeneous antiretroviral drug distribution and HIV/SHIV detection in the gut of three species. Sci. Transl. Med. 11. https://doi.org/10.1126/scitranslmed.aap8758.
Thompson, P.J., Macfarlan, T.S., Lorincz, M.C., 2016. Long terminal repeats: from parasitic elements to building blocks of the transcriptional regulatory repertoire. Mol. Cell 62, 766–776.
Truyen, U., Lochelt, M., 2006. Relevant oncogenic viruses in veterinary medicine: original pathogens and animal models for human disease. Contrib. Microbiol. 13, 101–117.
Van Linthout, S., Elsanhoury, A., Klein, O., Sosnowski, M., Miteva, K., Lassner, D., Abou-El-Enein, M., Pieske, B., Kuhl, U., Tschope, C., 2018. Telbivudine in chronic lymphocytic myocarditis and human parvovirus B19 transcriptional activity. ESC Heart Fail. 5, 818–829.
Van Linthout, S., Tschope, C., 2018. Viral myocarditis: a prime example for endomyocardial biopsy-guided diagnosis and therapy. Curr. Opin. Cardiol. 33, 325–333.
Verbeeck, N., Caprioli, R.M., van de Plas, R., 2019. Unsupervised machine learning for exploratory data analysis in imaging mass spectrometry. Mass Spectrom. Rev. 39, 245–291.
Waldchen, F., Spengler, B., Heiles, S., 2019. Reactive matrix-assisted laser desorption/ionization mass spectrometry imaging using an intrinsically photoreactive Paterno-Buchi matrix for double-bond localization in isomeric phospholipids. J. Am. Chem. Soc. 141, 11816–11820.
Wang, F., Flanagan, J., Su, N., Wang, L.C., Bui, S., Nielson, A., Wu, X., Vo, H.T., Ma, X.J., Luo, Y., 2012. RNAscope: a novel in situ RNA analysis platform for formalin-fixed, paraffin-embedded tissues. J. Mol. Diagn. 14, 22–29.
Wang, H.A., Grolimund, D., Giesen, C., Borca, C.N., Shaw-Stewart, J.R., Bodenmiller, B., Gunther, D., 2013. Fast chemical imaging at high spatial resolution by laser ablation inductively coupled plasma mass spectrometry. Anal. Chem. 85, 10107–10116.
Watrous, J.D., Dorrestein, P.C., 2011. Imaging mass spectrometry in microbiology. Nat. Rev. Microbiol. 9, 683–694.

Wee, S., Alli-Shaik, A., Kek, R., Swa, H.L.F., Tien, W.P., Lim, V.W., Leo, Y.S., Ng, L.C., Hapuarachchi, H.C., Gunaratne, J., 2019. Multiplex targeted mass spectrometry assay for one-shot flavivirus diagnosis. Proc. Natl. Acad. Sci. U. S. A. 116, 6754–6759.
Wen, Z., Song, H., Ming, G.L., 2017. How does Zika virus cause microcephaly? Genes Dev. 31, 849–861.
Wiegelmann, M., Dreisewerd, K., Soltwisch, J., 2016. Influence of the laser spot size, focal beam profile, and tissue type on the lipid signals obtained by MALDI-MS imaging in oversampling mode. J. Am. Soc. Mass Spectrom. 27, 1952–1964.
Wilson, W.H., Van Etten, J.L., Allen, M.J., 2009. The Phycodnaviridae: the story of how tiny giants rule the world. Curr. Top. Microbiol. Immunol. 328, 1–42.
Witte, R., Andriasyan, V., Georgi, F., Yakimovich, A., Greber, U.F., 2018. Concepts in light microscopy of viruses. Viruses 10, 202.
Yang, B., Tsui, T., Caprioli, R.M., Norris, J.L., 2020. Sample preparation and analysis of single cells using high performance MALDI FTICR mass spectrometry. Methods Mol. Biol. 2064, 125–134.
Zavalin, A., Yang, J., Hayden, K., Vestal, M., Caprioli, R.M., 2015. Tissue protein imaging at 1 μm laser spot diameter for high spatial resolution and high imaging speed using transmission geometry MALDI TOF MS. Anal. Bioanal. Chem. 407, 2337–2342.
Zhang, J., Lan, Y., Sanyal, S., 2017. Modulation of lipid droplet metabolism-a potential target for therapeutic intervention in Flaviviridae infections. Front. Microbiol. 8, 2286.
Zhang, Y., Liu, X., 2018. Machine learning techniques for mass spectrometry imaging data analysis and applications. Bioanalysis 10, 519–522.
Zheng, M., Huang, J., Tong, A., Yang, H., 2019. Oncolytic viruses for cancer therapy: barriers and recent advances. Mol. Ther. Oncolytics 15, 234–247.
Zhu, L., Yin, L., Xue, J., Wang, Z., Nie, Z., 2018. Mass spectrometry genotyping of human papillomavirus based on high-efficiency selective enrichment of nanoparticles. ACS Appl. Mater. Interfaces 10, 41178–41184.
Zhu, Z., Gorman, M.J., McKenzie, L.D., Chai, J.N., Hubert, C.G., Prager, B.C., Fernandez, E., Richner, J.M., Zhang, R., Shan, C., Tycksen, E., Wang, X., Shi, P.Y., Diamond, M.S., Rich, J.N., Chheda, M.G., 2017. Zika virus has oncolytic activity against glioblastoma stem cells. J. Exp. Med. 214, 2843–2857.
Zubair, F., Laibinis, P.E., Swisher, W.G., Yang, J., Spraggins, J.M., Norris, J.L., Caprioli, R.M., 2016. Trypsin and MALDI matrix pre-coated targets simplify sample preparation for mapping proteomic distributions within biological tissues by imaging mass spectrometry. J. Mass Spectrom. 51, 1168–1179.

CHAPTER THREE

Proximity labeling approaches to study protein complexes during virus infection

Francisco José Zapatero-Belinchón[a,b,c,d,*], Belén Carriquí-Madroñal[a,b], and Gisa Gerold[a,b,c,d,*]

[a]Department of Biochemistry & Research Center for Emerging Infections and Zoonoses (RIZ), University of Veterinary Medicine Hannover, Hannover, Germany
[b]Institute for Experimental Virology, TWINCORE, Centre for Experimental and Clinical Infection Research, a Joint Venture Between the Medical School Hannover and the Helmholtz Centre for Infection Research, Hannover, Germany
[c]Department of Clinical Microbiology, Virology, Umeå University, Umeå, Sweden
[d]Wallenberg Centre for Molecular Medicine (WCMM), Umeå University, Umeå, Sweden
*Corresponding authors: e-mail address: francisco.zapatero@tiho-hannover.de; gisa.gerold@tiho-hannover.de

Contents

Abstract

Cellular compartmentalization of proteins and protein complex formation allow cells to tightly control biological processes. Therefore, understanding the subcellular localization and interactions of a specific protein is crucial to uncover its biological function. The advent of proximity labeling (PL) has reshaped cellular proteomics in infection biology. PL utilizes a genetically modified enzyme that generates a "labeling cloud" by covalently labeling proteins in close proximity to the enzyme. Fusion of a PL enzyme

Advances in Virus Research, Volume 109
ISSN 0065-3527
https://doi.org/10.1016/bs.aivir.2021.02.001

to a specific antibody or a "bait" protein of interest in combination with affinity enrichment mass spectrometry (AE-MS) enables the isolation and identification of the cellular proximity proteome, or proxisome. This powerful methodology has been paramount for the mapping of membrane or membraneless organelles as well as for the understanding of hard-to-purify protein complexes, such as those of transmembrane proteins. Unsurprisingly, more and more infection biology research groups have recognized the potential of PL for the identification of host-pathogen interactions. In this chapter, we introduce the enzymes commonly used for PL labeling as well as recent promising advancements and summarize the major achievements in organelle mapping and nucleic acid PL. Moreover, we comprehensively describe the research on host-pathogen interactions using PL, giving special attention to studies in the field of virology.

1. Introduction

Cellular processes are compartmentalized within the cell. Hence, the same protein may perform different functions depending on its protein interaction partners in a given subcellular compartment. Consequently, uncovering the subcellular interactome of a protein of interest is key for the identification and characterization of its specific biological function.

A wide array of methods has been developed for isolation and quantification of cellular and subcellular proteomes. For instance, density gradient or liquid chromatography fractionation coupled with mass spectrometry (MS) can reveal the protein composition of cellular organelles or protein complexes, respectively (reviewed in Yates et al., 2005). A widely-used technique to isolate and quantify protein complexes is affinity enrichment coupled to MS (AE-MS). This technique relies either on the tagging of the protein of interest, or "bait," with a purification-compatible epitope tag (streptavidin-tag, histidine-tag, etc.) or on antibodies that bind the bait with high affinity (reviewed in Dunham et al., 2012). Although classical proteomics and AE-MS have been successfully used to identify host-virus interactions (reviewed in Gerold et al., 2016; Lasswitz et al., 2018; Lum and Cristea, 2016), all these techniques have important drawbacks. First, organelle capture is limited to those that can be isolated and preserved during cell lysis and purification, i.e., insoluble or membraneless organelles are excluded from this approach. Second, since strong protein interaction partners are more likely to be captured than weak and transient interaction partners, these techniques over-represent strong and stable protein interactions. Lastly, cell lysis mixes proteins that would normally not localize to the same subcellular compartment inside the cell and thus these methods can yield false positive interaction partners.

Since 2012, proximity labeling (PL) methods have been developed to circumvent the need to maintain organelle and protein complex integrity during purification, allowing harsher cell lysis and washing methods before protein isolation. PL involves fusion of the bait protein with a PL enzyme that promiscuously biotinylates proteins in close vicinity to the bait. Biotinylated proteins, or "preys" can then be captured through the covalently-bound biotin using streptavidin-based technology and quantified by MS (Fig. 1A). Importantly, PL can be performed in living cells and organisms and thus biotinylated proteins represent the physiological cellular proxisome of the bait.

In this chapter, we describe the discovery, evolution, and features of the most commonly used and recently developed PL enzymes as well as exemplary applications in molecular and cellular biology. Moreover, we summarize recent studies that have used PL for successful identification of host-pathogen interactions, with a special emphasis on virology.

2. Methodology

The vast majority of PL approaches modifies neighboring proteins with biotin. We will in the following give a short introduction to biotin and its occurrence in cells to then describe the three most commonly used PL techniques, namely biotin protein ligases, peroxidases and prokaryotic ubiquitin-like protein ligases.

2.1 Biotin

Biotin, or vitamin H, is an essential vitamin that can be synthesized by prokaryotes, bacteria and some fungi but not by vertebrates and has to be actively transported across the plasma membrane via the sodium dependent multivitamin transporter (SMVT) (Cronan, 2018; Prasad et al., 1998). In mammals and birds, biotin is a covalently-bound co-factor of biotin-dependent carboxylase enzymes that participate in the metabolism of carbohydrates, amino acids, and fatty acids (reviewed in Tong, 2013). Covalently-bound biotin both activates carboxylases and act as an intermediate carboxyl-acceptor before final transfer of the carboxyl group to its substrate. As each cell contains a number of endogenously biotinylated proteins, it is important to determine this cellular background when using PL techniques.

Biotin has become an excellent biotechnological tool due to its high affinity to the two tetrameric proteins egg-white avidin (dissociation

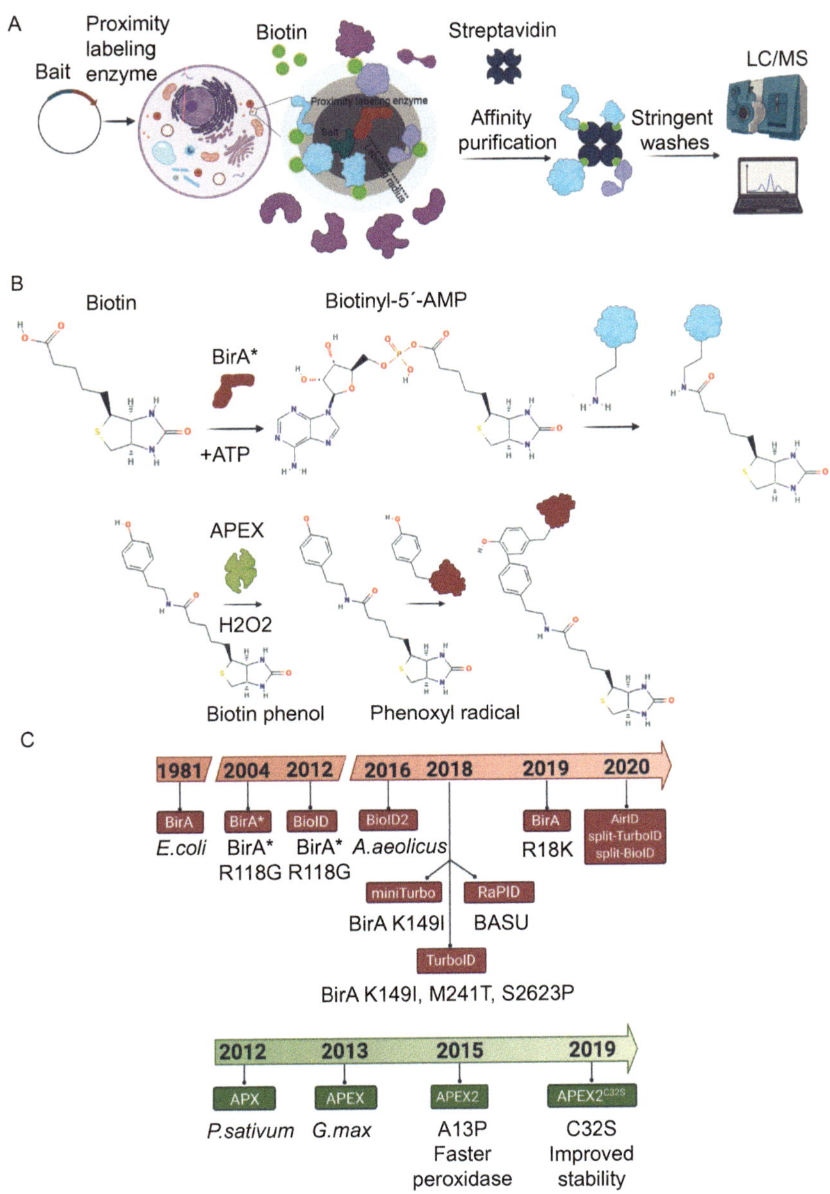

Fig. 1 Basic principles and evolution of proximity labeling approaches. (A) Schematic representation of proximity labeling workflow. The biotin protein ligase (BPL) enzyme is fused to a bait protein of interest. The BPL then catalyzes the synthesis of a reactive biotin intermediate, which labels neighboring proteins within a radius of 10–20 nm depending on the precise method (area shown in gray). The biotinylated proteins are then captured on streptavidin resins, washed under stringent conditions, and quantified using liquid chromatography coupled to mass spectrometry (LC/MS). (B) Labeling

constant (Kd) ≈ 10^{-15} M) and *Streptomyces* streptavidin (Kd of 4×10^{-14} M) (Chaiet and Wolf, 1964; Chaiet et al., 1963; Eakin et al., 1940; Michael Green, 1990). In fact, the non-covalent interactions between these proteins and biotin are among the strongest known in nature. The power of the biotin-avidin/streptavidin interactions has been extensively harnessed in biotechnological assays such as Western-blotting, enzyme-linked immunosorbent assay (ELISA), immunolabeling, and protein pull-down/purification (reviewed in Dundas et al., 2013). In summary, the specificity and affinity of biotin to avidin/streptavidin make biotin the ideal fishhook in the pond of cellular proteins.

2.2 Biotin protein ligases

In order to covalently bind biotin to carboxylases a third player, biotin protein ligases (BPLs), must catalyze the attachment (Chapman-Smith and Cronan, 1999). BPLs are highly evolutionary conserved and specific to their carboxylase substrates (Chapman-Smith et al., 2001).

BPL are classified into class I, II, and III (Sternicki et al., 2017). All of them possess a central catalytic domain that mediates biotin attachment as well as a C-terminal domain that regulates their enzymatic activity by a yet-to-identify mechanism (Chapman-Smith et al., 2001). Whereas class I BPLs do not have any domain at their N-terminus, class II BPLs possess a short DNA-binding domain that represses biotin attachment and class III BPLs have a large N-terminal domain of unknown function (Xu and Beckett, 1996). Prokaryotes, plants, and archaea express class I and II BPLs and insects, yeast and mammals express class III BPLs. Bacterial class II BPLs are among the best-studied and utilized proteins for biotechnological applications including PL.

---

mechanism mediated by BPL (upper reaction) and peroxidase (lower reaction) enzymes. In the presence of ATP, the BPL BirA* synthesizes a reactive intermediate (bio-AMP) using biotin as a precursor. This reactive intermediate interacts with the lysine residues of the neighboring proteins located in a radius of 10 nm, resulting in the addition of biotin to the protein. In the presence of H_2O_2, ascorbate peroxidase enzyme (APEX) transforms biotin-phenol into a short-lived intermediate biotin-phenoxyl radical which interacts with neighboring proteins in a radius of approx. 20 nm. This phenoxyl radical reacts with electron-rich amino acids (like tyrosine) present in the neighboring proteins resulting in addition of biotin. The chemical structures were obtained from PubChem database: Biotin (PubChem CID: 171548), AMP (PubChem CID: 6083), lysine (PubChem CID: 5962), biotin tyramide (PubChem CID: 21826423) and tyrosine (PubChem CID: 6057). (C) Timeline describing the evolution of both BPL (red) and peroxidase based (green) labeling techniques. Image created with BioRender.com.

The mechanism of BPL biotinylation is a multi-step process, which requires energy in the form of adenosine triphosphate (ATP). For the well-studied *Escherichia coli* class II 35 kDa BPL bifunctional ligase/repressor (BirA) the attachment starts by the binding of biotin to BirA (Fig. 1B). This interaction triggers conformational changes that allow binding of ATP to the catalytic pocket of BirA (Sternicki et al., 2017). Biotin then catalytically attacks the alpha phosphate of ATP, forming reactive biotinyl 5′ adenosyl monophosphate (biotynil-5′-AMP) (Lane et al., 1964). Importantly, this metabolite remains bound to BirA through hydrogen bonds with arginine 118 (R118). Lastly, an epsilon aminogroup of a lysine sidechain in the carboxylase substrate, namely biotin-carboxylase cargo protein (BCCP), attacks the biotin carboxyl group and thereby biotinylates the lysine and releases AMP.

While BirA could in principle be used as a PL enzyme in biotechnological applications, its main disadvantage is the binding strength between biotinyl-5′-AMP and BirA. The search for a more promiscuous BirA version resulted in the identification of a BirA variant, BirA*, with a R118G single-point mutation (Buoncristiani et al., 1986). This mutation not only destabilizes the complex between metabolite and protein (>400-fold higher dissociation rate for biotinyl-5′-AMP) but also leads to stronger binding of biotin to the substrate (K > 100-fold increase) (Kwon and Beckett, 2000; Kwon et al., 2000). Hence, it is better suited for labeling approaches than the endogenous *E. coli* enzyme. In 2004, Choi-Rhee et al. observed that BirA* promiscuously biotinylated proteins by proximity and hypothesized that this variant could be used as a tool for protein labeling and recovery (For a timeline summary of the development of BPL enzymes, refer to Fig. 1C) (Choi-Rhee et al., 2004). Eight years later, Roux et al. discovered the potential of BirA* and developed BioID, "a new technique for proximity-dependent labeling of proteins in eukaryotic cells" (Roux et al., 2012). BioID takes advantage of the higher dissociation rate of reactive biotinyl-5′-AMP. These metabolites generate a "biotinylation cloud" of ~10 nm radius that will promiscuously biotinylate primary amines of proteins in close proximity to BirA* in living cells (Kim et al., 2014; Roux et al., 2012). As a proof of concept, Roux et al. fused BirA* to the nuclear lamina protein lamin-A and identified both known and new members of the nuclear lamina by MS (Roux et al., 2012). An alternative PL method, termed BioID2, was developed in 2016 by introducing a R40G mutation in BPL of the bacterium *Aquifex aeolicus* (Kim et al., 2016). This BPL is smaller in size (26 KDa), lacks the N-terminal DNA-binding domain, and requires less biotin. Of note, the reduced need of exogenous biotin leads to BioID2 efficiently using biotin from serum in cell culture media and hence, an increased non-specific background. Finally, a screen of R118 BirA mutants identified R118K BirA in addition to R118G as new promiscuous BPL variant (Oostdyk et al., 2019).

The BioID and BioID2 methodologies require biotinylation to occur for as long as 18 h due to the slow enzyme kinetics of the BirA variants. To decrease labeling times, Branon et al. used yeast surface display in combination with error-prone PCR to generate BirA* variants with shorter labeling times (Branon et al., 2018). They generated TurboID, a 12-mutation variant, and Miniturbo, which harnesses 14 mutations and misses the first 63 aa of the DNA-binding domain. Both improved BPLs, TurboID and Miniturbo, have labeling times of ~10 min and a 23-fold increased activity over the original protein. Importantly, TurboID can use biotin from media and supplements like BioID2 leading to increased background and there have been reports of cellular toxicity (Branon et al., 2018; May et al., 2020).

In the search for a better BPL to label RNA, Ramanathan et al. searched for proteins with similar functional domains as BirA and identified a *Bacillus subtilis* BPL, which they modified by removing the DNA-binding domain (Ramanathan et al., 2018). The authors claim that the engineered protein, BASU, has >1000-fold increase kinetics and more than 30-fold stronger labeling activity than BirA* but consecutive studies could not confirm increased kinetics in comparison to other BPLs (Branon et al., 2018). Recently, a novel BPL has been identified by using ancestral genetic sequences from BirA (Kido et al., 2020). AirID (ancestral BirA for proximity-dependent biotin identification) has lower labeling times (3 h) and therefore lower risk of toxicity but further studies are required to assess the performance of this protein in the wide array of biotin ligases identified.

Finally in an effort to improve spatial-specific biotinylation of protein-protein interactions, three split-BioID versions (De Munter et al., 2017; Kwak et al., 2020; Schopp et al., 2017) and one split-TurboID (Cho et al., 2020a) version have been generated. These split variants required expression of two fusion proteins in order to allow for assembly of the active enzyme and may thus provide better specificity.

2.3 Peroxidases

Peroxidases belong to a large group of oxidoreductases, which catalyze redox reactions and have different biological functions. Peroxidases can be classified as heme or non-heme peroxidases based on the presence/absence of an iron-protoporphyrin IX prosthetic group and a catalytic histidine

(Koua et al., 2009). Heme peroxidases are present in all life kingdoms and can be further classified in five superfamilies with the peroxidase-catalase family being the largest (Savelli et al., 2019; Zámocký et al., 2015). Catalases use hydrogen peroxidase (H_2O_2) as the main electron donor forming short-lived free reactive hydroxyl radicals that oxidize a variety of electron-rich molecules like tyrosines. All members of the peroxidase-catalase superfamily possess a unique sequence, have a globular quaternary structure and can be subdivided into three classes (Welinder, 1992). Prokaryotic class I peroxidases scavenge excess intracellular H_2O_2 and important members are cytochrome C and ascorbate peroxidase like APX. Class II peroxidases are fungal peroxidases that degrade lignin in soils. Lastly, Class III peroxidases are plant secretory peroxidases that oxidize secondary metabolites, hormones like auxin, and lignin. Horseradish peroxidase (HRP) is a 44kDa calcium-dependent secreted glycoprotein of this class III that has been extensively used in biotechnological applications since its discovery in 1810 (for a review on the historical and molecular aspects of HRP, please refer to Veitch, 2004).

HRP has been used for chemiluminescence, electron microscopy (EM), and protein labeling applications because of its ability to oxidize luminol, 3,3′-diaminobenzidine, and phenolic substances (biotin-phenol), respectively (Bobrow et al., 1989; Hopkins et al., 2000; Thorpe and Kricka, 1986). HRP can be either conjugated to an antibody (ab) or genetically expressed on cells. Ab-conjugated HRP specifically targets the extracellular space of intact cells or the intra- and extracellular space of fixed cells (Honke et al., 2019; Kotani et al., 2008; Li et al., 2014). In the second approach, HRP is genetically fused to a bait protein or signal sequence and expressed in cells in a similar fashion as BioID or TurboID. HRP fusion proteins could successfully label ER and Golgi for EM visualization (Connolly et al., 1994). Interestingly, even though HRP retains activity in the lumen of ER and Golgi, this protein is inactive in the reducing environment of the cytosol (Connolly et al., 1994; Hopkins et al., 2000). The latter causes the loss of four disulfide bonds essential for the catalytic activity of HRP (Martell et al., 2012). Nevertheless, HRP could map the proteome of extracellular compartments such as the synaptic cleft and lipid rafts (Loh et al., 2016; Miyagawa-Yamaguchi et al., 2014). Furthermore, recent work describes enhanced HRP mutants. vHRP (v variant HPR) and enhanced HRP (eHRP) were generated by addition of stabilizing mutations or unbiased directed evolution, respectively (Cruz-Lopez et al., 2018; Yamagata and Sanes, 2018). These promising new variants await testing for their value as PL enzymes and future studies will shed light on their potential.

The Ting lab initiated the search and development of an alternate peroxidase that unlike HRP would be active in the cell cytosol and identified the pea ascorbate peroxidase APX (For a timeline summary of the development of ascorbate peroxidase enzymes, refer to Fig. 1C) (Martell et al., 2012). This peroxidase is a plant cytosolic 28 kDa class I peroxidase that oxidizes L-ascorbate. Unlike HRP, APX lacks disulfide bonds and calcium binding sites. APX is approx. 40% smaller than HRP but is naturally present as a homodimer. Therefore, Martel et al. engineered a monomeric version comprising the mutations K14D and E112K to reduce unwanted protein dimerization potential and the W41F mutation on the catalytic domain to restore catalytic activity. They named this engineered APX protein APEX (enhanced APX). It is worth mentioning that the current widely use APEX version comes from a soybean APX with identical mutations as the initial pea APX (Rhee et al., 2013). APEX catalyzes the H_2O_2-dependent oxidation of biotin-phenol generating short-lived (<1 ms), highly reactive biotin-phenoxyl radicals that covalently react to electron-rich aa (Tyr, Trp, His, Cys) in a radius of ~20 nm (Amini et al., 2002; Bendayan, 2001; Bhaskar et al., 2003; Mayer and Bendayan, 1997; Minamihata et al., 2011; Mortensen et al., 2001; Rogers et al., 2008) (Fig. 1B). Just like for the development of TurboID and miniTurbo (see Section 2.2), the Ting lab developed an enhanced version of APEX by yeast display direct evolution (Lam et al., 2015). Strikingly, unlike for their enhanced BPLs, their enhanced APEX2 only harbored a A134P mutation. Since this residue is commonly found in class II and III peroxidases (like HRP), they hypothesized that this mutant might be more resistant to H_20_2-induced inhibition. In addition, in 2019 an additional C32S mutation was identified to increase stability of APEX2-fused proteins (Huang et al., 2019).

Both HRP and APEX2 have also been split for protein-protein interaction mapping (Han et al., 2019; Martell et al., 2016; Xue et al., 2017). Whereas there is only one split-HRP developed, two groups generated split-APEX2 versions with the breakpoint at almost identical sites (E200/G201 or G201/L202), but with one of them performing additional directed evolution on the N-terminal fragment to increase its activity.

2.4 Prokaryotic ubiquitin-like protein (Pup) ligases

Pupylation is ubiquitination-like process that tags proteins for proteasomal degradation. Firstly observed in pathogenic *Mycobacterium tuberculosis* (Pearce et al., 2008), the Pup locus has been found in actinobacteria and some species of planctomycetes, verrucomicrobia, nitrospora proteobacteria and armatimonadetes (Iyer et al., 2008). The Pup locus encodes for Pup, the ubiquitin-like protein, and the proteasome accessory factor A (PafA), which is the Pup-ligase. Pup is a 64-aa small protein with a glutamine at its C-terminus. During pupylation, this peptide is deaminated resulting in a Pup with C-terminal glutamate designated Pup(E). Subsequently, the PafA ligase phosphorylates the glutamate γ-carboxylate in an ATP-dependent manner. Finally, PafA transfers the phosphorylated Pup(E) to ε-amine lysine residues of the target protein (Iyer et al., 2008; Özcelik et al., 2012).

PafA is a member of the glutamine synthetase (GS) family (Iyer et al., 2008; Özcelik et al., 2012). This GS family is subclassified in the carboxylate-amine/ammonia ligases superfamily that phosphorylates and ligates carboxyl groups to proteins or ammonia and the guanidino phosphotransferase superfamily that phosphorylates arginine or creatine. According to its structure and function PafA is classified as a member of the carboxylate-amine ligase superfamily (Iyer et al., 2008; Özcelik et al., 2012).

In 2018, Liu et al. showed that pupylation can be exploited for PL and termed the methodology PUP-IT (Liu et al., 2018b). In this system, both PafA and Pup(E) can be genetically expressed or exogenously supplemented. Pup(E) tags proteins in close proximity as well as PafA and the bait. Additionally, the authors further modified Pup(E) by adding a BCCP domain (see Section 2.2 for more information). The new peptide, bio-Pup(E), can be detected and pulled-down using streptavidin technology for detection and quantification of pupylated proteins. Using PUP-IT the authors identified known interactors of CD28 by MS revealing the potential of this methodology. In addition to Pup(E) (64 aa), DE28 (28 aa) and Peptide 4.1 (14 aa) peptide variants with pupylation potential have been identified (Sun et al., 2020). Peptide 4.1 is particularly interesting because it does not contain any lysines, which could cause pupylation branching. Although it seems that biotin ligase- and peroxidase-based PL approaches are more efficient and faster, the specificity and good cellular tolerability of Pup(E) makes PUP-IT an interesting tool in the arsenal of PL methods.

3. Cell biological applications of proximity labeling

3.1 Organelle mapping

The functionality of cellular proteins is both dependent on their subcellular localization and their proximal protein neighbors, so that cells can regulate biological function by spatial protein compartmentalization. Consequently, to understand a cellular function, it is imperative to characterize the subcellular protein fraction. The major advantage of PL in comparison to previously developed organelle mapping techniques is its ability to label proteins prior to cell lysis as well as its tolerability to harsh cell lysis and washing conditions before protein isolation, making it an exceptional tool for organelle mapping. In this section, we summarize studies using PL for organelle mapping and their major findings.

The nuclear lamina (NL) was the first organelle ever mapped by PL using BioID (Roux et al., 2012). By fusing the NL resident protein lamin A to BirA* the authors pulled down and identified known NL components such as nuclear protein complex 153 and 50 (Nup153, Nup50), as well as unknown components like the soluble lamina-associated protein of 75 kDa SLAP75 (Fig. 2). A year later, Rhee et al. mapped the mitochondrial matrix using APEX and identified 495 proteins, 31 of which were mitochondrial proteins (Rhee et al., 2013). These pioneering studies laid the ground for extensive organelle mapping in the last decade.

Among these more recent PL studies, the mitochondrion is the most heavily studied organelle. Using APEX2 the mitochondrial matrix, nucleoid, intermembrane space, and outer mitochondrial membrane (OMM) have been mapped (Fig. 2) (Han et al., 2017; Hung et al., 2014, 2017; Lam et al., 2015; Rhee et al., 2013). Likewise, BioID, BioID2, TurboID and Miniturbo have been used to map several sub-organelle compartments of mitochondria (Fig. 2) (Antonicka et al., 2017; Bagchi et al., 2020; Branon et al., 2018; Callegari et al., 2019; Chen et al., 2019; Cho et al., 2020a; Cole et al., 2015; Janer et al., 2016; Yoshinaka et al., 2019). Besides mitochondria, other membrane organelles such as the nucleus, ER, Golgi, lipid droplets, endosomes, autophagosome, peroxisome, and cytosol have been extensively characterized by PL (Fig. 2) (Bagchi et al., 2020; Bersuker et al., 2018; Cao et al., 2018; Chan et al., 2019; Cho et al., 2020a,b; Cross et al., 2020; De Munter et al., 2017; Del Olmo et al., 2019; Fazal et al., 2019; Guo et al., 2014; Han et al., 2019; Honke and Kotani, 2011; Hua et al., 2017; Hung et al., 2017; Kido et al., 2020; Kim et al., 2016; Kwak et al., 2020; Lam et al., 2015; Le Guerroué et al., 2017; Lee et al., 2016; Li et al., 2014; Liu et al., 2018a; Martell et al., 2016; Mirza et al., 2019; Pagac et al., 2016; Ramanathan et al., 2018; Roux et al., 2018; Van Itallie et al., 2014; Vidak et al., 2015). In addition, contact sites between organelles, known to be important for regulation of biological functions, were studied with PL. For instance, the ER membrane contacts mitochondria to regulate

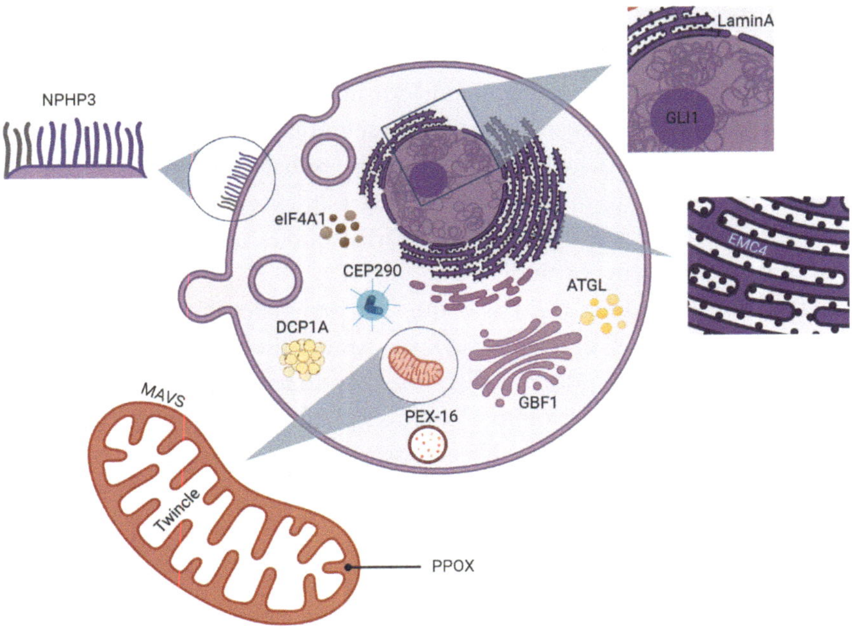

Fig. 2 Representative subcellular markers used in proximity labeling studies. Proximity labeling (PL) has been used for organelle mapping by fusing the labeling enzyme (BirA* or APEX) to a previously identified protein with known subcellular localization. Here, we depict some organelles and the markers used for subcellular mapping. LaminA and GLI1 are markers for nuclear lamina and nucleolus, respectively. EMC4 and GBF1 were used to characterize the ER and Golgi. Outer membrane, intermembrane space, and matrix from the mitochondria were mapped with MAVS, PPOX, and Twincle, respectively. PEX-16 and ATGL serve as markers of peroxisomes and lipid droplets, respectively. Finally, labeling of membraneless organelles including cilia (marker: NPHP3), centrosomes (CEP290), stress granules (eIF4A1) and P-bodies (CDP1A) gives the opportunity to study the protein composition of these otherwise hard to track subcellular structures. Image created with BioRender.com.

phospholipid synthesis, stress, cell cycle, and calcium responses, among others (Hamasaki et al., 2013; Paillusson et al., 2016; Rizzuto et al., 1998; Vance, 1990). A study mapping the cytosolic site of mitochondria and ER identified 94 proteins that were enriched in both datasets and are likely present at contact sites. Furthermore, the authors observed that the synaptojanin-2 binding protein SYNJ2BP induces contacts of mitochondria with the ER and interacts with the rough ER membrane protein ribosome binding protein 1 (RRBP1) (Hung et al., 2017). This mitochondria ER contact site was further studied using the split versions of APEX2, BioID and TurboID (Cho et al., 2020a; Han et al., 2019; Kwak et al., 2020).

Several studies address the extensive protein network of ER at contact sites with other organelles such as the plasma membrane (Jing et al., 2015; van Vliet et al., 2017), peroxisomes (Hua et al., 2017), autophagosomes (Hamasaki et al., 2013), endosomes (Hoyer et al., 2018), and lipid droplets (Bersuker et al., 2018).

Another major advancement in organelle mapping by PL is the characterization of membraneless compartments such as the base of cilia, the centrosome, stress granules, or the synaptic cleft. In a landmark study, Mick et al. fused APEX to ciliary target signals and identified the kinases liver kinase B1(LKB1), 5′ AMP-activated protein kinase (AMPK), and protein kinase A (PKA) as cilia-resident proteins (Fig. 2) (Mick et al., 2015). Interestingly, they found that PKA regulates the Hedgehog signaling pathway in this organelle. In the same year, Gupta et al. defined the proxisome of the cilium-centrosome interface using 58 BirA*-fused bait proteins (Fig. 2) (Gupta et al., 2015). They detected >1700 proteins and suggested that 112 newly identified proteins played a role in centriolar satellite biogenesis, centriole duplication, and ciliogenesis. Stress granules and processing bodies (P-bodies) are liquid-liquid phase-separated aggregations of ribonucleoproteins that regulate mRNA turnover and cellular stress protection, respectively (reviewed in Luo et al., 2018; Protter and Parker, 2016). A large prey-centric study involving 119 bait proteins targeting either P-bodies and/or stress granules captured 1972 preys which helped delineate a "core" P-body and stress granule proteasome as well as the overlapping interface between both organelles (Fig. 2) (Youn et al., 2018). Strikingly, Markmiller et al. showed almost simultaneously that the stress granule proteome is not only cell-type specific, but also changes according to different stresses and disease (Markmiller et al., 2018). This plasticity may also be observed during different viral infections, as there is growing evidence that viruses either block or highjack stress granules for successful replication (reviewed in Gaete-Argel et al., 2019). Until recently, the synaptic cleft was the only extracellular compartment mapped by HRP-based PL (Loh et al., 2016). However, new studies show that APEX2 can biotinylate the glycocalyx and extracellular vesicles (Joeh et al., 2020; Sanwald et al., 2020). Looking for glycan-binding partners of the extracellular bait protein galectin-3, Joeh et al. identified approx. 300 glycosylated proteins and hence proved that APEX2 can take "snapshots" of the extracellular space. On the other hand, Sanwald et al. revealed by APEX2 technology that the autophagy-related ATG8 protein GABA Type A Receptor-Associated Protein (GABARAP) is secreted in extracellular vesicles and together with other autophagy-related proteins. These groundbreaking studies showed that PL is not necessarily restricted to intracellular compartments or enclosed extracellular spaces like synaptic grooves and opens the door to mapping studies of other extracellular processes such as hormone release or immune signaling. Finally, a major advantage of APEX2-based PL, smartly exploited by the Kruse and Krogan labs, is its rapid labeling kinetics (1 min) for spatio-temporal resolution of protein signaling networks. Using G-protein-coupled receptor (GPCR) signaling as a model, the authors of these studies demonstrated that APEX2 can dissect spatio-temporal GPCR protein interaction dynamics after a stimulus (Lobingier et al., 2017; Paek et al., 2017).

In recent years, a large effort aimed at generating a cellular proxisome. The Varjosalo lab has set up a AP-MS/BioID dual system, coined MAC-tag, to comprehensively map subcellular localizations and protein interactions (Liu et al., 2018c). In their study, they used 18 proteins of different known localization to map each subcellular compartment and obtained a cellular "blueprint" including 1911 high-confidence interactions. Their study revealed that in general BioID identifies higher numbers of interactors than AE-MS but the overlap between AE-MS and BioID strongly depends on the protein bait and subcellular localization. The Gingras lab went one step further an generated a BioID-based HEK293 cellular map using 192 cellular markers spanning 32 different subcellular compartments (Go et al., 2019). This colossal effort accurately localized 4145 proteins and identified 35,902 unique high confidence proximity interactions. These outstanding studies provide a cellular map to confidently pinpoint the exact location of any protein of interest.

3.2 Nucleic acid proximity ligation

Proteins interact with both DNA and RNA to regulate transcription and translation. In recent years, PL methods have been developed to identify DNA and RNA-binding proteins as well as the genomic loci and RNA species they interact with.

Several methods help understand the protein and nucleotide structures responsible for DNA regulation. In 2017, Liu et al. developed CAPTURE (CRISPR affinity purification in situ of regulatory elements) to identify gene locus-specific chromatin regulatory proteins with PL enzymes (Liu et al., 2017). This method relies on the co-expression of a catalytically dead version of the gene editing endonuclease protein Cas9 (dCas9) containing a biotin acceptor peptide together with a gene-specific single-guide RNA (sgRNA) and wildtype BirA. BirA specifically biotinylates the DNA-bound dCas9 which can be captured and analyzed by MS. The authors of this study were able to identify regulatory proteins, long-range DNA interactions, and chromatin-associated RNA of different cis-regulatory DNA. An important caveat of this study was the need for crosslinking and AE for enrichment of the complexes. Schmidtmann et al. avoided these steps with CasID, a BioID-based approach in which BirA* is fused to dCas9 (Schmidtmann et al., 2016). CasID identified known and unknown chromatin regulatory factors of telomeric, major, and minor repetitive DNA regions. Likewise, the Sontheimer and Carr labs developed C-BERST (dCas9–APEX2 biotinylation at genomic elements by restricted spatial tagging) and GLoPro (genomic locus proteomics), two methods relying on the expression of locus-specific sgRNA and a dCas9-APEX2 fusion protein (Gao et al., 2018; Myers et al., 2018). Whereas C-BERST revealed proteins associated with telomere and centromere repeated sequences, GLoPro identified interactors of MYC proto-oncogene, bHLH transcription factor (MYC) and human telomerase reverse transcriptase (hTERT) promoter regions, thereby demonstrating that dCas9-APEX2 target single loci. In addition, Qiu et al. could show that CAPLOCUS (Combining CRISPR and peroxidase APEX2 system to identify local chromatin interactions), another dCas9-APEX2 method, could not only identify protein interactors of repetitive and single-locus genomic regions, but also long-range DNA interactors and chromatin-associated RNAs (Qiu et al., 2019). Finally, using a BASU ligase fused to specific chromatin regulator proteins Villaseñor et al. developed ChromID and identified proteins associated with specific chromatin modifications like methylation in a radius of approx. 35 nm (Villaseñor et al., 2020).

With respect to RNA-centric PL, methods to study RNA-protein interactions, the organelle-specific RNA transcriptome, and RNA labeling have been established. APEX-RIP (APEX RNA Immunoprecipitation) was the first methodology available to study protein-bound subcellular RNAs (Kaewsapsak et al., 2017), followed by Proximity-CLIP (Benhalevy et al., 2018; Ule et al., 2003). Both techniques use APEX2 biotinylation of proteins followed by either mild formaldehyde or photoreactive ribonucleoside-enhanced crosslinking, respectively. Whereas APEX-RIP performed well in the mitochondrial matrix, ER, and nucleus; Proximity-CLIP scored better in the cytosol and at cell-cell junctions. To avoid protein-RNA crosslinking and improve RNA mapping quality the Ting lab developed APEX-seq, a method to directly biotinylate and sequence

RNA transcripts (Fazal et al., 2019). They generated a RNA-atlas of nine different subcellular localizations (nucleus, nucleolus, nuclear lamina, nuclear pore, cytosol, ER cytosol-facing membrane and lumen, mitochondrial outer membrane and matrix), identified the transcript localization of over 32,000 genes and strikingly found that different RNA isoforms from the same gene have different localization in accordance to their protein localization. APEX-seq was soon used to study the RNA transcriptome of stress granules, finding major transcriptome differences depending on the type of cellular stress (Padrón et al., 2019). Replacement of biotin-phenol with biotin-aniline further improved APEX-seq (Zhou et al., 2019). An alternative approach to identify RNA specific proteomes is targeting the PL enzyme to a specific RNA sequence. In 2018, Ramanathan et al. developed RaPID (RNA-protein interaction detection). In RaPID, the RNA sequence of interest is flanked by two BoxB stem loops from the bacteriophage λ and co-expressed with an λN peptide fused to BirA* or BASU. The fusion protein will specifically bind BoxB stem loops and thereby enable PL of proteins associated with the target RNA sequence (Ramanathan et al., 2018). Using RaPID the authors correlate pathological human disorders with altered protein-RNA interactions as well as identify new protein interactors of zika virus (ZIKV) RNA. One year later, Mukherjee et al. developed RNA-BioID, a similar approach using bacteriophage MS2 loops and BirA*-MS2 coat protein (MCP), and investigated the β-actin mRNA interactome (Mukherjee et al., 2019). Finally, a hybrid study using both MS2-MCP and a dead Cas13 endonuclease in combination with APEX2 revealed new interactors of the human telomerase RNA hTR with better temporal resolution (Han et al., 2020). Taken together proximity labeling has advanced the field of protein-nucleic acid interactions immensely and the combination of different PL strategies with CRISPR technology led to a plethora of methods to address specific questions of molecular interactions between these major cellular biomolecules.

4. Proximity labeling to study host-pathogen interactions

Since the discovery of the principle of proximity biotinylation and its use, most of the studies have focused on better understanding the spatial proteome and interactome of eukaryotic cells. However, recently several labs have appreciated the value of this technology to study host-pathogen interactions. In this section we provide a non-exhaustive summary of major studies on host-pathogen interactions, with a special focus on viruses. As virology applications are just dawning, we start by describing the successful use of the technology to map bacteria—host interactions.

4.1 Bacteria

A pathogenic bacterium extensively studied by PL is *Chlamydia trachomatis*. *C. trachomatis* is a gram-negative obligate intracellular bacterium and causative agent of a major sexually transmitted disease worldwide with approx. 127 million annual cases. It can cause urethritis, cervicitis, pelvic inflammation, ectopic pregnancy, miscarriage, obstructive infertility, and blindness. Within the cell, *Chlamydia* spp. reside in vacuole-like organelles denominated inclusions (reviewed in Di Pietro et al., 2019). Inclusion formation is dependent on bacterial type III secretion system (T3SS) effector proteins such as Incs. These Inc proteins are anchored in the inclusion membrane and maintain the integrity of the inclusion by interacting with each other and with host proteins. However, due to major difficulties to purify chlamydial inclusions little was known about the interactome between Inc effectors and host cells. The first study characterizing inclusions by PL used the eukaryotic signal sequence, YGRL, to target proteins to the inclusion membrane (Kabeiseman et al., 2014). By fusing the BirA* to syntaxin 6 with and without the signal sequence, the authors observed that the fusion protein only localized and biotinylated the inclusion membrane in the presence of the YGRL signal peptide. In 2017, Rucks et al. fused APEX2 to the chlamydial effectors IncA and IncF (Rucks et al., 2017). Whereas IncA-APEX2 and IncF-APEX2 localized to and biotinylated inclusions, the expression of the fusion proteins had appreciable yet not drastic effect on the diameter, lipid acquisition and stability of the inclusions, suggesting that either APEX2 fusion alters Inc function or that Inc expression regulates homeostasis of inclusions. In 2019, two studies by Dickinson et al. and Olson et al. revealed the interactome of Incs at the inclusion membrane using APEX2 (Dickinson et al., 2019; Olson et al., 2019). In the first study, the authors identified more than 400 protein interactors of IncB-APEX2 at early, middle, and late stages of the bacterial intracellular life cycle. At early stages of infection, they found an association of the inclusion membrane with the ER exit site (ERES), which was critical for bacterial growth and generation of progeny. Using IncA-APEX2 and IncF-APEX2, the authors of the second study found significantly lower number of protein interactors. Among them, they found leucine-rich repeat flightless-interacting protein 1 (LRRF1), which specifically interacted with

Inc CT226. Due to discrepancies between both studies, Olson et al. performed a meta-analysis and suggested the study by Dickinson probably over-estimated high-confidence interactions, recommending Significance Analysis of INTeractome (SAINT) analysis as a more stringent and reliable analysis to identify true interactors (Olson et al., 2020).

Other PL-based research on pathogenic bacteria has identified protein interactors of *Chlamydia psittaci* and *Salmonella* Type-III secretion system effectors with BioID as well as crucial assembly modulators of enteroaggregative *E. coli* type-VI secretion system using APEX2 (D'Costa et al., 2019; Mojica et al., 2015; Santin et al., 2018). Thus, PL could successfully map bacteria induced subcellular compartments as well as large bacterial protein complexes in host cells.

4.2 Parasites

Trypanosoma species, *Plasmodium* species, and *Toxoplasma gondii* are among the most common human parasites (reviewed in Theel and Pritt, 2016) and cause sleeping sickness, malaria and toxoplasmosis, respectively. Morriswood et al. were the first to study the protozoan *Trypanosoma brucei* with PL techniques (Morriswood et al., 2013). The authors tagged TbMORN1, a protein found in a cytoskeletal structure of the parasite called bilobe, with BirA* and applied BioID. Nine previously unknown proteins were found, of which seven localized to the bilobe and two to the flagellum. Their study demonstrates that PL can be used to study unicellular organisms. In 2017, De Pablos et al. tagged two *T. brucei* RNA-binding proteins, RBP9 and 10, and performed BioID followed by RNA sequencing (De Pablos et al., 2017). Using the two baits, they identified over 200 high-confidence potential protein interactors of which 71 were related to mRNA metabolism, suggesting that those are likely functionally relevant interactors. Furthermore BioID identified proteins that regulate the polarization and motility of *T. brucei* (Dong et al., 2020; Hilton et al., 2018).

For *Plasmodium* spp. PL approaches elucidated the subcellular interactome during the intracellular stage of the parasite. *Plasmodium* parasites must disrupt a specialized parasitophorous vacuole and the red blood cell membrane to egress from cells and be taken up by mosquitoes (Tuteja, 2007). Parasite proteins contained in vesicles like osmiophilic bodies mediate the membrane lysis. To identify the *Plasmodium berghei* proteins involved in cell egress, Kehrer et al. fused BirA* to the osmiophilic body protein MDV1/PEG3 and performed BioID in human red blood cells (Kehrer et al., 2016). From the 34 parasite proteins identified, 22 were involved in cell egress and one, the Merozoite TRAP-like Protein (MTRAP), was essential for mosquito transmission. A subsequent BioID study on the composition of the parasitophorous vacuole membrane of *P. berghei* revealed 61 parasite proteins as potential neighbors of exported protein 1 (EXP1) (Schnider et al., 2018). From those 61, 15 had known parasitophorous vacuole membrane localization, seven novel candidates were further characterized and three of those were confirmed to be part of the parasitophorous vacuole membrane. Two more studies using BioID and BioID2 on the *P. falciparum* vacuole found proteins that were crucial for parasite growth (Khosh-Naucke et al., 2018) and revealed a new structural role for EXP1 in the vacuole integrity (Nessel et al., 2020). Finally, a study on the *P. falciparum* apicoplast, a semi-autonomous organelle of the parasite, using BioID identified 346 high-confidence potential interactors of which ATP-binding cassette protein ABCF1 was shown to be essential for blood-stage parasite survival (Boucher et al., 2018).

The most extensively studied parasite using PL methods is *Toxoplasma gondii*. This parasite harbors the so-called inner membrane complex, a highly-specialized peripheral membrane comprised of alveolar sacs that play a key role in parasite invasion, motility, and replication. Due to its importance, Chen et al. set out to identify the proteome of this membrane by generating recombinant parasites expressing different inner membrane complex proteins as bait for BioID and identified 19 novel parasite proteins with different sub-organellar localizations (Chen et al., 2015). In the same year, Gaji et al. showed with BioID that calcium dependent protein kinase TgCDPK3 interacted with and phosphorylated the unconventional motor myosin protein TgMyoA and this phosphorylation was critical for parasite motility (Gaji et al., 2015). Just like *Plasmodium* spp., *T. gondii* needs a parasitophorous vacuole for proper replication and protection within the host cell. To maintain it, *T. gondii* releases dense granules proteins (GRAs) that influence immune evasion, nutrient uptake, and vacuolar structure. To identify novel vacuole-secreted GRAs, Nadipuram et al. performed BioID on *T. gondii* infected fibroblasts using GRA17 as a bait and identified several parasite proteins of which 13 localized to the vacuole (Nadipuram et al., 2016). Out of the novel GRAs, GRA39 was important for virulence in mice as genetic ablation of GRA39 affected parasite growth and caused atypical lipid deposition in the vacuole. Two subsequent studies further elucidated the vacuolar proteome and identified additional GRAs. In the first study, the authors tagged GRA1 with both BirA* and APEX2 and found 46 proteins, of which 17 were common with the two methods, and 14 were GRAs (Pan et al., 2019). The second study focused on GRAs important for chronic infection and revealed five novel proteins by MAG1-BirA* BioID (Nadipuram et al., 2020). Importantly, they found that GRA55 was crucial for cyst formation and maintenance in mouse brain. Another important subcellular organelle of *Plasmodium* spp. is the conoid, a motile organelle associated with microtubules and essential for parasite motility (Graindorge et al., 2016). PL of the conoid proteome revealed CPH1 as an essential hub protein that stabilizes the conoid structure and AC9 and AC10 as essential proteins for the formation of the conoid (Long et al., 2017; Tosetti et al., 2020). Lastly, Engelberg et al. found in an organelle termed the apical annuli several signaling proteins and proteins containing coiled-coil domains. The latter are embedded in a pore-like shape in the inner membrane complex, which suggests that the apical annuli may originate from the centrosome (Engelberg et al., 2020). In sum, PL helped identify parasite proteins in subcellular organelles of the parasites and in host cell compartments in the context of infection.

4.3 Early studies on viruses using biotinylation methods

The use of PL methods in the field of virology is relatively new and few studies are available despite the potential of the method for host-pathogen interaction studies.

The first studies using PL enzymes to characterize viruses relied on WT BirA and a minimum biotin acceptor sequence (BAS) peptide. Thus, these are not PL studies per se but applications of WT BirA. Fusion of BAS to a viral protein specifically biotinylates the tagged protein in the presence of BirA and thereby allows isolation of proteins interacting with the BAS tagged protein by co-immunoprecipitation. The first study using this system was done by Belshan et al. and combined WT BirA with BAS to biotinylate the human immunodeficiency virus 1 (HIV-1) in vivo (Belshan et al., 2009). The BAS sequence was fused to either the matrix region of the structural protein gag or the integrase region of non-structural protein pol. The generated viruses were infectious but biotinylation of the integrase rendered progeny non-infectious. Few years later, the same lab generated immortalized T cells stably expressing BirA and replication competent HIV-2 viruses with a BAS-tag at the Vpx or Vpu accessory proteins (Belshan et al., 2011). Other groups followed tagging viruses such as influenza A virus (IAV) (Postel et al., 2011), rotavirus (De Lorenzo et al., 2012), herpesvirus (Lajko et al., 2015), and bacteriophage T7 (Wang et al., 2016) with BAS for BirA-dependent biotinylation. In 2017,

Santana et al. tagged the murine gammaherpesvirus 68 transcription factor ORF50, an important viral protein for lytic reactivation, and studied the chromatin localization of this protein in BirA-expressing transgenic mice (Santana et al., 2017). Using this approach, the authors identified a genomic locus important for viral reactivation in B cells. In an independent study published in 2017, Calton et al. investigated the nuclear translocation of viral DNA (vDNA) of the human papillomavirus 16 (HPV16) during infection (Calton et al., 2017). By tagging the viral L2 chaperone, known to transport the viral DNA from the Golgi to the nucleus, they found that the L2/vDNA complex only translocates to the nucleus at the beginning of mitosis, ensuring equal partitioning of vDNA between infected and daughter cell.

4.3.1 DNA viruses

The first virus studied with PL was Epstein-Barr virus (EBV). EBV is an enveloped double-stranded DNA (dsDNA) virus belonging to the *Herpesviridae* family and can cause mononucleosis, nasopharyngeal carcinoma, Burkitt lymphoma, Hodgkin lymphoma, post-transplant lymphoma, or gastric carcinoma and is the causative agent of approx. 1.5% of global carcinomas (reviewed in Farrell, 2019). The latent membrane protein 1 (LMP1) is a large multipass transmembrane oncoprotein that immortalizes B cells and non-lymphoid tissues by the constitutive activation of signaling pathways leading to survival of infected cells. In EBV infected cells, LMP1 localizes to detergent insoluble membrane microdomains, making the study of protein interactions by AE-MS extremely difficult. In 2015, Holthusen et al. set out to identify host factors interacting with LMP1 using a biomolecular fluorescence complementation (BiFC) and BioID (Holthusen et al., 2015). BiFC is used to validate protein interactions and relies on the reconstitution and fluorescent emission of a reporter protein that has been fragmented and fused to two proteins of interest. Whereas the main screen for interactors was performed with BiFC, the authors used a LMP1-BirA* fusion protein to verify the interaction of the cytoskeleton-associated protein actinin 1 to LMP1 and suggested that a specific post-translationally modified version of this protein was the major interactor of LMP1. Although this study did not use BioID for identification of neighboring proteins, it showed that BioID can reveal interactions of host cell proteins with hard-to-purify viral proteins and paved the way for future studies. Not much later, Rider et al. used BioID coupled with MS to investigate the interactome of LMP1 and identified 1862 proteins in seven independent experiments (Table 1) (Rider et al., 2018). The identified proteins were mainly involved in protein and

Table 1 Summary of PL studies in virology.

Year	Virus	Genome	Method	Bait	Identified proteins	Verified interactors	Major findings	References
2018	EBV	dsDNA	BioID	LMP1	1862	CD63, HSC70, ITGB1, syntenin-1, TSG101, and STAT3	The identified proteins were suggested to be involved in protein and vesicle trafficking as well as signaling transduction	Rider et al. (2018)
2020	HSV-1	dsDNA	BioID	Glycoprotein gM	2455	–	gM requires exportin 6 (XPO6) for the release from the nuclear membrane to the TGN	Boruchowicz et al. (2020)
2020	SV40	dsDNA	BioID2	EMC4, EMC7	–	Rab7	SV40 shuttles from late endosomes (Rab7+) to ER using EMC4 and 7 as molecular tethers	Bagchi et al. (2020)
2020	HBoV1	ssDNA	BioID	NP1	~300	DHX15,CPSF6	CPSF6 imports NP1 to the nucleus and thereby modulates viral mRNA processing and DNA replication	Wang et al. (2020)
2015	HIV-1	+ssRNA	BioID	Gag	50	–	Enriched proteins were annotated for membrane, cytoskeleton, nuclear transport, lipid metabolism, translation factor, and RNA processing	Ritchie et al. (2015)
2018	ZIKV	+ssRNA	BioID	10 viral proteins	1224	–	NS5A helicase disrupted nuclear cajal bodies and NS2A negatively impacts peroxisome density.	Coyaud et al. (2018)
2020	SARS-CoV-2	+ssRNA	Miniturbo	27 viral proteins	2242	–	To pinpoint the subcellular localization of the viral proteins, they mapped A549 cells using 17 host proteins of known localization	Samavarchi-Tehrani et al. (2020)
2019	IAV	-ssRNA	BioID	PA-X	156	NUDT21	Main interactors were involved in RNA splicing and RNA processing	Gaucherand et al. (2019)

vesicle trafficking as well as signal transduction. Interestingly, 1392 of the 1862 proteins were annotated in the vesiclepedia, a repository for extracellular proteins (Kalra et al., 2012). In line with this, some of the identified proteins (CD63, syntenin-1, ALIX, TSG101, Hrs, CHMPs, sorting nexins) are well-known regulators of exosome biogenesis. These findings are in accordance with previous studies showing that LMP1 is secreted from cells in exosomes. Indeed, downregulation of the exocytosis regulators ALIX and syntenin-1 diminished LMP1 secretion, supporting this notion. In 2020, the same group showed that LMP1 expression increases levels of exosome biogenesis and knockdown (KD) of ESCRT proteins CHMP4, 5, and 6 as well as exosome biogenesis modulators syntenin-1, ALIX, Hrs, and TSG101 decreases LMP1 packaging in these vesicles (Nkosi et al., 2020). Moreover, the authors showed that KD of syntenin-1 and Hrs alters the endolysosomal trafficking of LMP1, ultimately altering the metastatic properties of tumor cells. The work on EBV demonstrates that PL not only aids elucidating basic molecular details of virus host interactions but may also reveal protein interactions causative for disease phenotypes caused by viruses.

Herpesviruses human cytomegalovirus (HCMV) and herpes simplex 1 (HSV-1) have also been studied by BioID. Both viruses cause persistent infections, which mostly remain asymptomatic but can cause complications in immunocompromised individuals and neonates. During HCMV infection, the endosecretory machinery is remodeled forming the so-called cytoplasmic virion assembly complex (cVAC), which is important for viral tegument formation and secondary envelopment (reviewed in Alwine, 2012). This process is mediated by the viral protein pUL103. To identify host interactors of pUL103, Ortiz et al. generated viruses expressing either BirA*-pUL103 or pUL-V5-His and used BioID and AE-MS for identification of interacting viral and host proteins (Ortiz et al., 2016). AE-MS identified 49 viral proteins and 1745 cellular proteins and BioID 47 viral proteins and 1126 cellular proteins. After selecting for proteins identified by both methods, they described 13 viral and 18 cellular high-confidence interactors. The viral proteins identified were tegument, nuclear, and DNA-associated proteins and the cellular proteins were associated with Golgi transport, protein localization, and innate immune mechanisms. With regards to HSV-1, Boruchowicz et al. studied the interactome of the glycoprotein gM (Table 1) (Boruchowicz et al., 2020). This protein localizes to the nuclear membrane at 4 h post infection and to the trans-Golgi network (TGN) 12 h post infection. gM modulates the viral fusion machinery, molecular transport of viral proteins, and capsid secondary envelopment. Using gM-BirA*, the authors studied the interactome of gM at 4 and 12 h post infection and identified 2455 proteins of which 170 were found at both time points and with high abundance. Interestingly, 35% of the proteins were implicated in protein transport. Of these proteins, exportin 6 (XPO6) was found to be required for the release of gM from the nuclear membrane to the TGN.

Naked DNA viruses like the polyomavirus simian vacuolating factor 40 (SV40) and the parvovirus human bocavirus 1 (HBoV1) have recently been studied with PL. While the method was used to study the entry process of SV40, later stages of the viral life cycle were investigated for HBoV1. SV40 causes latent infection and may induce tumors. SV40 enters cells by endocytosis, transport to the ER, translocation to the cytosol and finally import into the nucleus (Schelhaas et al., 2007). To understand the ER complex subunit (EMC)-dependent transfer from late endosomes to ER, Bagchi et al. tagged ER membrane complex subunits EMC4 and 7 and applied BioID2 (Table 1) (Bagchi et al., 2020). They found that only the late endosomal marker Rab7 but not its early counterpart (Rab5) was biotinylated, strongly suggesting that late endosomes and ER might be in proximity to allow this transfer. HBoV1 can cause acute respiratory tract infection in children. Since the viral NP1 protein is crucial for capsid maturation and replication, Wang et al. studied the interactome of HBoV1 NP1 using BioID (Wang et al., 2020). They found that two mRNA processing factor proteins, DEAH-Box helicase 15 (DHX15) and cleavage and polyadenylation specific factor 6 (CPSF6) interact with NP1 (Table 1). Downregulation and upregulation of CPSF6 modulated capsid protein levels. Moreover, NP1 nuclear localization depended on CPSF6. Hence it is assumed that CPSF6 imports NP1 to the nucleus and thereby modulates viral mRNA processing and DNA replication.

Finally, host cell interactions of the fish iridovirus Singapore grouper iridovirus (SGIV), a major threat of grouper fish with up to 90% mortality, were recently studied using PL. The authors of the study used BioID to identify the interactome of ICP18, an immediate early protein that modulates viral replication and assembly by regulation of the cell cycle (Wu et al., 2020). Using a non-permissive HEK293T cell line, they found 112 proteins in proximity to ICP18, of which many modulated cell cycle and adhesion. Importantly, they verified the interaction of ICP18 with cyclin-dependent kinase 1 (CDK1), a key player in cell cycle regulation.

Taken together, although the first description of a PL approach to study DNA virus-host cell interactions dates back to 2015, only few studies applied the methodology to discover host factors involved in the virus life cycle or disease phenotypes of DNA viruses. Given the relatively large genomes and number of proteins of DNA viruses, it is expected that future PL studies will contribute significantly to the elucidation of protein networks during DNA virus infection.

4.3.2 Retroviruses

Proximity labeling also served to uncover interactions between retroviruses and cells. HIV-1 is an enveloped positive single-stranded RNA retrovirus that causes Acquired Immunodeficiency Syndrome (AIDS). An estimated 37.9 million people are infected with HIV-1 worldwide. Several cellular proteins have been identified to restrict HIV-1 infection (reviewed in Colomer-Lluch et al., 2018). Among them, tetherin potently hampers viral release and spread by tethering budding virions to the plasma membrane. Vpu is an HIV-1 accessory protein counteracting tetherin by targeting it to proteasomal degradation, downregulating it on the plasma membrane, and sequestering it in endosomal compartments. In 2015, Kueck et al. studied the molecular requirements underlying tetherin antagonism by Vpu (Kueck et al., 2015). Using Vpu-myc-BirA*, the authors show that the Vpu/tetherin complex interacts with the clathrin adaptor AP-1 only when Vpu is phosphorylated at a specific position and this interaction leads to endosomal trafficking and proteasomal degradation. Another protein studied with PL is the group-specific antigen (Gag), a polyprotein that matures into the structural proteins matrix (MA), capsid (CA), nucleocapsid (NC), and p6. This protein is necessary and sufficient to generate non-infectious viral particles. In 2015, two studies used BioID to identify protein interactors of Gag. In the study by Ritchie et al. the authors generated constructs with a BirA* fusion between MA and CA and a MA-deleted BirA*-Gag variant (Ritchie et al., 2015). They were able to identify 50 cellular proteins that were enriched for the gene ontology (GO) terms membrane, cytoskeleton, nuclear transport, lipid metabolism, translation factor, and RNA processing (Table 1). Interestingly, the authors observed that the presence of MA hampered biotinylation of the viral glycoprotein Env, suggesting that MA may sterically impede Env labeling. In the second study on Gag, Le Sage et al. used a Myc-BirA*-Gag for PL and found 47 proteins, among which they verified DEAD-Box helicase 17 (DDX17) and ribosomal protein s6 (RPS6) as Gag interaction partners (Le Sage et al., 2015). Overall, the authors identified many ribosomal proteins and the antiviral protein IFI16. Strikingly, only one protein found in this study, lyric, overlapped with the results from Ritchie et al. suggesting that the positioning of the

BirA* tagging strongly affects the proteins accessible to the biotinylation cloud. Other factors such as cell type, expression levels of the BirA* construct and the time post infection may contribute to the observed discrepancies.

Another retrovirus studied by PL is human T-lymphotropic virus (HTLV-1). HTLV-1 causes T-cell lymphoma in adults and induces cellular transformation and inflammation by constitutive activation of NF-κB signaling (Mori et al., 1999), through the viral protein Tax. In 2019, Schwob et al. performed BioID with BirA*-tagged Tax and identified p62 as a novel interactor. Since p62 silencing decreased and p62 overexpression increased NF-κB activity the authors concluded that p62 is a mediator of Tax-induced NF-κB activation (Schwob et al., 2019). In sum, PL methods aided in the discovery of innate factors and mechanisms during retroviral infection.

4.3.3 RNA viruses

The first RNA virus studied with PL was ZIKV. ZIKV is a positive single-stranded RNA (+ssRNA) virus of the family *Flaviviridae*. It causes Guillain-Barré syndrome in adults and microcephaly in newborns (reviewed in Musso and Gubler, 2016). In an extensive study, Coyaud et al. studied the global interactome of ZIKV by tagging the ZIKV 10 viral proteins with a Flag-BirA* at both the N- and C-terminus, expressing the proteins individually, and performing BioID and AE-MS (Coyaud et al., 2018). They identified 3033 high-confidence binary interactions between 1224 proteins, more than 60% of which were known to play a role in infection with other viruses (Table 1). Interestingly, BioID identified three times more membrane proteins that AE-MS, suggesting a better performance of BioID on transmembrane proteins. Proteins enriched in the combined dataset were involved in polypeptide processing and quality control, vesicle trafficking, RNA processing, and lipid metabolism. Furthermore, the authors used the dataset for hypothesis building and found that the viral capsid protein remodeled nuclear and ER membranes, the viral NS5A helicase disrupted nuclear cajal bodies, NS3-NS2B modified lysosome volumes, and NS2A negatively impacted peroxisome density. Although this study is limited to ectopically expressed viral proteins, it shows the massive potential of BioID to identify the interactome of viruses and link viral protein localization to plausible functions.

V'kovski et al. and Samavarchi-Tehrani et al. used BioID to study coronaviruses (CoV) (Samavarchi-Tehrani et al., 2020; V'kovski et al., 2019). CoVs are enveloped positive single stranded (+ssRNA) viruses of the family

Coronaviridae. Several human low pathogenic CoVs cause mild cold-like symptoms. However, emerging CoVs like severe acute respiratory CoV (SARS-CoV), middle east respiratory CoV (MERS) and the recent SARS-CoV-2 can cause serious respiratory disease and be lethal (reviewed in Fung and Liu, 2019). In 2019, the group of Volker Thiel used PL for the first time for the betacoronavirus murine hepatitis virus (MHV). By engineering a nsp2 fused in-frame to myc-tagged $BirA^{R118G}$ the authors revealed proteins interacting within the ER-derived coronavirus replication and transcription complex (RTC) using BioID (V'kovski et al., 2019). In total, 1381 host proteins were identified, 513 of which were significantly enriched. siRNA screening of all 513 proteins identified 53 proteins that reduced MHV infection. These proteins belong to the functional groups of vesicular transport, ESCRT, proteasome, and protein synthesis. Altogether, their results suggest that initiation of translation of viral proteins occurs near the RTC and thus host protein translation is spatially separated from viral protein synthesis. This may be a mechanism to ensure viral protein synthesis while shutting down protein synthesis. In 2020, the group of Anne-Claude Gingras published a preprint describing the global interactome of SARS-CoV-2 (Samavarchi-Tehrani et al., 2020). In this comprehensive study, the authors tagged 27 viral proteins with miniTurbo and expressed them individually in A549 lung carcinoma cells. In total, they found 2242 host proteins and 7810 binary interactions (Table 1). They mapped the subcellular localization of each viral protein in A549 cells using 17 miniTurbo fused host proteins of known localization. In subsequent imaging analysis they confirmed the protein localization determined by PL and suggested a possible role of the viral nucleoprotein in stress granule formation. In summary, the study is an invaluable source for data-driven hypothesis formulation regarding the underlying mechanism of betacoronavirus infection of lung epithelial cells and resulting pathogenesis.

Among enveloped negative-stranded (-ssRNA) viruses, host interaction partners of influenza A virus (IAV) and Ebola virus (EBOV) proteins have been identified using PL techniques. IAV is a member of the family *Orthomyxoviridae* that causes flu and EBOV belongs to the *Filoviridae* family and causes Ebola virus disease. To understand the mechanism behind IAV PA-X mediated host mRNA degradation, Gaucherand et al. performed BioID using PA-X as bait (Gaucherand et al., 2019). PA-X is fusion protein produced by ribosomal frameshifting that comprises the ribonuclease domain of the viral polymerase fused to the unique X-open reading frame X-ORF (Jagger et al., 2012). From the 156 potential interactors, 29 of them were of high-confidence (defined as statistically significant) and were involved in RNA splicing and RNA processing (Table 1). Among them were Nudix Hydrolase 21 (NUDT21) and CPSF6, both members of the cleavage factor I mammalian (CFIm) that is essential for maturation of pre-mRNAs. For EBOV, BioID delineated the interactome of matrix VP40, uncovering a functional role of the Rab14 GTPase in the plasma trafficking and cell secretion of the viral VP40 matrix protein (Fan et al., 2020).

Besides human pathogens, the plant tomato mosaic virus (TMV) interactome has been studied *in planta* using BioID. TMV is a naked +ssRNA virus of the family *Virgaviridae* that causes a characteristic "mosaic" pattern on leaves. Das et al. identified 18 significant protein interactors of the 162kDa replicase protein in tobacco plants. The authors showed that the interaction partner suppressor of G2 allele of SKP1 (SGT1) protects against infection (Das et al., 2019). In an independent study, Macharia et al. studied interactors of the autophagy protein ATG8 in TMV infected tobacco plants. The authors identified 67 proteins and discovered an autophagy-related protective role for the Huntington-interacting protein K-like (HYPK). In their study, HYPK silencing rendered plants more susceptible to TMV infection and pathology.

Taken together, BioID and miniTurbo PL approaches helped identify virus-host protein interactions in plants and animals for a broad spectrum of viruses. So far, no study used APEX2 to study virus host interactions, but several studies used this peroxidase to visualize viral proteins by EM (Eyre et al., 2016, 2017; Lin et al., 2017; Plescia et al., 2020; V'kovski et al., 2019). In addition, Sherwood and Hayhurst developed nanobodies fused to APEX2 instead of HRP for generation of affordable rapid detection assays for viral proteins (Sherwood and Hayhurst, 2019).

5. Future perspectives and limitations of the technology

Since the discovery and application of BPL and peroxidases for PL, this methodology has proven invaluable to characterize the cellular and subcellular proteome. The ability to label proteins in living cells or even in intact organisms such as mice and tobacco plants combined with the tolerance of stringent protein isolation techniques has allowed mapping of the protein composition of insoluble or membraneless organelles such as stress granules or centrosomes and greatly reduced the number of false negatives as compared to classical AE-MS methods. An important point to consider when using these techniques is that identified proteins are not necessarily direct interaction partners of the bait. Therefore, it is imperative to evaluate protein interactions by orthogonal assays such as ELISA or BiFC—also known as split reporter assays. Second, as for any given technique revealing proximities or interactions, PL does not reveal the biological relevance of proximity or interaction. Follow-up experimentation for instance RNA interference or CRISPR/Cas9 gene knock-out is mandatory to assess the biological relevance of the protein complexes identified.

The use of PL in host-pathogens interaction studies has started to bloom recently and will likely lead to groundbreaking studies in the near future. For the field of virology the technique is of particular interest as viruses have evolved to hijack cellular mechanisms and reshape cellular protein complexes to their advantage. As peroxidase PL allows kinetics studies, the technique is ideally suited to study the impact of virus infection on host cells and their protein complexes in a spatially and time resolver manner. Moreover, PL allows uncovering of the diverse interactions of viral gene products with each other and with host proteins. To that end, generation of viruses modified to express viral proteins fused to PL enzymes may allow analysis of viral and host proteins in the context of infected cells instead of cells expressing single viral proteins. In addition, split versions of PL enzymes will allow studies of viral protein-protein interactions in different subcellular localizations. Furthermore, we expect to see studies dissecting the intracellular and extracellular virion proteome, which would identify host factors crucial for virion integrity and infectivity. Obviously, in particular small viruses have proven difficult to modify and thus smart ways of introducing the BPL or peroxidase into virions are the prerequisite for such studies. Finally, we envision PL to foster studies on the subcellular viral proteome and transcriptome and on the spatiotemporal dissection of viral life steps including viral entry, intracellular trafficking, replication complex formation, assembly, and release of progeny particles from the cells.

As for all methods which ultimately involve MS to identify prey proteins, it is critical to carefully design experiments to produce biologically relevant and reproducible results. Key points of consideration are to determine the cellular background, e.g., from endogenous peroxidases with scrutiny, perform statistics on a reliable number of biological, not technical, replicates, choose an experimental system which is most physiological, and carefully assess if the genetically modified viruses still recapitulate the biological mechanisms of the wild type strains.

Given that these hurdles are overcome, we believe that the relative simplicity in conjunction with the reliability of PL techniques will encourage more virologist around the world to use PL as the primary choice for the study of host-pathogen interactions by proteomics.

Acknowledgments

We thank Barbora Veselková for critical reading of the manuscript. G.G was supported by the Knut and Alice Wallenberg Foundation, the Federal Ministry of Education and Research together with the Ministry of Science and Culture of Lower Saxony through the Professorinnen programm III and the Federal Ministry of Education and Research (project COVID-Protect, # 01KI20143C).

References

Alwine, J.C., 2012. The human cytomegalovirus assembly compartment: a masterpiece of viral manipulation of cellular processes that facilitates assembly and egress. PLoS Pathog. 8, e1002878.

Amini, F., Kodadek, T., Brown, K.C., 2002. Protein affinity labeling mediated by genetically encoded peptide tags. Angew. Chem. Int. Ed. 41, 356.

Antonicka, H., Choquet, K., Lin, Z.-Y., Gingras, A.-C., Kleinman, C.L., Shoubridge, E.A., 2017. A pseudouridine synthase module is essential for mitochondrial protein synthesis and cell viability. EMBO Rep. 18, 28–38.

Bagchi, P., Torres, M., Qi, L., Tsai, B., 2020. Selective EMC subunits act as molecular tethers of intracellular organelles exploited during viral entry. Nat. Commun. 11, 1127.

Belshan, M., Schweitzer, C.J., Donnellan, M.R., Lu, R., Engelman, A., 2009. In vivo biotinylation and capture of HIV-1 matrix and integrase proteins. J. Virol. Methods 159, 178–184.

Belshan, M., Matthews, J.M., Madson, C.J., 2011. Replication of biotinylated human immunodeficiency viruses. J. Virol. Methods 171, 299–302.

Bendayan, M., 2001. Tech.Sight. Worth its weight in gold. Science 291, 1363–1365.

Benhalevy, D., Anastasakis, D.G., Hafner, M., 2018. Proximity-CLIP provides a snapshot of protein-occupied RNA elements in subcellular compartments. Nat. Methods 15, 1074–1082.

Bersuker, K., Peterson, C.W.H., To, M., Sahl, S.J., Savikhin, V., Grossman, E.A., Nomura, D.K., Olzmann, J.A., 2018. A proximity labeling strategy provides insights into the composition and dynamics of lipid droplet proteomes. Dev. Cell 44, 97–112.e7.

Bhaskar, B., Immoos, C.E., Shimizu, H., Sulc, F., Farmer, P.J., Poulos, T.L., 2003. A novel heme and peroxide-dependent tryptophan-tyrosine cross-link in a mutant of cytochrome c peroxidase. J. Mol. Biol. 328, 157–166.

Bobrow, M.N., Harris, T.D., Shaughnessy, K.J., Litt, G.J., 1989. Catalyzed reporter deposition, a novel method of signal amplification. Application to immunoassays. J. Immunol. Methods 125, 279–285.

Boruchowicz, H., Hawkins, J., Cruz-Palomar, K., Lippé, R., 2020. The XPO6 exportin mediates herpes simplex virus 1 gm nuclear release late in infection. J. Virol. 94 (21), e00753–20.

Boucher, M.J., Ghosh, S., Zhang, L., Lal, A., Jang, S.W., Ju, A., Zhang, S., Wang, X., Ralph, S.A., Zou, J., Elias, J.E., Yeh, E., 2018. Integrative proteomics and bioinformatic prediction enable a high-confidence apicoplast proteome in malaria parasites. PLoS Biol. 16, e2005895.

Branon, T.C., Bosch, J.A., Sanchez, A.D., Udeshi, N.D., Svinkina, T., Carr, S.A., Feldman, J.L., Perrimon, N., Ting, A.Y., 2018. Efficient proximity labeling in living cells and organisms with TurboID. Nat. Biotechnol. 36, 880–887.

Buoncristiani, M.R., Howard, P.K., Otsuka, A.J., 1986. DNA-binding and enzymatic domains of the bifunctional biotin operon repressor (BirA) of Escherichia coli. Gene 44, 255–261.

Callegari, S., Müller, T., Schulz, C., Lenz, C., Jans, D.C., Wissel, M., Opazo, F., Rizzoli, S.O., Jakobs, S., Urlaub, H., Rehling, P., Deckers, M., 2019. A MICOS-TIM22 association promotes carrier import into human mitochondria. J. Mol. Biol. 431, 2835–2851.

Calton, C.M., Bronnimann, M.P., Manson, A.R., Li, S., Chapman, J.A., Suarez-Berumen, M., Williamson, T.R., Molugu, S.K., Bernal, R.A., Campos, S.K., 2017. Translocation of the papillomavirus L2/vDNA complex across the limiting membrane requires the onset of mitosis. PLoS Pathog. 13, e1006200.

Cao, Q., Wang, Z., Wan, H., Xu, L., You, X., Liao, L., Chen, Y., 2018. PAQR3 regulates endoplasmic reticulum-to-Golgi trafficking of COPII vesicle via interaction with Sec13/Sec31 coat proteins. iScience 9, 382–398.

Chaiet, L., Wolf, F.J., 1964. The properties of streptavidin, a biotin-binding protein produced by streptomycetes. Arch. Biochem. Biophys. 106, 1–5.

Chaiet, L., Miller, T.W., Tausig, F., Wolf, F.J., 1963. Antibiotic msd-235. ii. Separation and purification of synergistic components. Antimicrob. Agents Chemother. (Bethesda) 161, 28–32.

Chan, C.J., Le, R., Burns, K., Ahmed, K., Coyaud, E., Laurent, E.M.N., Raught, B., Melançon, P., 2019. Bioid performed on golgi enriched fractions identify c10orf76 as a GBF1 binding protein essential for golgi maintenance and secretion. Mol. Cell. Proteomics 18, 2285–2297.

Chapman-Smith, A., Cronan, J.E., 1999. Molecular biology of biotin attachment to proteins. J. Nutr. 129, 477S–484S.

Chapman-Smith, A., Mulhern, T.D., Whelan, F., Cronan, J.E., Wallace, J.C., 2001. The C-terminal domain of biotin protein ligase from E. coli is required for catalytic activity. Protein Sci. 10, 2608–2617.

Chen, A.L., Kim, E.W., Toh, J.Y., Vashisht, A.A., Rashoff, A.Q., Van, C., Huang, A.S., Moon, A.S., Bell, H.N., Bentolila, L.A., Wohlschlegel, J.A., Bradley, P.J., 2015. Novel components of the Toxoplasma inner membrane complex revealed by BioID. MBio 6, e02357–14.

Chen, Z., Lei, C., Wang, C., Li, N., Srivastava, M., Tang, M., Zhang, H., Choi, J.M., Jung, S.Y., Qin, J., Chen, J., 2019. Global phosphoproteomic analysis reveals ARMC10 as an AMPK substrate that regulates mitochondrial dynamics. Nat. Commun. 10, 104.

Cho, K.F., Branon, T.C., Rajeev, S., Svinkina, T., Udeshi, N.D., Thoudam, T., Kwak, C., Rhee, H.-W., Lee, I.-K., Carr, S.A., Ting, A.Y., 2020a. Split-TurboID enables contact-dependent proximity labeling in cells. Proc. Natl. Acad. Sci. U. S. A. 117, 12143–12154.

Cho, K.F., Branon, T.C., Udeshi, N.D., Myers, S.A., Carr, S.A., Ting, A.Y., 2020b. Proximity labeling in mammalian cells with TurboID and split-TurboID. Nat. Protoc. 15, 3971–3999.

Choi-Rhee, E., Schulman, H., Cronan, J.E., 2004. Promiscuous protein biotinylation by Escherichia coli biotin protein ligase. Protein Sci. 13, 3043–3050.

Cole, A., Wang, Z., Coyaud, E., Voisin, V., Gronda, M., Jitkova, Y., Mattson, R., Hurren, R., Babovic, S., Maclean, N., Restall, I., Wang, X., Jeyaraju, D.V., Sukhai, M.A., Prabha, S., Bashir, S., Ramakrishnan, A., Leung, E., Qia, Y.H., Zhang, N., Combes, K.R., Ketela, T., Lin, F., Houry, W.A., Aman, A., Al-Awar, R., Zheng, W., Wienholds, E., Xu, C.J., Dick, J., Wang, J.C.Y., Moffat, J., Minden, M.D., Eaves, C.J., Bader, G.D., Hao, Z., Kornblau, S.M., Raught, B., Schimmer, A.D., 2015. Inhibition of the mitochondrial protease clpp as a therapeutic strategy for human acute myeloid leukemia. Cancer Cell 27, 864–876.

Colomer-Lluch, M., Ruiz, A., Moris, A., Prado, J.G., 2018. Restriction factors: from intrinsic viral restriction to shaping cellular immunity against HIV-1. Front. Immunol. 9, 2876.

Connolly, C.N., Futter, C.E., Gibson, A., Hopkins, C.R., Cutler, D.F., 1994. Transport into and out of the Golgi complex studied by transfecting cells with cDNAs encoding horseradish peroxidase. J. Cell Biol. 127, 641–652.

Coyaud, E., Ranadheera, C., Cheng, D., Gonçalves, J., Dyakov, B.J.A., Laurent, E.M.N., St-Germain, J., Pelletier, L., Gingras, A.-C., Brumell, J.H., Kim, P.K., Safronetz, D., Raught, B., 2018. Global interactomics uncovers extensive organellar targeting by zika virus. Mol. Cell. Proteomics 17, 2242–2255.

Cronan, J.E., 2018. Advances in synthesis of biotin and assembly of lipoic acid. Curr. Opin. Chem. Biol. 47, 60–66.

Cross, S.H., Mckie, L., Hurd, T.W., Riley, S., Wills, J., Barnard, A.R., Young, F., MacLaren, R.E., Jackson, I.J., 2020. The nanophthalmos protein TMEM98 inhibits MYRF self-cleavage and is required for eye size specification. PLoS Genet. 16, e1008583.

Cruz-Lopez, D., Ramos, D., Castilloveitia, G., Schikorski, T., 2018. Quintuple labeling in the electron microscope with genetically encoded enhanced horseradish peroxidase. PLoS One 13, e0200693.

Das, P.P., Macharia, M.W., Lin, Q., Wong, S.-M., 2019. In planta proximity-dependent biotin identification (BioID) identifies a TMV replication co-chaperone NbSGT1 in the vicinity of 126 kDa replicase. J. Proteomics 204, 103402.

D'Costa, V.M., Coyaud, E., Boddy, K.C., Laurent, E.M.N., St-Germain, J., Li, T., Grinstein, S., Raught, B., Brumell, J.H., 2019. BioID screen of Salmonella type 3 secreted effectors reveals host factors involved in vacuole positioning and stability during infection. Nat. Microbiol. 4, 2511–2522.

De Lorenzo, G., Eichwald, C., Schraner, E.M., Nicolin, V., Bortul, R., Mano, M., Burrone, O.R., Arnoldi, F., 2012. Production of in vivo-biotinylated rotavirus particles. J. Gen. Virol. 93, 1474–1482.

De Munter, S., Görnemann, J., Derua, R., Lesage, B., Qian, J., Heroes, E., Waelkens, E., Van Eynde, A., Beullens, M., Bollen, M., 2017. Split-BioID: a proximity biotinylation assay for dimerization-dependent protein interactions. FEBS Lett. 591, 415–424.

De Pablos, L.M., Kelly, S., de Freitas Nascimento, J., Sunter, J., Carrington, M., 2017. Characterization of RBP9 and RBP10, two developmentally regulated RNA-binding proteins in *Trypanosoma brucei*. Open Biol. 7 (4), 160159. https://doi.org/10.1098/rsob.160159.

Del Olmo, T., Lauzier, A., Normandin, C., Larcher, R., Lecours, M., Jean, D., Lessard, L., Steinberg, F., Boisvert, F.-M., Jean, S., 2019. APEX2-mediated RAB proximity labeling identifies a role for RAB21 in clathrin-independent cargo sorting. EMBO Rep. 20 (2), e47192. https://doi.org/10.15252/embr.201847192.

Di Pietro, M., Filardo, S., Romano, S., Sessa, R., 2019. *Chlamydia trachomatis* and *Chlamydia pneumoniae* interaction with the host: latest advances and future prospective. Microorganisms 7 (5), 140. https://doi.org/10.3390/microorganisms7050140.

Dickinson, M.S., Anderson, L.N., Webb-Robertson, B.-J.M., Hansen, J.R., Smith, R.D., Wright, A.T., Hybiske, K., 2019. Proximity-dependent proteomics of the chlamydia trachomatis inclusion membrane reveals functional interactions with endoplasmic reticulum exit sites. PLoS Pathog. 15, e1007698.

Dong, X., Lim, T.K., Lin, Q., He, C.Y., 2020. Basal body protein TbSAF1 is required for microtubule quartet Anchorage to the basal bodies in *Trypanosoma brucei*. mBio 11 (3), e00668–20. https://doi.org/10.1128/mBio.00668-20.

Dundas, C.M., Demonte, D., Park, S., 2013. Streptavidin-biotin technology: improvements and innovations in chemical and biological applications. Appl. Microbiol. Biotechnol. 97, 9343–9353.

Dunham, W.H., Mullin, M., Gingras, A.-C., 2012. Affinity-purification coupled to mass spectrometry: basic principles and strategies. Proteomics 12, 1576–1590.

Eakin, R.E., McKinley, W.A., Williams, R.J., 1940. Egg-white injury in chicks and its relationship to a deficiency of vitamin h (biotin). Science 92, 224–225.

Engelberg, K., Chen, C.-T., Bechtel, T., Sánchez Guzmán, V., Drozda, A.A., Chavan, S., Weerapana, E., Gubbels, M.-J., 2020. The apical annuli of Toxoplasma gondii are composed of coiled-coil and signalling proteins embedded in the inner membrane complex sutures. Cell. Microbiol. 22, e13112.

Eyre, N.S., Hampton-Smith, R.J., Aloia, A.L., Eddes, J.S., Simpson, K.J., Hoffmann, P., Beard, M.R., 2016. Phosphorylation of NS5A Serine-235 is essential to hepatitis C virus RNA replication and normal replication compartment formation. Virology 491, 27–44.

Eyre, N.S., Johnson, S.M., Eltahla, A.A., Aloi, M., Aloia, A.L., McDevitt, C.A., Bull, R.A., Beard, M.R., 2017. Genome-wide mutagenesis of dengue virus reveals plasticity of the NS1 protein and enables generation of infectious tagged reporter viruses. J. Virol. 91 (23). https://doi.org/10.1128/JVI.01455-17, e01455–17.

Fan, J., Liu, X., Mao, F., Yue, X., Lee, I., Xu, Y., 2020. Proximity proteomics identifies novel function of Rab14 in trafficking of Ebola virus matrix protein VP40. Biochem. Biophys. Res. Commun. 527, 387–392.

Farrell, P.J., 2019. Epstein-Barr virus and cancer. Annu. Rev. Pathol. 14, 29–53.

Fazal, F.M., Han, S., Parker, K.R., Kaewsapsak, P., Xu, J., Boettiger, A.N., Chang, H.Y., Ting, A.Y., 2019. Atlas of subcellular RNA localization revealed by APEX-Seq. Cell 178, 473–490.e26.

Fung, T.S., Liu, D.X., 2019. Human coronavirus: host-pathogen interaction. Annu. Rev. Microbiol. 73, 529–557.

Gaete-Argel, A., Márquez, C.L., Barriga, G.P., Soto-Rifo, R., Valiente-Echeverría, F., 2019. Strategies for success. viral infections and membraneless organelles. Front. Cell. Infect. Microbiol. 9, 336.

Gaji, R.Y., Johnson, D.E., Treeck, M., Wang, M., Hudmon, A., Arrizabalaga, G., 2015. Phosphorylation of a myosin motor by TgCDPK3 facilitates rapid initiation of motility during Toxoplasma gondii egress. PLoS Pathog. 11, e1005268.

Gao, X.D., Tu, L.-C., Mir, A., Rodriguez, T., Ding, Y., Leszyk, J., Dekker, J., Shaffer, S.A., Zhu, L.J., Wolfe, S.A., Sontheimer, E.J., 2018. C-BERST: defining subnuclear proteomic landscapes at genomic elements with dCas9-APEX2. Nat. Methods 15, 433–436.

Gaucherand, L., Porter, B.K., Levene, R.E., Price, E.L., Schmaling, S.K., Rycroft, C.H., Kevorkian, Y., McCormick, C., Khaperskyy, D.A., Gaglia, M.M., 2019. The influenza a virus endoribonuclease PA-X usurps host mRNA processing machinery to limit host gene expression. Cell Rep. 27, 776–792.e7.

Gerold, G., Bruening, J., Pietschmann, T., 2016. Decoding protein networks during virus entry by quantitative proteomics. Virus Res. 218, 25–39. https://doi.org/10.1016/j.virusres.2015.09.006.

Go, C.D., Knight, J.D.R., Rajasekharan, A., Rathod, B., Hesketh, G.G., Abe, K.T., Youn, J.-Y., Samavarchi-Tehrani, P., Zhang, H., Zhu, L.Y., Popiel, E., Lambert, J.-P., Coyaud, É., Cheung, S.W.T., Rajendran, D., Wong, C.J., Antonicka, H., Pelletier, L., Raught, B., Palazzo, A.F., Shoubridge, E.A., Gingras, A.-C., 2019. A proximity biotinylation map of a human cell. BioRxiv.

Graindorge, A., Frénal, K., Jacot, D., Salamun, J., Marq, J.B., Soldati-Favre, D., 2016. The conoid associated motor MyoH is indispensable for Toxoplasma gondii entry and exit from host cells. PLoS Pathog. 12, e1005388.

Guo, Z., Neilson, L.J., Zhong, H., Murray, P.S., Zanivan, S., Zaidel-Bar, R., 2014. E-cadherin interactome complexity and robustness resolved by quantitative proteomics. Sci. Signal. 7, rs7.

Gupta, G.D., Coyaud, É., Gonçalves, J., Mojarad, B.A., Liu, Y., Wu, Q., Gheiratmand, L., Comartin, D., Tkach, J.M., Cheung, S.W.T., Bashkurov, M., Hasegan, M.,

Knight, J.D., Lin, Z.-Y., Schueler, M., Hildebrandt, F., Moffat, J., Gingras, A.-C., Raught, B., Pelletier, L., 2015. A dynamic protein interaction landscape of the human centrosome-cilium Interface. Cell 163, 1484–1499.

Hamasaki, M., Furuta, N., Matsuda, A., Nezu, A., Yamamoto, A., Fujita, N., Oomori, H., Noda, T., Haraguchi, T., Hiraoka, Y., Amano, A., Yoshimori, T., 2013. Autophagosomes form at ER-mitochondria contact sites. Nature 495, 389–393.

Han, S., Udeshi, N.D., Deerinck, T.J., Svinkina, T., Ellisman, M.H., Carr, S.A., Ting, A.Y., 2017. Proximity biotinylation as a method for mapping proteins associated with mtDNA in living cells. Cell Chem. Biol. 24, 404–414.

Han, Y., Branon, T.C., Martell, J.D., Boassa, D., Shechner, D., Ellisman, M.H., Ting, A., 2019. Directed evolution of split APEX2 peroxidase. ACS Chem. Biol. 14, 619–635.

Han, S., Zhao, B.S., Myers, S.A., Carr, S.A., He, C., Ting, A.Y., 2020. RNA-protein interaction mapping via MS2- or Cas13-based APEX targeting. Proc. Natl. Acad. Sci. U. S. A. 117, 22068–22079.

Hilton, N.A., Sladewski, T.E., Perry, J.A., Pataki, Z., Sinclair-Davis, A.N., Muniz, R.S., Tran, H.L., Wurster, J.I., Seo, J., de Graffenried, C.L., 2018. Identification of TOEFAZ1-interacting proteins reveals key regulators of Trypanosoma brucei cytokinesis. Mol. Microbiol. 109, 306–326.

Holthusen, K., Talaty, P., Everly, D.N., 2015. Regulation of latent membrane protein 1 signaling through interaction with cytoskeletal proteins. J. Virol. 89, 7277–7290.

Honke, K., Kotani, N., 2011. The enzyme-mediated activation of radical source reaction: a new approach to identify partners of a given molecule in membrane microdomains. J. Neurochem. 116, 690–695.

Honke, K., Miyagawa-Yamaguchi, A., Kotani, N., 2019. The EMARS reaction for proximity labeling. Methods Mol. Biol. 2008, 1–12.

Hopkins, C., Gibson, A., Stinchcombe, J., Futter, C., 2000. Chimeric molecules employing horseradish peroxidase as reporter enzyme for protein localization in the electron microscope. In: Applications of Chimeric Genes and Hybrid Proteins—Part B: Cell Biology and Physiology. Methods in Enzymology, Elsevier, pp. 35–45.

Hoyer, M.J., Chitwood, P.J., Ebmeier, C.C., Striepen, J.F., Qi, R.Z., Old, W.M., Voeltz, G.K., 2018. A novel class of ER membrane proteins regulates ER-associated endosome fission. Cell 175, 254–265.e14.

Hua, R., Cheng, D., Coyaud, É., Freeman, S., Di Pietro, E., Wang, Y., Vissa, A., Yip, C.M., Fairn, G.D., Braverman, N., Brumell, J.H., Trimble, W.S., Raught, B., Kim, P.K., 2017. VAPs and ACBD5 tether peroxisomes to the ER for peroxisome maintenance and lipid homeostasis. J. Cell Biol. 216, 367–377.

Huang, M.-S., Lin, W.-C., Chang, J.-H., Cheng, C.-H., Wang, H.Y., Mou, K.Y., 2019. The cysteine-free single mutant C32S of APEX2 is a highly expressed and active fusion tag for proximity labeling applications. Protein Sci. 28, 1703–1712.

Hung, V., Zou, P., Rhee, H.-W., Udeshi, N.D., Cracan, V., Svinkina, T., Carr, S.A., Mootha, V.K., Ting, A.Y., 2014. Proteomic mapping of the human mitochondrial intermembrane space in live cells via ratiometric APEX tagging. Mol. Cell 55, 332–341.

Hung, V., Lam, S.S., Udeshi, N.D., Svinkina, T., Guzman, G., Mootha, V.K., Carr, S.A., Ting, A.Y., 2017. Proteomic mapping of cytosol-facing outer mitochondrial and ER membranes in living human cells by proximity biotinylation. Elife 6, e24463. https://doi.org/10.7554/eLife.24463.

Iyer, L.M., Burroughs, A.M., Aravind, L., 2008. Unraveling the biochemistry and provenance of pupylation: a prokaryotic analog of ubiquitination. Biol. Direct 3, 45.

Jagger, B.W., Wise, H.M., Kash, J.C., Walters, K.A., Wills, N.M., Xiao, Y.L., Dunfee, R.L., Schwartzman, L.M., Ozinsky, A., Bell, G.L., Dalton, R.M., Lo, A., Efstathiou, S., Atkins, J.F., Firth, A.E., Taubenberger, J.K., Digard, P., 2012. An overlapping protein-coding region in influenza a virus segment 3 modulates the host response. Science 337, 199–204.

Janer, A., Prudent, J., Paupe, V., Fahiminiya, S., Majewski, J., Sgarioto, N., Des Rosiers, C., Forest, A., Lin, Z.-Y., Gingras, A.-C., Mitchell, G., McBride, H.M., Shoubridge, E.A., 2016. SLC25A46 is required for mitochondrial lipid homeostasis and cristae maintenance and is responsible for Leigh syndrome. EMBO Mol. Med. 8, 1019–1038.

Jing, J., He, L., Sun, A., Quintana, A., Ding, Y., Ma, G., Tan, P., Liang, X., Zheng, X., Chen, L., Shi, X., Zhang, S.L., Zhong, L., Huang, Y., Dong, M.-Q., Walker, C.L., Hogan, P.G., Wang, Y., Zhou, Y., 2015. Proteomic mapping of ER-PM junctions identifies STIMATE as a regulator of Ca^{2+} influx. Nat. Cell Biol. 17, 1339–1347.

Joeh, E., O'Leary, T., Li, W., Hawkins, R., Hung, J.R., Parker, C.G., Huang, M.L., 2020. Mapping glycan-mediated galectin-3 interactions by live cell proximity labeling. Proc. Natl. Acad. Sci. U. S. A. 117 (44), 27329–27338. https://doi.org/10.1073/pnas.2009206117.

Kabeiseman, E.J., Cichos, K.H., Moore, E.R., 2014. The eukaryotic signal sequence, YGRL, targets the chlamydial inclusion. Front. Cell. Infect. Microbiol. 4, 129.

Kaewsapsak, P., Shechner, D.M., Mallard, W., Rinn, J.L., Ting, A.Y., 2017. Live-cell mapping of organelle-associated RNAs via proximity biotinylation combined with protein-RNA crosslinking. Elife 6, e29224. https://doi.org/10.7554/eLife.29224.

Kalra, H., Simpson, R.J., Ji, H., Aikawa, E., Altevogt, P., et al., 2012. Vesiclepedia: a compendium for extracellular vesicles with continuous community annotation. PLoS Biol. 10, e1001450.

Kehrer, J., Frischknecht, F., Mair, G.R., 2016. Proteomic analysis of the plasmodium berghei gametocyte egressome and vesicular bioID of osmiophilic body proteins identifies merozoite TRAP-like protein (MTRAP) as an essential factor for parasite transmission. Mol. Cell. Proteomics 15, 2852–2862.

Khosh-Naucke, M., Becker, J., Mesén-Ramírez, P., Kiani, P., Birnbaum, J., Fröhlke, U., Jonscher, E., Schlüter, H., Spielmann, T., 2018. Identification of novel parasitophorous vacuole proteins in P. falciparum parasites using BioID. Int. J. Med. Microbiol. 308, 13–24.

Kido, K., Yamanaka, S., Nakano, S., Motani, K., Shinohara, S., Nozawa, A., Kosako, H., Ito, S., Sawasaki, T., 2020. AirID, a novel proximity biotinylation enzyme, for analysis of protein-protein interactions. Elife 9, e54983. https://doi.org/10.7554/eLife.54983.

Kim, D.I., Birendra, K.C., Zhu, W., Motamedchaboki, K., Doye, V., Roux, K.J., 2014. Probing nuclear pore complex architecture with proximity-dependent biotinylation. Proc. Natl. Acad. Sci. U. S. A. 111, E2453–E2461.

Kim, D.I., Jensen, S.C., Noble, K.A., Kc, B., Roux, K.H., Motamedchaboki, K., Roux, K.J., 2016. An improved smaller biotin ligase for BioID proximity labeling. Mol. Biol. Cell 27, 1188–1196.

Kotani, N., Gu, J., Isaji, T., Udaka, K., Taniguchi, N., Honke, K., 2008. Biochemical visualization of cell surface molecular clustering in living cells. Proc. Natl. Acad. Sci. U. S. A. 105, 7405–7409.

Koua, D., Cerutti, L., Falquet, L., Sigrist, C.J.A., Theiler, G., Hulo, N., Dunand, C., 2009. PeroxiBase: a database with new tools for peroxidase family classification. Nucleic Acids Res. 37, D261–D266.

Kueck, T., Foster, T.L., Weinelt, J., Sumner, J.C., Pickering, S., Neil, S.J.D., 2015. Serine phosphorylation of HIV-1 Vpu and its binding to tetherin regulates interaction with clathrin adaptors. PLoS Pathog. 11, e1005141.

Kwak, C., Shin, S., Park, J.-S., Jung, M., Nhung, T.T.M., Kang, M.-G., Lee, C., Kwon, T.-H., Park, S.K., Mun, J.Y., Kim, J.-S., Rhee, H.-W., 2020. Contact-ID, a tool for profiling organelle contact sites, reveals regulatory proteins of mitochondrial-associated membrane formation. Proc. Natl. Acad. Sci. U. S. A. 117, 12109–12120.

Kwon, K., Beckett, D., 2000. Function of a conserved sequence motif in biotin holoenzyme synthetases. Protein Sci. 9, 1530–1539.

Kwon, K., Streaker, E.D., Ruparelia, S., Beckett, D., 2000. Multiple disordered loops function in corepressor-induced dimerization of the biotin repressor. J. Mol. Biol. 304, 821–833.

Lajko, M., Haddad, A.F., Robinson, C.A., Connolly, S.A., 2015. Using proximity biotinylation to detect herpesvirus entry glycoprotein interactions: limitations for integral membrane glycoproteins. J. Virol. Methods 221, 81–89.

Lam, S.S., Martell, J.D., Kamer, K.J., Deerinck, T.J., Ellisman, M.H., Mootha, V.K., Ting, A.Y., 2015. Directed evolution of APEX2 for electron microscopy and proximity labeling. Nat. Methods 12, 51–54.

Lane, M.D., Rominger, K.L., Young, D.L., Lynen, F., 1964. The enzymatic synthesis of holotranscarboxylase from apotranscarboxylase and (+)-biotin. ii. investigation of the reaction mechanism. J. Biol. Chem. 239, 2865–2871.

Lasswitz, L., Chandra, N., Arnberg, N., Gerold, G., 2018. Glycomics and proteomics approaches to investigate early adenovirus-host cell interactions. J. Mol. Biol. 430, 1863–1882.

Le Guerroué, F., Eck, F., Jung, J., Starzetz, T., Mittelbronn, M., Kaulich, M., Behrends, C., 2017. Autophagosomal content profiling reveals an LC3C-dependent piecemeal mitophagy pathway. Mol. Cell 68, 786–796.e6.

Le Sage, V., Cinti, A., Valiente-Echeverría, F., Mouland, A.J., 2015. Proteomic analysis of HIV-1 Gag interacting partners using proximity-dependent biotinylation. Virol. J. 12, 138.

Lee, S.-Y., Kang, M.-G., Park, J.-S., Lee, G., Ting, A.Y., Rhee, H.-W., 2016. APEX fingerprinting reveals the subcellular localization of proteins of interest. Cell Rep. 15, 1837–1847.

Li, X.-W., Rees, J.S., Xue, P., Zhang, H., Hamaia, S.W., Sanderson, B., Funk, P.E., Farndale, R.W., Lilley, K.S., Perrett, S., Jackson, A.P., 2014. New insights into the DT40 B cell receptor cluster using a proteomic proximity labeling assay. J. Biol. Chem. 289, 14434–14447.

Lin, D.L., Cherepanova, N.A., Bozzacco, L., MacDonald, M.R., Gilmore, R., Tai, A.W., 2017. Dengue virus hijacks a noncanonical oxidoreductase function of a cellular oligosaccharyltransferase complex. mBio 8 (4), e00939–17. https://doi.org/10.1128/mBio.00939-17.

Liu, X., Zhang, Y., Chen, Y., Li, M., Zhou, F., Li, K., Cao, H., Ni, M., Liu, Y., Gu, Z., Dickerson, K.E., Xie, S., Hon, G.C., Xuan, Z., Zhang, M.Q., Shao, Z., Xu, J., 2017. In situ capture of chromatin interactions by biotinylated dCas9. Cell 170, 1028–1043.e19.

Liu, L., Doray, B., Kornfeld, S., 2018a. Recycling of Golgi glycosyltransferases requires direct binding to coatomer. Proc. Natl. Acad. Sci. U. S. A. 115, 8984–8989.

Liu, Q., Zheng, J., Sun, W., Huo, Y., Zhang, L., Hao, P., Wang, H., Zhuang, M., 2018b. A proximity-tagging system to identify membrane protein-protein interactions. Nat. Methods 15, 715–722.

Liu, X., Salokas, K., Tamene, F., Jiu, Y., Weldatsadik, R.G., Öhman, T., Varjosalo, M., 2018c. An AP-MS- and BioID-compatible MAC-tag enables comprehensive mapping of protein interactions and subcellular localizations. Nat. Commun. 9, 1188.

Lobingier, B.T., Hüttenhain, R., Eichel, K., Miller, K.B., Ting, A.Y., von Zastrow, M., Krogan, N.J., 2017. An approach to spatiotemporally resolve protein interaction networks in living cells. Cell 169, 350–360.e12.

Loh, K.H., Stawski, P.S., Draycott, A.S., Udeshi, N.D., Lehrman, E.K., Wilton, D.K., Svinkina, T., Deerinck, T.J., Ellisman, M.H., Stevens, B., Carr, S.A., Ting, A.Y., 2016. Proteomic analysis of unbounded cellular compartments: synaptic clefts. Cell 166, 1295–1307.e21.

Long, S., Anthony, B., Drewry, L.L., Sibley, L.D., 2017. A conserved ankyrin repeat-containing protein regulates conoid stability, motility and cell invasion in Toxoplasma gondii. Nat. Commun. 8, 2236.

Lum, K.K., Cristea, I.M., 2016. Proteomic approaches to uncovering virus-host protein interactions during the progression of viral infection. Expert Rev. Proteomics 13, 325–340.

Luo, Y., Na, Z., Slavoff, S.A., 2018. P-bodies: composition, properties, and functions. Biochemistry 57, 2424–2431.

Markmiller, S., Soltanieh, S., Server, K.L., Mak, R., Jin, W., Fang, M.Y., Luo, E.-C., Krach, F., Yang, D., Sen, A., Fulzele, A., Wozniak, J.M., Gonzalez, D.J., Kankel, M.W., Gao, F.-B., Bennett, E.J., Lécuyer, E., Yeo, G.W., 2018. Context-dependent and disease-specific diversity in protein interactions within stress granules. Cell 172, 590–604.e13.

Martell, J.D., Deerinck, T.J., Sancak, Y., Poulos, T.L., Mootha, V.K., Sosinsky, G.E., Ellisman, M.H., Ting, A.Y., 2012. Engineered ascorbate peroxidase as a genetically encoded reporter for electron microscopy. Nat. Biotechnol. 30, 1143–1148.

Martell, J.D., Yamagata, M., Deerinck, T.J., Phan, S., Kwa, C.G., Ellisman, M.H., Sanes, J.R., Ting, A.Y., 2016. A split horseradish peroxidase for the detection of intercellular protein-protein interactions and sensitive visualization of synapses. Nat. Biotechnol. 34, 774–780.

May, D.G., Scott, K.L., Campos, A.R., Roux, K.J., 2020. Comparative application of BioID and TurboID for protein-proximity biotinylation. Cells 9 (5), 1070. https://doi.org/10.3390/cells9051070.

Mayer, G., Bendayan, M., 1997. Biotinyl-tyramide: a novel approach for electron microscopic immunocytochemistry. J. Histochem. Cytochem. 45, 1449–1454.

Michael Green, N., 1990. [5] Avidin and streptavidin. In: Avidin-Biotin Technology. Methods in Enzymology, Elsevier, pp. 51–67.

Mick, D.U., Rodrigues, R.B., Leib, R.D., Adams, C.M., Chien, A.S., Gygi, S.P., Nachury, M.V., 2015. Proteomics of primary cilia by proximity labeling. Dev. Cell 35, 497–512.

Minamihata, K., Goto, M., Kamiya, N., 2011. Protein heteroconjugation by the peroxidase-catalyzed tyrosine coupling reaction. Bioconjug. Chem. 22, 2332–2338.

Mirza, A.N., McKellar, S.A., Urman, N.M., Brown, A.S., Hollmig, T., Aasi, S.Z., Oro, A.E., 2019. LAP2 proteins chaperone GLI1 movement between the lamina and chromatin to regulate transcription. Cell 176, 198–212.e15.

Miyagawa-Yamaguchi, A., Kotani, N., Honke, K., 2014. Expressed glycosylphosphatidylinositol-anchored horseradish peroxidase identifies co-clustering molecules in individual lipid raft domains. PLoS One 9, e93054.

Mojica, S.A., Hovis, K.M., Frieman, M.B., Tran, B., Hsia, R., Ravel, J., Jenkins-Houk, C., Wilson, K.L., Bavoil, P.M., 2015. SINC, a type III secreted protein of chlamydia psittaci, targets the inner nuclear membrane of infected cells and uninfected neighbors. Mol. Biol. Cell 26, 1918–1934.

Mori, N., Fujii, M., Ikeda, S., Yamada, Y., Tomonaga, M., Ballard, D.W., Yamamoto, N., 1999. Constitutive activation of NF-κB in primary adult T-cell Leukemia cells. Blood 93 (7), 2360–2368.

Morriswood, B., Havlicek, K., Demmel, L., Yavuz, S., Sealey-Cardona, M., Vidilaseris, K., Anrather, D., Kostan, J., Djinovic-Carugo, K., Roux, K.J., Warren, G., 2013. Novel bilobe components in Trypanosoma brucei identified using proximity-dependent biotinylation. Eukaryot. Cell 12, 356–367.

Mortensen, A., Skibsted, L.H., Truscott, T.G., 2001. The interaction of dietary carotenoids with radical species. Arch. Biochem. Biophys. 385, 13–19.

Mukherjee, J., Hermesh, O., Eliscovich, C., Nalpas, N., Franz-Wachtel, M., Maček, B., Jansen, R.-P., 2019. β-Actin mRNA interactome mapping by proximity biotinylation. Proc. Natl. Acad. Sci. U. S. A. 116, 12863–12872.

Musso, D., Gubler, D.J., 2016. Zika virus. Clin. Microbiol. Rev. 29, 487–524.

Myers, S.A., Wright, J., Peckner, R., Kalish, B.T., Zhang, F., Carr, S.A., 2018. Discovery of proteins associated with a predefined genomic locus via dCas9-APEX-mediated proximity labeling. Nat. Methods 15, 437–439.

Nadipuram, S.M., Kim, E.W., Vashisht, A.A., Lin, A.H., Bell, H.N., Coppens, I., Wohlschlegel, J.A., Bradley, P.J., 2016. In vivo biotinylation of the toxoplasma parasitophorous vacuole reveals novel dense granule proteins important for parasite growth and pathogenesis. mBio 7 (4), e00808–16. https://doi.org/10.1128/mBio.00808-16.

Nadipuram, S.M., Thind, A.C., Rayatpisheh, S., Wohlschlegel, J.A., Bradley, P.J., 2020. Proximity biotinylation reveals novel secreted dense granule proteins of Toxoplasma gondii bradyzoites. PLoS One 15, e0232552.

Nessel, T., Beck, J.M., Rayatpisheh, S., Jami-Alahmadi, Y., Wohlschlegel, J.A., Goldberg, D.E., Beck, J.R., 2020. EXP1 is required for organisation of EXP2 in the intraerythrocytic malaria parasite vacuole. Cell. Microbiol. 22, e13168.

Nkosi, D., Sun, L., Duke, L.C., Patel, N., Surapaneni, S.K., Singh, M., Meckes, D.G., 2020. Epstein-Barr virus LMP1 promotes syntenin-1- and Hrs-induced extracellular vesicle formation for its own secretion to increase cell proliferation and migration. mBio 11 (3), e00589–20. https://doi.org/10.1128/mBio.00589-20.

Olson, M.G., Widner, R.E., Jorgenson, L.M., Lawrence, A., Lagundzin, D., Woods, N.T., Ouellette, S.P., Rucks, E.A., 2019. Proximity labeling to map host-pathogen interactions at the membrane of a bacterium-containing vacuole in *Chlamydia trachomatis*-infected human cells. Infect. Immun. 87 (11), e00537–19. https://doi.org/10.1128/IAI.00537-19.

Olson, M.G., Ouellette, S.P., Rucks, E.A., 2020. A meta-analysis of affinity purification-mass spectrometry experimental systems used to identify eukaryotic and chlamydial proteins at the chlamydia trachomatis inclusion membrane. J. Proteomics 212, 103595.

Oostdyk, L.T., Shank, L., Jividen, K., Dworak, N., Sherman, N.E., Paschal, B.M., 2019. Towards improving proximity labeling by the biotin ligase BirA. Methods 157, 66–79.

Ortiz, D.A., Glassbrook, J.E., Pellett, P.E., 2016. Protein-protein interactions suggest novel activities of human cytomegalovirus tegument protein pUL103. J. Virol. 90, 7798–7810.

Özcelik, D., Barandun, J., Schmitz, N., Sutter, M., Guth, E., Damberger, F.F., Allain, F.H.-T., Ban, N., Weber-Ban, E., 2012. Structures of pup ligase PafA and depupylase Dop from the prokaryotic ubiquitin-like modification pathway. Nat. Commun. 3, 1014.

Padrón, A., Iwasaki, S., Ingolia, N.T., 2019. Proximity RNA labeling by APEX-Seq reveals the organization of translation initiation complexes and repressive RNA granules. Mol. Cell 75, 875–887.e5.

Paek, J., Kalocsay, M., Staus, D.P., Wingler, L., Pascolutti, R., Paulo, J.A., Gygi, S.P., Kruse, A.C., 2017. Multidimensional tracking of GPCR signaling via peroxidase-catalyzed proximity labeling. Cell 169, 338–349.e11.

Pagac, M., Cooper, D.E., Qi, Y., Lukmantara, I.E., Mak, H.Y., Wu, Z., Tian, Y., Liu, Z., Lei, M., Du, X., Ferguson, C., Kotevski, D., Sadowski, P., Chen, W., Boroda, S., Harris, T.E., Liu, G., Parton, R.G., Huang, X., Coleman, R.A., Yang, H., 2016. SEIPIN regulates lipid droplet expansion and adipocyte development by modulating the activity of glycerol-3-phosphate acyltransferase. Cell Rep. 17, 1546–1559.

Paillusson, S., Stoica, R., Gomez-Suaga, P., Lau, D.H.W., Mueller, S., Miller, T., Miller, C.C.J., 2016. There's something wrong with my MAM; the ER-mitochondria axis and neurodegenerative diseases. Trends Neurosci. 39, 146–157.

Pan, M., Li, M., Li, L., Song, Y., Hou, L., Zhao, J., Shen, B., 2019. Identification of novel dense-granule proteins in toxoplasma gondii by two proximity-based biotinylation approaches. J. Proteome Res. 18, 319–330.

Pearce, M.J., Mintseris, J., Ferreyra, J., Gygi, S.P., Darwin, K.H., 2008. Ubiquitin-like protein involved in the proteasome pathway of Mycobacterium tuberculosis. Science 322, 1104–1107.

Plescia, C.B., David, E.A., Patra, D., Sengupta, R., Amiar, S., Su, Y., Stahelin, R.V., 2020. SARS-CoV-2 viral budding and entry can be modeled using BSL-2 level virus-like particles. J. Biol. Chem. 296, 100103. https://doi.org/10.1074/jbc.RA120.016148.

Postel, A., Letzel, T., Müller, F., Ehricht, R., Pourquier, P., Dauber, M., Grund, C., Beer, M., Harder, T.C., 2011. In vivo biotinylated recombinant influenza A virus hemagglutinin for use in subtype-specific serodiagnostic assays. Anal. Biochem. 411, 22–31.

Prasad, P.D., Wang, H., Kekuda, R., Fujita, T., Fei, Y.J., Devoe, L.D., Leibach, F.H., Ganapathy, V., 1998. Cloning and functional expression of a cDNA encoding a mammalian sodium-dependent vitamin transporter mediating the uptake of pantothenate, biotin, and lipoate. J. Biol. Chem. 273, 7501–7506.

Protter, D.S.W., Parker, R., 2016. Principles and properties of stress granules. Trends Cell Biol. 26, 668–679.

Qiu, W., Xu, Z., Zhang, M., Zhang, D., Fan, H., Li, T., Wang, Q., Liu, P., Zhu, Z., Du, D., Tan, M., Wen, B., Liu, Y., 2019. Determination of local chromatin interactions using a combined CRISPR and peroxidase APEX2 system. Nucleic Acids Res. 47, e52.

Ramanathan, M., Majzoub, K., Rao, D.S., Neela, P.H., Zarnegar, B.J., Mondal, S., Roth, J.G., Gai, H., Kovalski, J.R., Siprashvili, Z., Palmer, T.D., Carette, J.E., Khavari, P.A., 2018. RNA-protein interaction detection in living cells. Nat. Methods 15, 207–212.

Rhee, H.-W., Zou, P., Udeshi, N.D., Martell, J.D., Mootha, V.K., Carr, S.A., Ting, A.Y., 2013. Proteomic mapping of mitochondria in living cells via spatially restricted enzymatic tagging. Science 339, 1328–1331.

Rider, M.A., Cheerathodi, M.R., Hurwitz, S.N., Nkosi, D., Howell, L.A., Tremblay, D.C., Liu, X., Zhu, F., Meckes, D.G., 2018. The interactome of EBV LMP1 evaluated by proximity-based BioID approach. Virology 516, 55–70.

Ritchie, C., Cylinder, I., Platt, E.J., Barklis, E., 2015. Analysis of HIV-1 Gag protein interactions via biotin ligase tagging. J. Virol. 89, 3988–4001.

Rizzuto, R., Pinton, P., Carrington, W., Fay, F.S., Fogarty, K.E., Lifshitz, L.M., Tuft, R.A., Pozzan, T., 1998. Close contacts with the endoplasmic reticulum as determinants of mitochondrial Ca2+ responses. Science 280, 1763–1766.

Rogers, M.S., Hurtado-Guerrero, R., Firbank, S.J., Halcrow, M.A., Dooley, D.M., Phillips, S.E.V., Knowles, P.F., McPherson, M.J., 2008. Cross-link formation of the cysteine 228-tyrosine 272 catalytic cofactor of galactose oxidase does not require dioxygen. Biochemistry 47, 10428–10439.

Roux, K.J., Kim, D.I., Raida, M., Burke, B., 2012. A promiscuous biotin ligase fusion protein identifies proximal and interacting proteins in mammalian cells. J. Cell Biol. 196, 801–810.

Roux, K.J., Kim, D.I., Burke, B., May, D.G., 2018. BioID: a screen for protein-protein interactions. Curr. Protoc. Protein Sci. 91, 19.23.1–19.23.15.

Rucks, E.A., Olson, M.G., Jorgenson, L.M., Srinivasan, R.R., Ouellette, S.P., 2017. Development of a proximity labeling system to map the chlamydia trachomatis inclusion membrane. Front. Cell. Infect. Microbiol. 7, 40.

Samavarchi-Tehrani, P., Abdouni, H., Knight, J.D.R., Astori, A., Samson, R., Lin, Z.-Y., Kim, D.-K., Knapp, J.J., St-Germain, J., Go, C.D., Larsen, B., Wong, C.J., Cassonnet, P., Demeret, C., Jacob, Y., Roth, F.P., Raught, B., Gingras, A.-C., 2020. A SARS-CoV-2—host proximity interactome. bioRxiv. https://doi.org/10.1101/2020.09.03.282103.

Santana, A.L., Oldenburg, D.G., Kirillov, V., Malik, L., Dong, Q., Sinayev, R., Marcu, K.B., White, D.W., Krug, L.T., 2017. RTA occupancy of the origin of lytic replication during murine gammaherpesvirus 68 reactivation from B cell latency. Pathogens 6 (1), 9. https://doi.org/10.3390/pathogens6010009.

Santin, Y.G., Doan, T., Lebrun, R., Espinosa, L., Journet, L., Cascales, E., 2018. In vivo TssA proximity labelling during type VI secretion biogenesis reveals TagA as a protein that stops and holds the sheath. Nat. Microbiol. 3, 1304–1313.

Sanwald, J.L., Poschmann, G., Stühler, K., Behrends, C., Hoffmann, S., Willbold, D., 2020. The GABARAP co-secretome identified by APEX2-GABARAP proximity labelling of extracellular vesicles. Cells 9 (6), 1468. https://doi.org/10.3390/cells9061468.

Savelli, B., Li, Q., Webber, M., Jemmat, A.M., Robitaille, A., Zamocky, M., Mathé, C., Dunand, C., 2019. RedoxiBase: a database for ROS homeostasis regulated proteins. Redox Biol. 26, 101247.

Schelhaas, M., Malmström, J., Pelkmans, L., Haugstetter, J., Ellgaard, L., Grünewald, K., Helenius, A., 2007. Simian virus 40 depends on ER protein folding and quality control factors for entry into host cells. Cell 131, 516–529.

Schmidtmann, E., Anton, T., Rombaut, P., Herzog, F., Leonhardt, H., 2016. Determination of local chromatin composition by CasID. Nucleus 7, 476–484.

Schnider, C.B., Bausch-Fluck, D., Brühlmann, F., Heussler, V.T., Burda, P.-C., 2018. BioID reveals novel proteins of the *Plasmodium* parasitophorous vacuole membrane. mSphere 3 (1), e00522–17. https://doi.org/10.1128/mSphere.00522-17.

Schopp, I.M., Amaya Ramirez, C.C., Debeljak, J., Kreibich, E., Skribbe, M., Wild, K., Béthune, J., 2017. Split-BioID a conditional proteomics approach to monitor the composition of spatiotemporally defined protein complexes. Nat. Commun. 8, 15690.

Schwob, A., Teruel, E., Dubuisson, L., Lormières, F., Verlhac, P., Abudu, Y.P., Gauthier, J., Naoumenko, M., Cloarec-Ung, F.-M., Faure, M., Johansen, T., Dutartre, H., Mahieux, R., Journo, C., 2019. SQSTM-1/p62 potentiates HTLV-1 tax-mediated NF-κB activation through its ubiquitin binding function. Sci. Rep. 9, 16014.

Sherwood, L.J., Hayhurst, A., 2019. Periplasmic nanobody-APEX2 fusions enable facile visualization of Ebola, Marburg, and Mĕnglà virus nucleoproteins, alluding to similar antigenic landscapes among *Marburgvirus* and *Dianlovirus*. Viruses 11 (4), 364. https://doi.org/10.3390/v11040364.

Sternicki, L.M., Wegener, K.L., Bruning, J.B., Booker, G.W., Polyak, S.W., 2017. Mechanisms governing precise protein biotinylation. Trends Biochem. Sci. 42, 383–394.

Sun, W., Huo, Y., Mei, Y., Zhou, Q., Zhao, S., Zhuang, M., 2020. Identification of a small probe that can be conjugated to proteins by proximity labeling. ACS Chem. Biol. 15, 39–43.

Theel, E.S., Pritt, B.S., 2016. Parasites. Microbiol. Spectr. 4 (4). https://doi.org/10.1128/microbiolspe.

Thorpe, G.H.G., Kricka, L.J., 1986. [29] Enhanced chemiluminescent reactions catalyzed by horseradish peroxidase. In: Bioluminescence and Chemiluminescence Part B. Methods in Enzymology, Elsevier, pp. 331–353.

Tong, L., 2013. Structure and function of biotin-dependent carboxylases. Cell. Mol. Life Sci. 70, 863–891.

Tosetti, N., Dos Santos Pacheco, N., Bertiaux, E., Maco, B., Bournonville, L., Hamel, V., Guichard, P., Soldati-Favre, D., 2020. Essential function of the alveolin network in the subpellicular microtubules and conoid assembly in *Toxoplasma gondii*. Elife 9, e56635. https://doi.org/10.7554/eLife.56635.

Tuteja, R., 2007. Malaria—an overview. FEBS J. 274, 4670–4679.

Ule, J., Jensen, K.B., Ruggiu, M., Mele, A., Ule, A., Darnell, R.B., 2003. CLIP identifies Nova-regulated RNA networks in the brain. Science 302, 1212–1215.

Van Itallie, C.M., Tietgens, A.J., Aponte, A., Fredriksson, K., Fanning, A.S., Gucek, M., Anderson, J.M., 2014. Biotin ligase tagging identifies proteins proximal to E-cadherin, including lipoma preferred partner, a regulator of epithelial cell-cell and cell-substrate adhesion. J. Cell Sci. 127, 885–895.

van Vliet, A.R., Giordano, F., Gerlo, S., Segura, I., Van Eygen, S., Molenberghs, G., Rocha, S., Houcine, A., Derua, R., Verfaillie, T., Vangindertael, J., De Keersmaecker, H., Waelkens, E., Tavernier, J., Hofkens, J., Annaert, W., Carmeliet, P., Samali, A., Mizuno, H., Agostinis, P., 2017. The ER stress sensor PERK coordinates ER-plasma membrane contact site formation through interaction with filamin-A and F-actin remodeling. Mol. Cell 65, 885–899.e6.

Vance, J.E., 1990. Phospholipid synthesis in a membrane fraction associated with mitochondria. J. Biol. Chem. 265, 7248–7256.

Veitch, N.C., 2004. Horseradish peroxidase: a modern view of a classic enzyme. Phytochemistry 65, 249–259.

Vidak, S., Kubben, N., Dechat, T., Foisner, R., 2015. Proliferation of progeria cells is enhanced by lamina-associated polypeptide 2α (LAP2α) through expression of extracellular matrix proteins. Genes Dev. 29, 2022–2036.

Villaseñor, R., Pfaendler, R., Ambrosi, C., Butz, S., Giuliani, S., Bryan, E., Sheahan, T.W., Gable, A.L., Schmolka, N., Manzo, M., Wirz, J., Feller, C., von Mering, C., Aebersold, R., Voigt, P., Baubec, T., 2020. ChromID identifies the protein interactome at chromatin marks. Nat. Biotechnol. 38, 728–736.

V'kovski, P., Gerber, M., Kelly, J., Pfaender, S., Ebert, N., Braga Lagache, S., Simillion, C., Portmann, J., Stalder, H., Gaschen, V., Bruggmann, R., Stoffel, M.H., Heller, M., Dijkman, R., Thiel, V., 2019. Determination of host proteins composing the microenvironment of coronavirus replicase complexes by proximity-labeling. Elife 8, e42037. https://doi.org/10.7554/eLife.42037.

Wang, Z., Wang, D., Chen, J., Sela, D.A., Nugen, S.R., 2016. Development of a novel bacteriophage based biomagnetic separation method as an aid for sensitive detection of viable Escherichia coli. Analyst 141, 1009–1016.

Wang, X., Xu, P., Cheng, F., Li, Y., Wang, Z., Hao, S., Wang, J., Ning, K., Ganaie, S.S., Engelhardt, J.F., Yan, Z., Qiu, J., 2020. Cellular cleavage and polyadenylation specificity factor 6 (CPSF6) mediates nuclear import of human bocavirus 1 NP1 protein and modulates viral capsid protein expression. J. Virol. 94 (2), e01444–19. https://doi.org/10.1128/JVI.01444-19.

Welinder, K.G., 1992. Superfamily of plant, fungal and bacterial peroxidases. Curr. Opin. Struct. Biol. 2, 388–393.

Wu, G., Lin, Q., Lim, T.K., Zhang, Y., Aweya, J.J., Zhu, J., Yao, D., 2020. The interactome of Singapore grouper iridovirus protein ICP18 as revealed by proximity-dependent BioID approach. Virus Res. 291, 198218.

Xu, Y., Beckett, D., 1996. Evidence for interdomain interaction in the Escherichia coli repressor of biotin biosynthesis from studies of an N-terminal domain deletion mutant. Biochemistry 35, 1783–1792.

Xue, M., Hou, J., Wang, L., Cheng, D., Lu, J., Zheng, L., Xu, T., 2017. Optimizing the fragment complementation of APEX2 for detection of specific protein-protein interactions in live cells. Sci. Rep. 7, 12039.

Yamagata, M., Sanes, J.R., 2018. Reporter-nanobody fusions (RANbodies) as versatile, small, sensitive immunohistochemical reagents. Proc. Natl. Acad. Sci. U. S. A. 115, 2126–2131.

Yates, J.R., Gilchrist, A., Howell, K.E., Bergeron, J.J.M., 2005. Proteomics of organelles and large cellular structures. Nat. Rev. Mol. Cell Biol. 6, 702–714.

Yoshinaka, T., Kosako, H., Yoshizumi, T., Furukawa, R., Hirano, Y., Kuge, O., Tamada, T., Koshiba, T., 2019. Structural basis of mitochondrial scaffolds by prohibitin complexes: insight into a role of the coiled-coil region. iScience 19, 1065–1078.

Youn, J.-Y., Dunham, W.H., Hong, S.J., Knight, J.D.R., Bashkurov, M., Chen, G.I., Bagci, H., Rathod, B., MacLeod, G., Eng, S.W.M., Angers, S., Morris, Q., Fabian, M., Côté, J.-F., Gingras, A.-C., 2018. High-density proximity mapping reveals the subcellular organization of mRNA-associated granules and bodies. Mol. Cell 69, 517–532.e11.

Zámocký, M., Hofbauer, S., Schaffner, I., Gasselhuber, B., Nicolussi, A., Soudi, M., Pirker, K.F., Furtmüller, P.G., Obinger, C., 2015. Independent evolution of four heme peroxidase superfamilies. Arch. Biochem. Biophys. 574, 108–119.

Zhou, Y., Wang, G., Wang, P., Li, Z., Yue, T., Wang, J., Zou, P., 2019. Expanding APEX2 substrates for proximity-dependent labeling of nucleic acids and proteins in living cells. Angew. Chem. Int. Ed. Engl. 58, 11763–11767.

CHAPTER FOUR

Elucidation of host-virus surfaceome interactions using spatial proteotyping

Fabian Wendt[a,b,†], Emanuela S. Milani[a,b,†], and Bernd Wollscheid[a,b,*]
[a]Department of Health Sciences and Technology (D-HEST), ETH Zurich, Institute of Translational Medicine (ITM), Zurich, Switzerland
[b]Swiss Institute of Bioinformatics (SIB), Lausanne, Switzerland
*Corresponding author: e-mail address: bernd.wollscheid@hest.ethz.ch

Contents

Abstract

The cellular surfaceome and its residing extracellularly exposed proteins are involved in a multitude of molecular signaling processes across the viral infection cycle. Successful viral propagation, including viral entry, immune evasion, virion release and viral spread rely on dynamic molecular interactions with the surfaceome. Decoding of these viral-host surfaceome interactions using advanced technologies enabled the discovery of fundamental new functional insights into cellular and viral biology. In this review, we highlight recently developed experimental strategies, with a focus on spatial proteotyping technologies, aiding in the rational design of theranostic strategies to combat viral infections.

† These authors contributed equally.

Advances in Virus Research, Volume 109
ISSN 0065-3527
https://doi.org/10.1016/bs.aivir.2021.03.002

Abbreviations

ACE2	angiotensin-converting enzyme 2
ADCC	antibody-dependent cell-mediated cytotoxicity
AIDS	acquired immune deficiency syndrome
APC	antigen-presenting cell
APEX	engineered ascorbate peroxidase
APMS	affinity purification mass spectrometry
BAR	biotinylation by antibody recognition
BST2	bone marrow stromal antigen 2
CADA	cyclotriazadisulfonamide
CCR5	C-C chemokine receptor type 5
CD	cluster of differentiation
CITE-seq	cellular indexing of transcriptomes and epitopes by sequencing
Co-IPMS	co-immunoprecipitation mass spectrometry
COVID-19	coronavirus disease 2019
CR	complement receptor
CXCR4	C-X-C chemokine receptor type 4
DC	dendritic cell
DC-SIGN	DC-specific intercellular adhesion molecule-3-grabbing non-integrin
ECM	extracellular matrix
EGFR	epidermal growth factor receptor
EMARS	enzyme-mediated activation of radical sources
ER	endoplasmic reticulum
Fc	fragment crystallizable
FDA	Food and Drug Administration
GPI	glycosylphosphatidylinositol
HAdV	human adenovirus
HBV	hepatitis B virus
HCMV	human cytomegalovirus
HDV	hepatitis D virus
HIV	human immunodeficiency virus
HLA	human leukocyte antigen
HRP	horseradish peroxidase
HTLV-1	human T-lymphotropic virus-1
IAV	influenza A virus
ICAM1	intercellular adhesion molecule 1
IFN-γR	interferon-γ receptor
KIR	killer cell immunoglobulin-like receptors
LC-MS	liquid-chromatography mass spectrometry
LFA1	lymphocyte function-associated antigen 1
LRC	ligand-receptor capture
LRP1	low-density lipoprotein receptor-related protein 1
MHC	major histocompatibility complex
MIP	macrophage inflammatory proteins
MS	mass spectrometry
MTOC	microtubule organization center
NCAM1	neural cell adhesion molecule 1

NK	natural killer
NKG2A	NKG2-A/NKG2-B type II integral membrane protein
NTB-A	NK-T-B-antigen
NTCP	sodium taurocholate cotransporting polypeptide
PD-L1	programmed-death ligand 1
PTPRJ	receptor-type tyrosine-protein phosphatase eta
QTV	quantitative temporal viromics
REAP-seq	RNA expression and protein sequencing
SARS-CoV	severe acute respiratory syndrome coronavirus
Siglec-1	sialic acid-binding Ig-like lectin 1
SNAT1	sodium-coupled neutral amino acid transporter 1
SOG	singlet oxygen generator
SPPLAT	selective proteomic proximity labeling assay using tyramide
UV	ultraviolet
Vpu	viral protein U
ZIKV	Zika virus

1. The surfaceome is the host's cellular interface enabling viral entry, immune evasion, and viral spread

Proteins exposed at the exterior of the plasma membrane orchestrate the cellular communication with the extracellular environment. Cell surface residing receptors, transporters, channels, cell-adhesion proteins, and ectoenzymes are together referred to as the cell surface proteome or surfaceome (Almén et al., 2009; da Cunha et al., 2009). The surfaceome is involved in many fundamental cellular functions such as nutrient absorption, cell adhesion and signaling. Particularly, cell surface proteins mediate information transduction from the outside environment to intracellular signaling networks (Bausch-Fluck et al., 2019). As cellular gateway keepers cell surface proteins interact in trans with soluble ligands (small molecules, peptides, proteins, microorganisms), extracellular matrix (ECM) proteins and surfaceome members of different cells enabling cell-to-cell communication (Armingol et al., 2021). Moreover, the transmission of extracellular signals across the plasma membrane depends on the lateral interaction of receptors in cis with the adjacent surfaceome on the same cell, and eventually downstream intracellular effectors (Jacobson et al., 2019). Cell surface proteins are not randomly distributed within the plasma membrane and rather dynamically organized as monomers, in nanoscale domains, and in microscale higher-order structures. Although not fully understood yet, intrinsic membrane factors such as lipid rafts and tetraspanin nanodomains as well as extrinsic factors such as cortical actin and galectin orchestrate this lateral protein organization and the underlying signaling capacity (van Deventer et al., 2021). Lateral interactions and protein networks range from promiscuous heterodimerization, e.g., within the epidermal growth factor receptor (EGFR) family (Kennedy et al., 2016), co-receptor dependencies, e.g., dictating human immunodeficiency virus (HIV) tropism (Bozek et al., 2012), clustering of receptors in nanodomains, e.g., affecting interferon-γ receptor (IFN-γR) 2 signaling (Blouin et al., 2016), to the formation of functional microscale structures in immunological synapses (Maity et al., 2015). Fluctuating external conditions cause cells to dynamically alter cell surface protein abundances that in turn lead to a context-specific interactome, and adapted surface signaling capacity to allow subsequently for a rapid and efficient response (Bausch-Fluck et al., 2019; Washburn, 2016).

All plasma membrane proteins with at least one amino acid facing the extracellular space are considered part of the surfaceome (Bausch-Fluck et al., 2019). This includes integral and extracellular peripheral membrane proteins, and also exterior lipid-anchored proteins such as glycosylphosphatidylinositol (GPI)-linked proteins. A mammalian cell-type specific screen cataloged over 1500 cell surface proteins (Bausch-Fluck et al., 2015). This experimentally defined, carefully curated dataset enabled through feature extraction in combination with machine learning the prediction that approximately 2900 proteins, from the roughly 20,000 UniProt annotated human protein entries, belong to the human surfaceome (Bausch-Fluck et al., 2018). Around 90% of the surfaceome members are N-linked glycosylated within their extracellular domain; which plays an important role in immunity (Maverakis et al., 2015). Moreover, the surfaceome and its organization reflects the identity and functional state of a cell and thus represents a valuable resource as a diagnostic and therapeutic target pool (Overington et al., 2006; Yin and Flynn, 2016).

As the plasma membrane delimits the cell, obligate intracellular viruses need to breach this barrier. For this task, viruses take advantage of the exposed localization of cell surface proteins and non-proteinaceous, membrane-bound structures for initial attachment (Yamauchi and Helenius, 2013) (Fig. 1). Subsequently, the surfaceome is a critical interface for viral entry into the cell, immune evasion, virion release and spread (Doms and Trono, 2000). During these steps of the productive infection cycle, the abundance, and organizational integrity of host cell surface proteins are modulated. Furthermore, viral proteins eventually also localize at the plasma membrane and become

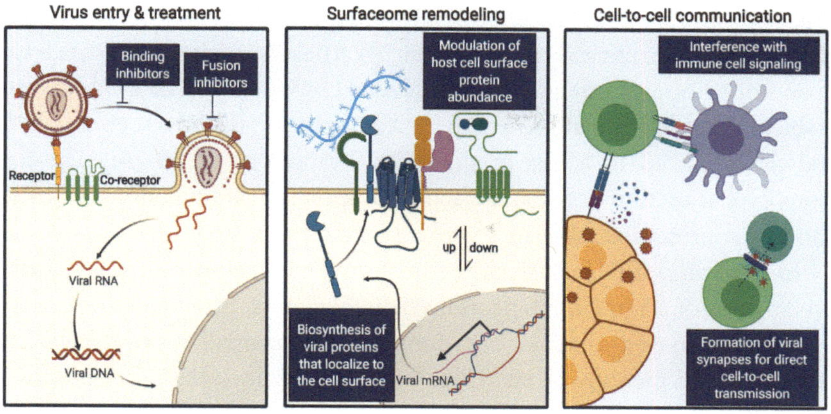

Fig. 1 The surfaceome is the cellular interface enabling viral entry, immune evasion, and viral spread. The surfaceome is a critical molecular interface during the entire span of the viral life cycle. As an initial step during the infection cycle, viruses attach and bind cell surface proteins to invade host cells, which is targeted by antiviral treatments to inhibit the infection. During the course of infection (post virus entry), viruses modulate the abundance and the organizational integrity of host cell surface proteins to counteract the host response. Additionally, viruses synthetize and localize viral encoded proteins at the plasma membrane becoming "foreign" surfaceome members, changing the surfaceome functionality. Overall, viruses induce a reorganization of cell surface proteins in order to alter intercellular communication in immune signaling as well as promoting their spread through direct-cell-to-cell transmission.

"foreign" surfaceome members, rewiring the cell surface signaling network leading to novel surfaceome functionality to promote viral propagation (Rodriguez Boulan and Pendergast, 1980). Thus, the surfaceome plays a key role during viral infection, ultimately defining pathogenesis (Fig. 1).

In this review, we describe recent technological advances that provided new insights into virus-host surfaceome interactions with a focus on mass spectrometry (MS)-based spatial proteotyping strategies. The highlighted strategies indicate great progress in the field of host-pathogen biology and show how information about molecular nanoscale organization opens up new theranostic opportunities to combat viral infection.

2. Mapping the surfaceome and its interaction network

Over the past decade, MS-based proteotyping strategies made significant strides forward facilitating the elucidation of fundamental biology and clinical phenotypes (Aebersold and Mann, 2003, 2016; Bantscheff et al., 2007;

Doll et al., 2019; Röst et al., 2015). Substantial technological improvements in sensitivity and quantitative accuracy, but also the advancements in data acquisition strategies, and machine-learning-based computational data analysis have enabled comprehensive proteotype measurements (Kelstrup et al., 2018; Ludwig et al., 2018; Meier et al., 2020; Tyanova et al., 2016). The term proteotype is used to describe the acute composition and functional organization of the proteome associated with a specific phenotype.

In relation to the analysis of the proteotype composition, the quantitative analysis of up to 10,000 proteins from *in vitro* cultured cell lines, as well as quantitation of around 1000 proteins from mammalian single cells have been reported recently (Kelly, 2020; Muntel et al., 2019). Coupled with sophisticated sample preparation methodologies, MS-based proteotyping is a powerful tool-box enabling the investigation of infectious diseases on the molecular level. Insights into virus-host biology were gained through the elucidation of the functional organization of the proteotype, including post-translational modifications, protein-protein interactions, and classification of subcellular proteome pools including spatial and temporal resolved surfaceome dynamics (Gordon et al., 2020; Greco and Cristea, 2017; Jean Beltran et al., 2016; Novy et al., 2018; Weekes et al., 2014).

The MS-based spatially restricted elucidation of the cell surface protein pool is challenging due the fact that the plasma membrane represents only 1–2% of cellular membranes, the hydrophobicity of transmembrane proteins, their low abundance in contrast to intracellular proteins and the fact that cell surface proteins are present at the cell surface, but can also reside in intracellular pools. Thus, whole cell lysate-based proteotyping studies often only cover a small subset of the surfaceome and do not allow for making statements about the quantity and functional organization of proteins in a particular location, in this case the cell surface. Transcriptomic approaches are also utilized to impute protein quantities based on the measurable mRNA pool, but both strategies still fall short of reporting the actual identity and quantity of the surfaceome-residing pool of receptors, as these strategies cannot report on spatially and functionally distinct receptor abundance information. Indeed, studies based on combined transcriptome sequencing with DNA-barcoded antibody mediated identification of proteins (CITE-seq and REAP-seq) have shown that single cell relative mRNA abundances and corresponding surface protein abundances correlate weakly (Liu et al., 2016; Peterson et al., 2017; Stoeckius et al., 2017). However, contrasting studies have been published for selected surfaceome subgroups (Nusinow et al., 2020). Neural network-based predictors of surfaceome abundance, which are based on imputation from single cell RNA sequencing data, showed early promising results for a set of 24 immunophenotypic markers, but did not prove system-wide prediction competence (Zhou et al., 2020). This shortcoming in the prediction accuracy is currently mainly due, as reported by the authors, to stochasticity in RNA processing, and lack of detailed knowledge/data about protein translation, protein transport, and proteostasis in very general terms. Therefore, tailored surfaceome enrichment strategies and direct measurements are required to report accurately on the functionally relevant surfaceome proteotype across the viral infection process.

The spatial characterization of the surfaceome encompasses the mapping of the global cell surface receptor repertoire including receptor abundances, as well as the locally confined cell surface signaling microenvironments the receptors reside in, including the organization of extracellular protein networks, and extracellular receptor-ligand interactions—which are eventually utilized during the viral life cycle (Fig. 2). In the past, methodologies such as proteolytic-cleavage of cell surface proteins by trypsin or proteinase K (Olaya-Abril et al., 2014), one-step plasma membrane purification by centrifugation (Lund et al., 2009), and silica bead coating-based affinity purification (Durr et al., 2004) were utilized in order to gain information about the surfaceome proteotype. However, these strategies typically suffer from protein contaminations, originating from cytosolic, intracellular plasma membrane attached or intracellular organelle membrane compartments. Thus, such strategies are only partially suited for the *bona fide* identification and quantification of only the functionally relevant pool of cell surface residing proteins. Alternatively, enrichment of cell surface proteins can be achieved by surfaceome-wide protein labeling/tagging methodologies (Fig. 2A). These strategies typically rely on bi-functional, membrane-impermeable small molecules. Next to the affinity moiety (usually biotin), these reagents harbor a reactive group to covalently label/tag cell surface proteins. Extracellularly exposed proteins can be labeled with, e.g., NHS-based chemistry on primary amines on accessible lysine residues or N-termini (Karhemo et al., 2012), or with hydrazide- or aminooxy-based reactions at their glycan structure (Weekes et al., 2010; Wollscheid et al., 2009). NHS-based labeling is usually fast, but may suffer from labeled cytosolic proteins originating from leaking cells. In addition, stochastic tagging of lysine residues with NHS blocks tryptic digestion at these tagged sites, typically leading to longer peptides upon tryptic cleavage. To more specifically

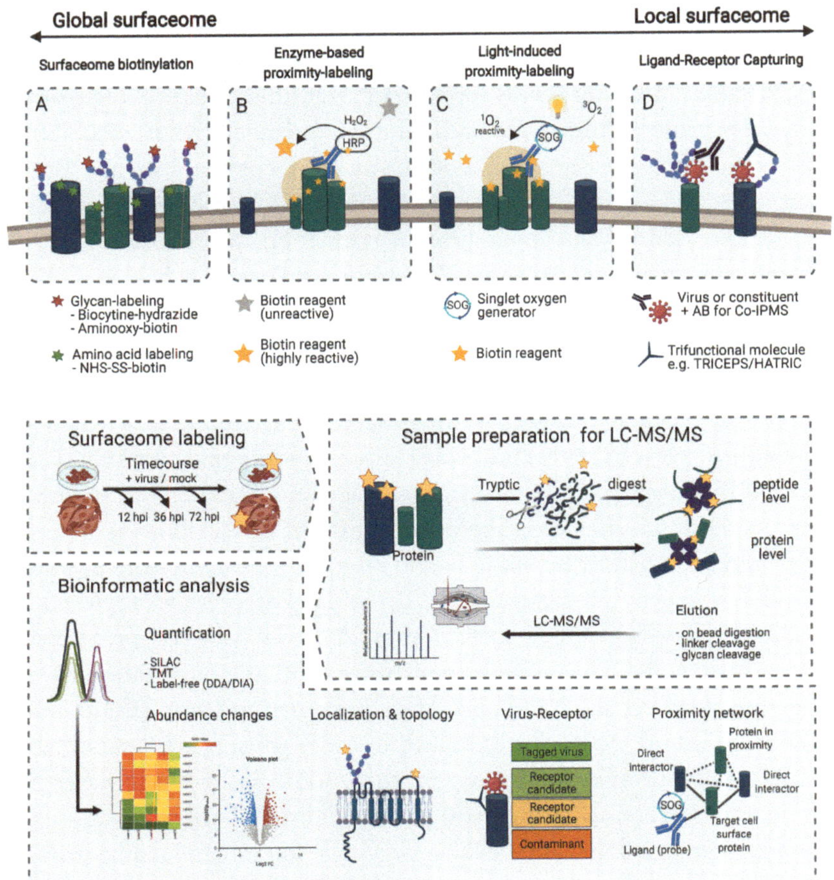

Fig. 2 Molecular characterization of virus-induced surfaceome dynamics, interaction and organization using spatial proteotyping strategies. The top panel illustrates exemplary labeling/tagging strategies (A) for global surfaceome biotinylation, (B and C) for proximity-based labeling of molecular microenvironments including interaction networks and (D) for the locally confined surfaceome analysis, e.g., direct ligand-receptor interactions. (A) Extracellularly exposed proteins can be tagged at selected protein amino acid residues with NHS-based chemistry or at their N- and O-glycosylation structures with hydrazide/aminooxy-biotin tags. (B) Extracellular enzyme-based proximity labeling strategies often utilize HRP-coupled antibodies against the target cell surface protein of interest. HRP catalyzes the generation of highly reactive biotin-probes upon addition of H_2O_2. Subsequently, the HRP-AB conjugate, protein of interest and adjacent proteins are labeled with biotin-tags. (C) Light-induced proximity labeling termed LUX-MS uses singlet oxygen generators (SOGs) coupled to antibody or ligands against the known or unknown target cell surface protein. Upon illumination, SOGs catalyze the formation of highly reactive singlet oxygen ($^1O_2$). Singlet oxygen oxidizes biomolecules in close vicinity and activates amino acid residues for labeling. Subsequently, biocytin-hydrazide covalently reacts with these activated amino acid

label surfaceome-residing proteins, the prominently occurring glycosylation of extracellular protein domains is targeted in glycan-based labeling approaches. However, glycan-based strategies exclude non-glycosylated cell surface proteins from analysis. Despite this, since around 90% of the surface proteins are N-linked glycosylated within their extracellular domain this strategy is successfully used in surfaceome studies, reporting a high enrichment of cell surface proteins. Chemical protein labeling approaches, partly combined with miniaturization and automation, were applied for surfaceome mapping of low-abundant receptors during cell differentiation, rare immune cell populations, and time-resolved dynamics in immune and neuronal signaling (Kalxdorf et al., 2017; Ravenhill et al., 2020; van Oostrum et al., 2019, 2020). Cell surface protein enrichment strategies and their molecular details were reviewed extensively in Elia (2008), Elschenbroich et al. (2010), Kuhlmann et al. (2018) and Li et al. (2020b).

Whereas the described chemical labeling/tagging methodologies enable the global analysis of the surfaceome, these strategies do not allow for the investigation of the functional organization of cell surface proteins, e.g., the nanoscale organization of immunological synapses upon viral infection. Here, proximity labeling methodologies can facilitate extracellular

residues and biotinylates proteins in proximity. (D) Extracellular receptors can be captured by Co-IPMS or with trifunctional crosslinkers, that (i) couple the ligand, (ii) crosslink with the receptors upon ligand binding on live cells, and (iii) enable affinity purification through a third moiety such as biotin. The bottom panel shows schematically the proteotyping strategy for analyzing the labeled/tagged cell surface proteins. Initially, the surfaceome of cells or tissue is tagged by the labeling strategies described above. Proteins are then extracted from the biological specimen and subsequently prepared for further proteotype analysis. Affinity purification of labeled/biotinylated species can be carried out at the protein- or peptide-level. Protein level: Undigested proteins are affinity purified exploiting the strong biotin/streptavidin binding. Upon stringent washing, proteins are eluted either by on-bead digestion, or through dissociation of cleavable linker molecules followed by in-solution digestion. Peptide-level: Proteins are digested by trypsin and biotinylated peptides are affinity purified with streptavidin. In case of NHS-based labeling, peptides can be eluted with harsh buffer conditions, and heating from streptavidin. Biotinylated N-glycopeptides can be enzymatically released with PNGase F. Release of formerly N-glycosylated peptides results in the deamidation of asparagine within the NXS/T consensus sequence (where X can be any amino acid except proline), indicating surface localization and glycosylation site. Eluted peptides are then analyzed with liquid-chromatography mass spectrometry (LC-MS). Relative quantification of measured peptides/proteins can be computed based on the chosen quantitative strategy (TMT, SILAC, label-free DDA or DIA). Obtained quantitative matrices finally shed light on virus-induced surfaceome dynamics, interaction and organization changes upon infection.

proteotyping in confined local areas at the cell surface (Bausch-Fluck et al., 2019; Chen and Perrimon, 2017). Delineating the molecular nanoscale organization of the surfaceome is a prerequisite for understanding cellular signaling. The developed proximity-tagging strategies can be classified in enzyme-based and light-induced labeling technologies (Fig. 2B and C) and enable tagging of a cell surface residing protein of interest, as well as its direct interaction partners, transient interactors and proteins localizing in close proximity. The tagged cell surface proteins can then be affinity purified under stringent conditions, as the affinity tag is covalently attached to the proteins. Extracellular enzyme-based proximity labeling strategies predominantly utilize horseradish peroxidase (HRP) as biocatalyst. HRP catalyzes under consumption of H_2O_2 the formation of reactive intermediates such as phenoxyl radicals, which freely diffuse to proteins in proximity and target electron-rich amino acid side chains of tyrosine, tryptophan, histidine, and cysteine. Of note is that the necessary usage of H_2O_2 induces oxidative cellular stress and activates kinases upon deactivation of phosphatases. The implication is that only activated cellular states can be analyzed. Nevertheless, HRP exhibits a high activity with short reaction times, and stability in the extracellular (oxidizing) environment, which is exploited in a variety of technologies including EMARS (Jiang et al., 2012), split HRP (Martell et al., 2016), SPPLAT (Rees et al., 2015), and BAR (Bar et al., 2018). HRP-based approaches have been applied, among others, to interrogate the lipid-raft composition, and B-cell receptor architecture (Jiang et al., 2012; Li et al., 2014).

Moreover, engineered ascorbate peroxidase (APEX), and Pup ligase PafA were used for example to elucidate the interactome of growth factor signaling domains (Zhen et al., 2018) and extracellular Interleukin-2 mediated receptor labeling (Liu et al., 2018), respectively. In contrast to HRP- and the similar APEX-based labeling, PafA-mediated labeling modifies lysine residues of proteins residing in very close proximity, but requires longer reaction times due to the lower activity of the enzyme.

Recently, two light-inducible optoproteomic proximity labeling technologies, termed μMap and LUX-MS, were developed enabling the decoding of surfaceome signaling architectures (Geri et al., 2020; Müller et al., 2020). These proximity-tagging strategies utilize, instead of the larger enzymes mentioned before, small-molecule catalysts, having likely also a smaller impact on the structure-function relationship upon coupling to the probe/bait (Fig. 2C). These catalysts were coupled to antibodies, ligands or in general terms affinity binders, targeting the cell surface protein of interest. Upon short illumination using selected light sources, these catalysts induce, through chemical reactions, modifications in amino acids of cell surface proteins surrounding the target protein. The modified proteins can be subsequently identified upon affinity enrichment using a MS-based technology. The μMap strategy developed by the MacMillan laboratory is utilizing a Dexter energy transfer-based mechanism to spatially restrict the generation of reactive carbene species around the protein of interest (Arias-Rotondo and McCusker, 2016). The formed carbene species exhibit a shorter half-life than HRP-generated phenoxyl radicals and thus confine the labeling radius further to a few nanometers, ultimately enhancing the labeling specificity. They applied this principle successfully for the elucidation of protein interaction networks, such as shown for CD45, CD47 and CD20 on Jurkat cells, identifying both known and previously unknown interaction partners. Furthermore, μMap enabled the dissection of the programmed-death ligand 1 (PD-L1) microenvironment on the surface of living lymphocytes (Geri et al., 2020). LUX-MS technology exploits, in contrast, singlet oxygen generators (SOG) for the light-triggered identification of acute protein interactions on living cells (Müller et al., 2020). Using SOG-coupled antibodies, small molecule-drugs, biologics and intact viral particles, LUX-MS was shown not only to decode ligand-receptor interactions across organisms, but also to decipher the surfaceome receptor nanoscale organization engaged by ligands. A spatial organization that has direct implications for drug action. Furthermore, investigation of functional immune synapses revealed that intercellular signaling between antigen-presenting cells (APCs) and CD8+ T cells can be mapped now providing insights into T cell activation with spatio-temporal resolution.

Proximity labeling strategies evolved into widely applicable technologies whose application space can be chosen and targeted to a particular microenvironment by the selected probe/affinity binder carrying the tagging functionality. Spatially confined probes such as antibodies or ligands enable the deciphering of cell surfaceome nanoscale signaling interaction networks, and widely distributing generic probes even facilitate the global *in vivo* surfaceome analysis (Li et al., 2020a).

Identified proteins in proximity labeling methodologies are close to the targeted protein at the cell surface, but might not necessarily be direct interactors and/or be of functional relevance in the infection process (Liu et al., 2020). Nevertheless, to decode the direct ligand-receptor interaction, e.g., for the identification of the cognate virus receptor, several strategies were developed (Fig. 2D). Well-established protein-protein interaction mapping technologies such as co-immunoprecipitation mass spectrometry (Co-IPMS) can help to decipher extracellular ligand-receptor interactions by antibody-based isolation of the whole virus or a virus constituent bound to its receptor (Gerold et al., 2016). However, Co-IPMS for extracellular targets remains challenging, as the necessity of detergents to extract proteins from the plasma membrane can disrupt the often-labile ligand-receptor interactions. The transient character of interactions and, typically unknown, on and off rates can reduce the success of identifying such ligand receptor interactions at the surface of living cells (Martinez-Martin, 2017). To overcome these drawbacks, the ligand receptor pair can be crosslinked on intact cells to stabilize the interaction prior to enrichment (Fig. 2D). Technologies such as the TRICEPS- and HATRIC-based ligand receptor capture (LRC) technologies, as well as the photo-crosslinking-based approach by Srivastava et al. utilize this principle by exploiting three-headed small molecules that (i) bind the ligand, (ii) crosslink with the receptors, and (iii) enable affinity purification through a third moiety (Frei et al., 2012; Sobotzki et al., 2018; Srivastava et al., 2020). HATRIC- and TRICEPS LRC are leveraging quantitative chemical reaction differences in order to identify unknown receptor(s) of known ligands while Srivastava et al. utilize UV exposure at different time points to generate quantitative differences, which enable the MS-based identification of the cognate receptors. Identification of ligand-receptor interactions on live cells by affinity purification takes into account the native functional environment of the receptor, but can be a delicate task, due to the variety of biophysical properties of ligands, and thus restricted to a single or few ligands per experiment. Recently established large-scale platforms enable testing of binary extracellular ligand-receptor interactions in a high-throughput manner, but require prior genetic engineering (Galaway and Wright, 2020; Martinez-Martin et al., 2016).

The above-described technologies and strategies allow for the investigation of diverse spatially confined surfaceome. These applications show furthermore that mapping of global surfaceome dynamics during viral infection, decoding of extracellular microenvironments, such as immunological or viral synapses and identification of locally confined ligand-receptor interactions, e.g., cognate virus receptors can be in principle approached successfully.

3. Deciphering virus binding and entry strategies to guide rational design of antiviral strategies

The interaction of viral particles with the cellular surfaceome plays an essential role during the first steps of the viral life cycle. It is the first contact/interaction point between the viral particle and the host cell, which sets the stage for viral entry, the initial step of the infection. Virus entry is initiated by the attachment of the virus to cell-surface receptors and leads ultimately to the delivery of the viral genetic content into the cell. The identification of host cell surface receptors is, therefore, fundamental for understanding the molecular processes involved in viral infection and virus tropism, as well crucial for the development of antiviral treatment to inhibit or interfere with successful viral infection.

With the recent evolution of MS-based proteotyping strategies, novel approaches have been developed to study different aspects of interactions between viruses and host cells including surfaceome interactions. A wide range of strategies have been developed over the years for the identification of viral receptors, extending from early biochemical and immunological strategies (Bass and Greenberg, 1992) to genomic strategies, protein microarrays, high-throughput screening, gain-of-function and loss-of-function strategies, and MS-based strategies (Barrass and Butcher, 2020). An extensive review describing the use of MS-based proteotyping technologies to map the host receptor space was written by Gerold et al. (2016). Here, we will describe a few prime examples highlighting the use of past and current MS-based approaches to identify host cell surface receptors for viral particles.

Co-IPMS and affinity purification mass spectrometry (APMS) are technologies widely used for the identification of protein-protein interactions. Co-IPMS was used for example to elucidate the receptor for severe acute respiratory syndrome coronavirus (SARS-CoV) during the first SARS-CoV pandemic in 2003 (Li et al., 2003). The truncated soluble S1-domain of the viral spike glycoprotein was fused to the Fc domain of human immunoglobulin-γ1. Immunoprecipitation allowed for the enrichment of the cognate receptor, which was subsequently identified by MS as the angiotensin-converting enzyme 2 (ACE2). Since SARS-CoV has very high homology with SARS-CoV-2, the cause of the current coronavirus disease 2019 (COVID-19) pandemic, ACE2 was also recently confirmed to play an essential role in SARS-CoV-2 entry (Hoffmann et al., 2020; Ou et al., 2020).

In combination with the above-mentioned affinity purification strategies a chemical cross-linking strategy was utilized to discover the receptor of hepatitis B virus (HBV). In 2012, the Li group used a "zero-distance photo-cross-linking" approach coupled to tandem affinity purification and MS to identify sodium taurocholate cotransporting polypeptide (NTCP), a multi-pass transmembrane transporter predominantly expressed in the liver, as the HBV receptor (Yan et al., 2012). In this study, a synthetic viral peptide based on a viral protein modified with photoreactive amino acid analogs and biotin was used. Irradiation with ultraviolet (UV) light enabled the direct covalent crosslinking to amino acids within the corresponding HBV-binding receptor, which was then purified using the biotin moiety. Crosslinking was essential to stabilize the HBV-receptor interaction and only enabled in this way the subsequent identification of NTCP by MS. Further experimental validation of the results showed that inhibition of NTCP expression blocked both HBV and hepatitis D virus (HDV) infection of liver cells. The discovery of NTCP as HBV entry receptor not only allowed the development of new HBV-sensitive cell lines that better phenocopied the complete life cycle of the virus but also enabled the further development of myrcludex B (Volz et al., 2013), a synthetic N-acylated preS1 lipopeptide-based inhibitor of HBV and HDV entry, currently used for the treatment of chronic HDV (Bogomolov et al., 2016) or dual infection of HBV and HDV (Blank et al., 2016; Donkers et al., 2017).

An alternative affinity capture approach is the above-mentioned HATRIC-LRC, which was implemented to shed light on the complex interaction between influenza A virus (IAV) and the surface of the host cell (Sobotzki et al., 2018). This chemoproteomic technology employs a trifunctional linker, termed HATRIC, which was first coupled to the intact human IAV H3N2 virus. Upon binding to the A549 cell surface, receptors in close proximity were then crosslinked to the virus and enriched through the azide affinity handle. MS was used to identify candidate receptors, and the list was further refined to six potential host entry facilitators for IAV by validating their impact on viral entry.

Srivastrava et al. used likewise a trifunctional chemical probe, but with different chemical properties to identify the entry receptor for Zika virus (ZIKV). The multifunctional chemical probe was first conjugated to the surface of the intact ZIKV particle, the photoreactive second group crosslinked to the cell surface receptor on the host cell upon UV exposure, and the third moiety, a biotin, enabled the purification of the cross-linked complex. The use of the whole viral particle, instead of a truncated soluble domain of a viral protein, to capture natural *bona fide* virus entry, enabled the identification of neural cell-adhesion molecule 1 (NCAM1) as the ZIKV receptor (Srivastava et al., 2020).

The progress achieved in the field of quantitative MS-based proteotyping has enabled the discovery of virus entry receptors shedding light on the mechanisms viruses exploit to enter the cells. The elucidation of viral entry mechanisms raises the opportunity for targeted development of therapeutic interventions that block the viral entry process. Currently, around 66% of all pharmaceutical drugs in the DrugBank target cell-surface proteins (Bausch-Fluck et al., 2019). In the field of antiviral treatments an opposite trend has been observed. Classically, host-targeting antiviral strategies have focused on inhibiting intracellular proteins crucial for virus replication and/or immune modulation. However, new classes of antiviral drugs affecting cell-surface proteins have gained a foothold as an antiviral strategy.

The first antiviral agent of this class to be approved by the United States Food and Drug Administration (FDA) in 2007 was maraviroc, a prime example for the importance of spatial knowledge about the lateral interaction network of receptors. Maraviroc inhibits the interaction between the host C-C chemokine receptor type 5 (CCR5) and the HIV-1 glycoprotein gp120, thereby inhibiting the entry of HIV into the host cell (Dorr et al., 2005; MacArthur and Novak, 2008). CCR5 was discovered by Alkhatib et al. as a co-receptor essential for HIV entry into CD4+ T lymphocytes (Alkhatib et al., 1996). Treating cells with the C-C chemokines MIP-1α, MIP-1β, or RANTES, the natural ligands of CCR5, significantly inhibited HIV virus entry. This discovery raised great interest in the pharmaceutical industry, and the efforts to screen for drugs targeting CCR5 led to the development of maraviroc.

Once the cell surface receptors exploited by viruses are identified, antibodies can also be developed against these specific receptors to block viral binding. A strategy established along with these recent developments is immunochemical engineering, a technology based on anchoring receptor specific antibodies to the plasma membrane in order to enhance local concentration and efficiency with which virus attachment to the surfaceome is blocked. This strategy was used to effectively render cells insensitive to human rhinovirus and HIV-1 (Xie et al., 2017). In the context of rhinovirus, a phage-displayed combinatorial human antibody library was screened for antibodies that bind to intercellular adhesion molecule 1 (ICAM1), the known receptor of rhinovirus (Greve et al., 1989). Selected ICAM1 specific antibodies were anchored to the plasma membrane to inhibit binding between the rhinovirus and its receptor at a higher efficiency compared to free circulating antibodies. The same surfaceome-targeting approach was used to render cells resistant to HIV infection by targeting the CD4 receptor, thereby opening the possibility of curing the infection. An alternative treatment approach is based on the principle that the inhibition of the entry receptor synthesis will result in a reduction of the abundance of the receptor at the plasma membrane. A strategy that hinders virus attachment and subsequent infection. A prime example for such an approach is the antiviral agent cyclotriazadisulfonamide (CADA), which lowers CD4 abundance and subsequently leads to reduced binding and infection by HIV-1 and human herpesvirus 7 (Vermeire et al., 2002).

Mapping of the cell surface receptor repertoire, and more generally the complex host-virus nanoscale interactions, is of critical importance for understanding the mechanisms by which viruses infect host cells. The progress achieved in the MS-based proteotyping field has enabled the discovery of cell surface receptors with high sensitivity in smaller subpopulations of cells. In the future, further elucidation of virus-induced surface remodeling is expected to guide the development of new antiviral strategies that will not only block virus entry and reduce the spread of the virus, but also selectively target cells that are already infected for the treatment of persistent infections.

4. Elucidation of virus-induced surfaceome dynamics

Invading viruses are under constant pressure to escape the host immune system. Among the viral countermeasures, viruses modulate cell surface residing and secreted proteins to their advantage (Alcami and Koszinowski, 2000). For example, HIV infection of T cells or APCs leads to a decrease in the cell surface abundance of complement receptors (CRs) CR1, CR2, CR3, C5aR, major histocompatibility complex (MHC) class I and II proteins, entry receptors CD4, CXCR4, CCR5, and T cell activating CD3 and Interleukin-2 receptor (Landi et al., 2011; Speth and Dierich, 1999). These abundance modulations of selected cellular receptors, and in turn their signaling functions, enable the virus eventually to evade the host immune response. The well-studied effects of HIV infection illustrate the diversity of modulated processes belonging to the innate and adaptive immune system occurring during virus infection. Interestingly, many virus families have evolved similar strategies to target immune signaling. To avoid clearance of infected cells by T cell-mediated cytotoxicity viruses such as HIV, human cytomegalovirus (HCMV), IAV and bovine papillomavirus modulate MHC class I and II antigen presentation. Viruses suppress the cell surface abundance of antigen-presenting proteins by altering their expression, their localization at the cell surface, or their turnover (Forsyth and Eisenlohr, 2016; Koutsakos et al., 2019; Petersen et al., 2003). As a counter effect, this loss renders cells susceptible to cytotoxicity mediated by natural killer (NK) cells (Kärre et al., 1986). NK cells are among the first line of defenders and function prior to the induction of the adaptive immune response. NK cells express the two major inhibitory receptor classes killer cell immunoglobulin-like receptors (KIRs) and the CD94-NKG2A heterodimers, that sense the presence of MHC class I proteins on cells (Pegram et al., 2011). Deficiency in MHC class I proteins prompts activation of NK cells, ultimately leading to pathogen clearance (Vivier et al., 2008). Viruses have developed numerous mechanisms to evade NK-based immune surveillance (Orange et al., 2002). Members of the retrovirus, poxvirus, herpesvirus, papillomavirus, and flavivirus families evade the NK cell response mainly by modulation of surface receptors and/or ligand secretion to interfere with NK cell receptor recognition. HCMV encodes a MHC class I homolog named UL18 (Beck and Barrell, 1988) that blocks NK cytotoxicity despite the absence of host MHC class I (Cosman et al., 1997; Reyburn et al., 1997). As UL18 has been reported to have both inhibitory and activating effects on the NK cell response (Leong et al., 1998; Prod'homme et al., 2007), further experiments are needed to elucidate its function in more detail. Studies based on murine CMV support the notion that multiple class I MHC viral homologs exist (Smith et al., 2002). Other viruses overcome the cytotoxic responses emanating from T and NK cells by specifically modulating only the expression of certain MHC class I alleles. Viruses reduce surface human leukocyte antigen (HLA) A and HLA-B molecules, which are efficient at presenting viral peptides to cytotoxic T lymphocytes, whereas they spare HLA-C and HLA-E, well-known ligands for NK cell inhibitory receptors (Lopez-Botet et al., 2001; Schust et al., 1998).

Apart from its immune modulatory function, the host surfaceome also plays a role in confining the spread of virions. For instance, the interferon (INF)-induced antiviral host restriction factor Bone marrow stromal antigen 2 (BST2, also named Tetherin) blocks the release of diverse mammalian enveloped viruses, e.g., HIV (Homann et al., 2011), filoviruses (Jouvenet et al., 2009), arenaviruses (Sakuma et al., 2009) and herpesviruses (Mansouri et al., 2009), by tethering virions to the plasma membrane. This mechanism hinders virion release toward the extracellular space and therefore limits the spread of the infection (Neil et al., 2008; Van Damme et al., 2008). This strategy is counteracted by the HIV protein Viral protein U (Vpu), which promotes in turn the downregulation of BST2. Interestingly, downregulation of BST2 by Vpu also protects HIV-infected cells from antibody-dependent cell-mediated cytotoxicity (ADCC) (Alvarez et al., 2014; Arias et al., 2014). However, it was reported that INF-α and/or small-molecule inhibitors of BST2 antagonists promote elimination of productively infected cells by broadly neutralizing antibodies (Pham et al., 2016). Therefore, BST2 antagonists are explored toward the effective treatment of infections caused by enveloped viruses.

The selected examples described above represent only a few of numerous mechanisms, which viruses developed during co-evolution with cellular hosts to evade the immune system. Nevertheless, the examples also illustrate the importance of studying the surfaceome and its virus-induced qualitative, quantitative and organizational changes. The analysis of surfaceome perturbations that occur upon infection is, therefore, imperative to shed light on the ever-improving strategies viruses use to escape the host immune system, as well as the rational basis for the development of new antiviral therapeutics. In this regard, several MS-based studies provided qualitative and quantitative data on surfaceome dynamics upon virus infection (Berro et al., 2007; Ersing et al., 2017; Gudleski-O'Regan et al., 2012; Hsu et al., 2015; Matheson et al., 2015; Soh et al., 2020; Stergiou et al., 2013; Viswanathan et al., 2017; Weekes et al., 2014; Zhang et al., 2019). Here, we will focus and describe in more detail two viruses and associated studies exploring surfaceome dynamics, in the context of HCMV infection, as a prime example for DNA viruses, and a second study focusing on HIV infection, as a prime example for RNA viruses.

HCMV is an enveloped, double stranded DNA virus belonging to the herpesvirus family. A high percentage of the world population is actually infected with HCMV, albeit this pathogenic virus has the ability to lie dormant, causing typically life-long latency. Nevertheless, HCMV infection can lead to severe issues in immunosuppressed individuals, and newborns (Crough and Khanna, 2009). Gudleski-O'Regan et al. analyzed the surfaceome of HCMV infected fibroblasts in order to gain insights into HCMV molecular pathogenesis (Gudleski-O'Regan et al., 2012). Cell surface residing proteins were tagged with sulfo-NHS-SS-biotin, a primary amine-reactive reagent enabling tagging of lysine residues and N-termini of proteins, which can subsequently be enriched by affinity purification strategies using the biotin handle with subsequent analysis by MS-based quantitation. Over the course of the viral infection, 500 annotated cell surface proteins were quantified at 6, 24 and 72 h post infection. While in the early phase only 8% of the surfaceome was classified as differentially abundant, the ratio of regulated to unregulated cell surface proteins increased with progression of infection to 24% at 72 h post infection. These cell surface proteins were involved in apoptosis (e.g., CD99), cell adhesion (CSPG4, VCAM1), immune response (CD55, CD59), metabolism

(LRP1, GLUT4), transport (Na+2/K-transporting ATPase-subunits), and signaling (ephrins). Further validation of the low-density lipoprotein receptor-related protein 1 (LRP1) demonstrated that its increased abundance during the early phases of infection reduced intracellular cholesterol levels and altered the lipid composition of virion envelopes, resulting in lower infectivity due to inefficient fusion of the virion envelope with the plasma membrane.

In a second study, Weekes et al. applied a quantitative temporal viromics (QTV) (Weekes et al., 2014) approach to characterize the surfaceome and in parallel the cellular proteotype during HCMV infection. Cell surface proteins were enriched through chemical labeling using aminooxy-biotin, which can be used to covalently attach biotin to aldehyde or ketone groups on polysaccharides/glycoproteins, with subsequent affinity purification. In total, about 1200 cell surface annotated proteins and >8000 intracellular proteins were quantified by using an MS3-based isobaric tandem mass tag (TMT) data acquisition strategy with seven time points starting from 6 to 96 h post infection. By integrating both datasets the regulation of several known, as well as potentially new cell surface residing NK and T cell ligands, were reported. Among these, protocadherins were exceptionally strongly reduced in their cell surface abundance as protein family, indicating a potential role as immunomodulators. Besides the quantitation of host proteins, 29 viral glycoproteins were identified as surfaceome members. Apart from known members of the virion envelope, which appear only at the plasma membrane during later stages of infection, a high correlation between expression and cell surface abundance for viral proteins was observed for many glycoproteins. The temporal resolution enabled the classification of the delayed appearing UL119 as a potentially new envelope protein (Weekes et al., 2014). The dataset from Weekes et al. was subsequently used to validate the isolation quality of the plasma membrane fraction in an elegant cell-wide organelle separation approach by Jean Beltran et al. enabling insights into molecular infection processes on the subcellular level upon HCMV infection of human fibroblasts (Jean Beltran et al., 2016). Cellular organelles were separated in a two-step centrifugation approach up to 120 h post infection with five time points in order to establish a spatio-temporal resolved proteotype during infection with compartmental/subcellular resolution. Isolated proteins were classified in spatially distinct plasma membrane-residing, cytosol, endoplasmic reticulum (ER), Golgi complex, lysosome, mitochondria, and peroxisome clusters. The plasma membrane fraction reached >80% agreement in the localization assignment with the QTV analysis (Weekes et al., 2014). Jean Beltran et al. further reported an early reduction of plasma membrane abundance of proteins involved in adhesion, cell junction, and antigen presentation, together with a global reorganization across the secretory pathway, plasma membrane, and mitochondria. Interestingly, they could show the dynamic localization switch of the viral protein UL13 between the plasma membrane/cytoplasm at 24 h post infection, to the mitochondria at 72 h post infection, and again to the plasma membrane and ER/Golgi late in infection (Jean Beltran et al., 2016). The multi-localizing viral protein UL13 may also have different functional relevance in its distinct environments. In addition to the temporal analysis of surfaceome dynamics, MS-based proteotyping can be used as a tool to dissect the contribution of individual proteins to molecular mechanisms. The viral HCMV proteins US3, US6, US11 and US2 were individually overexpressed in THP-1 cells. Subsequently, the surfaceome was analyzed by aminooxy-biotin-based chemical labeling. The HCMV proteins US3, US6 and US11 were found to specifically downregulate MHC class I proteins, as predicted from the literature. Besides that, expression of US2 in THP-1 had a broader impact and played the role of a mediator. Through triggering the proteasomal degradation of five distinct integrin chains, CD112, the interleukin-12 receptor, and the receptor tyrosine-protein phosphatase PTPRJ via the cellular E3 ligase TRC8 (translocation in renal carcinoma on chromosome 8 protein), US2 ultimately affected integrin signaling, cell adhesion and migration (Hsu et al., 2015).

HIV causes, since the early 1980s, an ongoing pandemic and has infected an estimated 65 million people worldwide. HIV infection eventually leads to the acquired immunodeficiency syndrome (AIDS), which resulted in approximately 25 million deaths to date according to the Joint United Nations Program on HIV/AIDS (UNAIDS) 2006 report on the global AIDS epidemic. Over the course of an HIV infection, the surfaceome is described as a "combat zone" (Doms and Trono, 2000). The surfaceome was shown to play multiple critical roles in the infection of HIV beyond the initial binding and uptake of the HIV particle itself (Abbas and Herbein, 2014). Viral adhesion, viral spreading and cellular apoptosis are additional aspects in which the surfaceome plays a role in HIV pathogenesis (Speth and Dierich, 1999).

In an interesting study by Matheson et al. the HIV infection of T cells was temporally analyzed during the first 72 h of the infection using the above-described QTV approach (Matheson et al., 2015). The proteomic data confirmed the previously reported reduction in cell surface abundance of CD4, HLA-A, CCR7, CD28, NTB-A, SELL, and the tetraspanins CD37, CD53, CD63. CD81, CD82, as well as controversially discussed proteins such as CD71, CXCR4, and CCR5. In total, more than one hundred host surfaceome members were differentially regulated in their cell surface abundance post infection. Many of these detected quantitative protein abundance changes were associated with cell surface receptors involved in immune functions, but others such as in case of sodium-coupled neutral amino acid transporter 1 (SNAT1), were interestingly associated with amino acid transport and therefore further validated. It was subsequently found that HIV protein Vpu mediates the depletion of amino acid transporter SNAT1 from the surfaceome pool by triggering its proteasomal degradation. The resulting data revealed a mechanism showing that SNAT1 depletion results in limited alanine uptake and reduced intracellular free alanine pools in CD4+ T cells. This mechanism ultimately leads to decreased alanine-dependent CD4+ mitogenesis and consequently to reduced T cell activation. The authors propose therefore a unique paradigm of HIV interference with immunometabolism (Matheson et al., 2015; Sugden and Cohen, 2015).

In conclusion, the above-described MS-based surfaceome analysis strategies enabled the characterization of dynamic surfaceome changes during the course of viral infection and, more importantly, contributed to a deeper mechanistic understanding of the molecular details underlying the interactions between viruses, the host surfaceome and the immune system.

5. The surfaceome and its role in viral spreading

During the final stage of the viral life cycle, newly generated viral particles egress the cell at spatially confined cell membrane locations. Viruses typically spread in between host cells through the release of viral particles in the extracellular space surrounding the initially infected cell. However, several viruses have developed alternative mechanisms to spread in a direct cell-to-cell fashion. These alternative strategies facilitate a rapid viral dissemination and avoid immune recognition of extracellular virions. Viruses achieve cell-to-cell dissemination through a variety of mechanisms as reviewed by Sattentau (2008): some viruses hijack intracellular protein complexes, like for example the recruitment of actin- or tubulin-containing structures that project the virus toward neighboring cell, strategy commonly used by poxviruses, herpesviruses and retroviruses, whereas others induce a surfaceome reorganization in order to engage cell-to-cell contacts. An illustration is the formation of virological synapses at the surface of immune cells.

Virological synapses are highly molecular organized cellular surfaceome junctions formed between communicating infected and uninfected cells, with a partial functional homology to immunological synapses (Bayliss and Piguet, 2018; Igakura et al., 2003). The formation of virological synapses between infected and uninfected T cells has been observed for numerous viruses, including the human T-lymphotropic virus-1 (HTLV-1) and HIV-1. HTLV-1 takes advantage of the additional engagement of ICAM1 with the ligand lymphocyte function-associated antigen 1 (LFA1) on the infected cells to stabilize virological synapses. This process initiates an intracellular relocation of the microtubule organization center (MTOC) in a Tax protein-dependent fashion inducing a polarization of viral assembly and budding toward the synaptic cleft of the virological synapses (Barnard et al., 2005; Nejmeddine and Bangham, 2010; Nejmeddine et al., 2005). Additional studies have shown that HIV-1 exploits the same strategy to form virological surfaceome synapses between infected and uninfected T cells to facilitate cell-to-cell transmission (Jolly et al., 2007, 2011; Vasiliver-Shamis et al., 2009). Beside cell-to-cell transmission between T cells, HIV-1 also takes advantage of the interaction between dendritic cells (DC) and T cells during lymphocyte activation for the early transmission of viral particles between the mucosal tissue and the secondary lymphoid tissue. HIV-1 virions are "trapped" at the surface of DCs by alternative HIV-1 receptors. Upon interaction with the T-cell a virological synapse is formed which recruits virions and the cognate receptors CD4, CXCR4 and CCR5 on the T-cell facilitating cell-to-cell spread. Several alternative receptors have been identified as HIV-1 binding factors, including: DC-specific intercellular adhesion molecule-3-grabbing non-integrin (DC-SIGN) which binds the viral glycoprotein gp120 (Arrighi et al., 2004; Curtis et al., 1992; Geijtenbeek et al., 2000) and sialic acid-binding Ig-like lectin 1 (Siglec-1, CD169), which mediates HIV-1 trans-infection by interacting with viral membrane gangliosides. The molecular, spatial and temporal organization of virological synapses and their molecular mode of action is currently not completely understood and poorly investigated by spatial MS-based strategies.

Viral synapses, and functionally similar mechanisms supporting cell-to-cell (or trans) infections are significant strategies for viruses to achieve high infectivity and at the same time escape the immune system response. Therefore, it is essential to understand at the molecular level the nanoscale reorganization of the surfaceome of infected cells and their respective communicating cells in order to be able to inhibit further viral dissemination. With the recent advancements in spatial MS-based proteotyping strategies described above the more detailed elucidation of such processes within virological synapses becomes feasible now.

6. Perspective

Proteotyping technologies considerably contributed to the identification of viral receptors and the elucidation of virus-induced surfaceome changes, thereby providing mechanistic insights and expanding our knowledge about viral pathogenicity. Temporally-resolved analysis of surfaceome abundance changes provide direct insights into the cellular host response toward viral intruders and enable the determination of virus-hijacked signaling networks. Surfaceome abundance changes can also be tell-tale signs as extracellular markers for intracellular signaling network changes. Moreover, they can provide additional mechanistic insights into the molecular mode of action during the stages of successful infection and viral propagation. 20 years after the human genome was sequenced, new technologies enable virologists now to go beyond genomics and elucidate the spatial molecular (surfaceome) organization influencing viral propagation. The analysis of protein-protein interactions, functional nanoclusters and receptor cross-talk are taking center stage in recognition that these submembrane signaling domains are playing key roles in viral propagation, immune signaling and pathogen clearance (Mattila et al., 2016). Recently established large-scale platforms enable now high-throughput testing of binary extracellular receptor interactions in the context of host-pathogen interactions (Galaway and Wright, 2020; Martinez-Martin et al., 2016). Martinez-Martin et al. mapped the extracellular host-virus interactome of recombinant human adenovirus (HAdV) E3 proteins versus 1500 host single transmembrane proteins. The Genentech team identified 51 new interaction pairs and validated them using complementary technologies (Martinez-Martin et al., 2016). While these approaches have great potential to decode extracellular interactions, complementary technologies are needed to also probe the native environment of receptors and their nanoscale organization. Here, recently developed proximity-tagging strategies can fill this gap and together these technologies can provide a systems virology understanding and perspective on the molecular nanoscale organization necessary for successful viral propagation. Among these new proximity-tagging strategies for decoding surfaceome nanoscale organization are the recently developed light-induced proximity labeling approaches μMap and LUX-MS. Both MS-based optoproteomic technologies were shown to enable the discovery of molecular nanoscale architecture in a spatio-temporal manner from receptor interaction networks to intercellular communication within immunological synapses (Geri et al., 2020; Müller et al., 2020). The application of light-induced proximity-tagging strategies in the context of viral infection will likely lead to new molecular mechanistic insights about the spatial organization and development of virological synapses. Moving from a protein-centric view in the context of viral infection to a proteoform-centric view will further be critical to bridge the current gap and quest in determining the cellular phenotype upon infection from the genotype. As scientists/virologists, we have the unique opportunity now to appreciate the complexity which resides within co-existing (surfaceome) proteoforms decorated with specific subsets of post- and co-translational modifications guiding functional molecular interactions with viruses uncovered by the latest (MS-)technology advancements. This newly gained knowledge can in turn be leveraged now to combat viral infections using innovative pharmaceutical strategies, targeting not only proteins residing within the drug-accessible surfaceome, but rather host proteoform-centric complexes—as the mediators of viral infection.

Acknowledgments

B.W. acknowledges generous support from ETH Zurich and D-HEST (BMPP). This work was also supported by a Swiss National Science Foundation Grant (grant 31003A_160259) to B.W. and by the Personalized Health and Related Technologies (PHRT) strategic focus area of ETH (to B.W.). Figures were created with a licensed version BioRender.com (to B.W.), partially by adaptation of BioRender templates, retrieved from https://app.biorender.com/biorender-templates. The authors would also like to acknowledge the Wollscheid lab (wollscheidlab.org) for critical feedback on the manuscript.

Conflict of interest statement

The authors declare no conflicts of interest, competing interests or financial interests.

CRediT authorship contribution statement

Fabian Wendt: Writing—original draft, review and editing
Emanuela Sara Milani: Writing—original draft, review and editing
Bernd Wollscheid: Funding acquisition, Supervision, Writing—review and editing

References

Abbas, W., Herbein, G., 2014. Plasma membrane signaling in HIV-1 infection. Biochim. Biophys. Acta 1838 (4), 1132–1142.

Aebersold, R., Mann, M., 2003. Mass spectrometry-based proteomics. Nature 422 (6928), 198–207.

Aebersold, R., Mann, M., 2016. Mass-spectrometric exploration of proteome structure and function. Nature 537 (7620), 347–355.
Alcami, A., Koszinowski, U.H., 2000. Viral mechanisms of immune evasion. Trends Microbiol. 8 (9), 410–418.
Alkhatib, G., et al., 1996. CC CKR5: a RANTES, MIP-1alpha, MIP-1beta receptor as a fusion cofactor for macrophage-tropic HIV-1. Science 272 (5270), 1955–1958.
Almén, M.S., et al., 2009. Mapping the human membrane proteome: a majority of the human membrane proteins can be classified according to function and evolutionary origin. BMC Biol. 7, 50.
Alvarez, R.A., et al., 2014. HIV-1 Vpu antagonism of tetherin inhibits antibody-dependent cellular cytotoxic responses by natural killer cells. J. Virol. 88 (11), 6031–6046.
Arias, J.F., et al., 2014. Tetherin antagonism by Vpu protects HIV-infected cells from antibody-dependent cell-mediated cytotoxicity. Proc. Natl. Acad. Sci. U. S. A. 111 (17), 6425–6430.
Arias-Rotondo, D.M., McCusker, J.K., 2016. The photophysics of photoredox catalysis: a roadmap for catalyst design. Chem. Soc. Rev. 45 (21), 5803–5820.
Armingol, E., et al., 2021. Deciphering cell-cell interactions and communication from gene expression. Nat. Rev. Genet. 22 (2), 71–88.
Arrighi, J.-F., et al., 2004. DC-SIGN–mediated infectious synapse formation enhances X4 HIV-1 transmission from dendritic cells to T cells. J. Exp. Med. 200 (10), 1279–1288.
Bantscheff, M., et al., 2007. Quantitative mass spectrometry in proteomics: a critical review. Anal. Bioanal. Chem. 389 (4), 1017–1031.
Bar, D.Z., et al., 2018. Biotinylation by antibody recognition—a method for proximity labeling. Nat. Methods 15 (2), 127.
Barnard, A.L., et al., 2005. Engagement of specific T-cell surface molecules regulates cytoskeletal polarization in HTLV-1—infected lymphocytes. Blood 106 (3), 988–995.
Barrass, S.V., Butcher, S.J., 2020. Advances in high-throughput methods for the identification of virus receptors. Med. Microbiol. Immunol. 209 (3), 309–323.
Bass, D.M., Greenberg, H.B., 1992. Strategies for the identification of icosahedral virus receptors. J. Clin. Invest. 89 (1), 3–9.
Bausch-Fluck, D., et al., 2015. A mass spectrometric-derived cell surface protein atlas. PLoS One 10 (3), e0121314.
Bausch-Fluck, D., et al., 2018. The in silico human surfaceome. Proc. Natl. Acad. Sci. U. S. A. 115 (46), E10988–E10997.
Bausch-Fluck, D., Milani, E.S., Wollscheid, B., 2019. Surfaceome nanoscale organization and extracellular interaction networks. Curr. Opin. Chem. Biol. 48, 26–33.
Bayliss, R.J., Piguet, V., 2018. Masters of manipulation: viral modulation of the immunological synapse. Cell. Microbiol. 20 (10), e12944.
Beck, S., Barrell, B.G., 1988. Human cytomegalovirus encodes a glycoprotein homologous to MHC class-I antigens. Nature 331 (6153), 269–272.
Berro, R., et al., 2007. Identifying the membrane proteome of HIV-1 latently infected cells. J. Biol. Chem. 282 (11), 8207–8218.
Blank, A., et al., 2016. First-in-human application of the novel hepatitis B and hepatitis D virus entry inhibitor myrcludex B. J. Hepatol. 65, 483–489. https://doi.org/10.1016/j.jhep.2016.04.013.
Blouin, C.M., et al., 2016. Glycosylation-dependent IFN-γR partitioning in lipid and actin nanodomains is critical for JAK activation. Cell 166 (4), 920–934.
Bogomolov, P., et al., 2016. Treatment of chronic hepatitis D with the entry inhibitor myrcludex B: first results of a phase Ib/IIa study. J. Hepatol. 65 (3), 490–498.
Bozek, K., et al., 2012. An expanded model of HIV cell entry phenotype based on multi-parameter single-cell data. Retrovirology 9, 60.

Chen, C.-L., Perrimon, N., 2017. Proximity-dependent labeling methods for proteomic profiling in living cells. Wiley Interdiscip. Rev. Dev. Biol. 6 (4). https://doi.org/10.1002/wdev.272.

Cosman, D., et al., 1997. A novel immunoglobulin superfamily receptor for cellular and viral MHC class I molecules. Immunity 7 (2), 273–282.

Crough, T., Khanna, R., 2009. Immunobiology of human cytomegalovirus: from bench to bedside. Clin. Microbiol. Rev. 22 (1), 76–98. https://doi.org/10.1128/CMR.00034-08.

Curtis, B.M., Scharnowske, S., Watson, A.J., 1992. Sequence and expression of a membrane-associated C-type lectin that exhibits CD4-independent binding of human immunodeficiency virus envelope glycoprotein gp120. Proc. Natl. Acad. Sci. U. S. A. 89 (17), 8356–8360.

da Cunha, J.P.C., et al., 2009. Bioinformatics construction of the human cell surfaceome. Proc. Natl. Acad. Sci. U. S. A. 106 (39), 16752–16757.

Doll, S., Gnad, F., Mann, M., 2019. The case for proteomics and phospho-proteomics in personalized cancer medicine. Proteomics Clin. Appl. 13 (2), e1800113.

Doms, R.W., Trono, D., 2000. The plasma membrane as a combat zone in the HIV battlefield. Genes Dev. 14 (21), 2677–2688.

Donkers, J.M., et al., 2017. Reduced hepatitis B and D viral entry using clinically applied drugs as novel inhibitors of the bile acid transporter NTCP. Sci. Rep. 7 (1), 15307.

Dorr, P., et al., 2005. Maraviroc (UK-427,857), a potent, orally bioavailable, and selective small-molecule inhibitor of chemokine receptor CCR5 with broad-spectrum anti-human immunodeficiency virus type 1 activity. Antimicrob. Agents Chemother. 49, 4721–4732.

Durr, E., et al., 2004. Direct proteomic mapping of the lung microvascular endothelial cell surface in vivo and in cell culture. Nat. Biotechnol. 22 (8), 985–992.

Elia, G., 2008. Biotinylation reagents for the study of cell surface proteins. Proteomics 8 (19), 4012–4024.

Elschenbroich, S., et al., 2010. Isolation of cell surface proteins for mass spectrometry-based proteomics. Expert Rev. Proteomics 7 (1), 141–154.

Ersing, I., et al., 2017. A temporal proteomic map of Epstein-Barr virus lytic replication in B cells. Cell Rep. 19 (7), 1479–1493.

Forsyth, K.S., Eisenlohr, L.C., 2016. Giving CD4+ T cells the slip: viral interference with MHC class II-restricted antigen processing and presentation. Curr. Opin. Immunol. 40, 123–129.

Frei, A.P., et al., 2012. Direct identification of ligand-receptor interactions on living cells and tissues. Nat. Biotechnol. 30 (10), 997–1001.

Galaway, F., Wright, G.J., 2020. 'Rapid and sensitive large-scale screening of low affinity extracellular receptor protein interactions by using reaction induced inhibition of Gaussia luciferase. Sci. Rep. 10 (1), 10522.

Geijtenbeek, T.B.H., et al., 2000. DC-SIGN, a dendritic cell–specific HIV-1-binding protein that enhances trans-infection of T cells. Cell 100 (5), 587–597.

Geri, J.B., et al., 2020. Microenvironment mapping via Dexter energy transfer on immune cells. Science 367 (6482), 1091–1097.

Gerold, G., Bruening, J., Pietschmann, T., 2016. Decoding protein networks during virus entry by quantitative proteomics. Virus Res. 218, 25–39.

Gordon, D.E., et al., 2020. Comparative host-coronavirus protein interaction networks reveal pan-viral disease mechanisms. Science 370 (6521). https://doi.org/10.1126/science.abe9403.

Greco, T.M., Cristea, I.M., 2017. Proteomics tracing the footsteps of infectious disease. Mol. Cell. Proteomics 16 (4 Suppl. 1), S5–S14.

Greve, J.M., et al., 1989. The major human rhinovirus receptor is ICAM-1. Cell 56 (5), 839–847.

Gudleski-O'Regan, N., et al., 2012. Increased expression of LDL receptor-related protein 1 during human cytomegalovirus infection reduces virion cholesterol and infectivity. Cell Host Microbe 12 (1), 86–96.

Hoffmann, M., et al., 2020. SARS-CoV-2 cell entry depends on ACE2 and TMPRSS2 and is blocked by a clinically proven protease inhibitor. Cell 181 (2), 271–280.e8.

Homann, S., et al., 2011. Upregulation of BST-2/Tetherin by HIV infection in vivo. J. Virol. 85 (20), 10659–10668.

Hsu, J.-L., et al., 2015. Plasma membrane profiling defines an expanded class of cell surface proteins selectively targeted for degradation by HCMV US2 in cooperation with UL141. PLoS Pathog. 11 (4), e1004811.

Igakura, T., et al., 2003. Spread of HTLV-I between lymphocytes by virus-induced polarization of the cytoskeleton. Science 299 (5613), 1713–1716.

Jacobson, K., Liu, P., Lagerholm, B.C., 2019. The lateral organization and mobility of plasma membrane components. Cell 177 (4), 806–819.

Jean Beltran, P.M., Mathias, R.A., Cristea, I.M., 2016. A portrait of the human organelle proteome in space and time during cytomegalovirus infection. Cell Syst. 3 (4), 361–373.e6.

Jiang, S., et al., 2012. A proteomics approach to the cell-surface interactome using the enzyme-mediated activation of radical sources reaction. Proteomics 12 (1), 54–62.

Jolly, C., Mitar, I., Sattentau, Q.J., 2007. Requirement for an intact T-cell actin and tubulin cytoskeleton for efficient assembly and spread of human immunodeficiency virus type 1. J. Virol. 81 (11), 5547–5560.

Jolly, C., et al., 2011. The regulated secretory pathway in CD4(+) T cells contributes to human immunodeficiency virus type-1 cell-to-cell spread at the virological synapse. PLoS Pathog. 7 (9), e1002226.

Jouvenet, N., et al., 2009. Broad-spectrum inhibition of retroviral and filoviral particle release by tetherin. J. Virol. 83 (4), 1837–1844.

Kalxdorf, M., et al., 2017. Monitoring cell-surface N-glycoproteome dynamics by quantitative proteomics reveals mechanistic insights into macrophage differentiation. Mol. Cell. Proteomics 16 (5), 770–785.

Karhemo, P.-R., et al., 2012. An optimized isolation of biotinylated cell surface proteins reveals novel players in cancer metastasis. J. Proteomics 77, 87–100.

Kärre, K., et al., 1986. Selective rejection of H–2-deficient lymphoma variants suggests alternative immune defence strategy. Nature 319 (6055), 675–678.

Kelly, R.T., 2020. Single-cell proteomics: progress and prospects. Mol. Cell. Proteomics 19 (11), 1739–1748.

Kelstrup, C.D., et al., 2018. Performance evaluation of the Q exactive HF-X for shotgun proteomics. J. Proteome Res. 17 (1), 727–738.

Kennedy, S.P., et al., 2016. The under-appreciated promiscuity of the epidermal growth factor receptor family. Front. Cell Dev. Biol. 4, 88.

Koutsakos, M., et al., 2019. Downregulation of MHC class I expression by influenza A and B viruses. Front. Immunol. 10, 1158.

Kuhlmann, L., et al., 2018. Cell-surface proteomics for the identification of novel therapeutic targets in cancer. Expert Rev. Proteomics 15 (3), 259–275.

Landi, A., et al., 2011. One protein to rule them all: modulation of cell surface receptors and molecules by HIV Nef. Curr. HIV Res. 9 (7), 496–504.

Leong, C.C., et al., 1998. Modulation of natural killer cell cytotoxicity in human cytomegalovirus infection: the role of endogenous class I major histocompatibility complex and a viral class I homolog. J. Exp. Med. 187 (10), 1681–1687.

Li, W., et al., 2003. Angiotensin-converting enzyme 2 is a functional receptor for the SARS coronavirus. Nature 426 (6965), 450–454.

Li, X.-W., et al., 2014. New insights into the DT40 B cell receptor cluster using a proteomic proximity labeling assay. J. Biol. Chem. 289 (21), 14434–14447.

Li, J., et al., 2020a. Cell-surface proteomic profiling in the fly brain uncovers wiring regulators. Cell 180 (2), 373–386.e15.
Li, Y., Qin, H., Ye, M., 2020b. An overview on enrichment methods for cell surface proteome profiling. J. Sep. Sci. 43 (1), 292–312.
Liu, Y., Beyer, A., Aebersold, R., 2016. On the dependency of cellular protein levels on mRNA abundance. Cell 165 (3), 535–550.
Liu, Q., et al., 2018. A proximity-tagging system to identify membrane protein-protein interactions. Nat. Methods 15 (9), 715–722.
Liu, X., et al., 2020. Combined proximity labeling and affinity purification-mass spectrometry workflow for mapping and visualizing protein interaction networks. Nat. Protoc. 15 (10), 3182–3211.
Lopez-Botet, M., Llano, M., Ortega, M., 2001. Human cytomegalovirus and natural killer-mediated surveillance of HLA class I expression: a paradigm of host-pathogen adaptation. Immunol. Rev. 181, 193–202. https://doi.org/10.1034/j.1600-065x.2001.1810116.x.
Ludwig, C., et al., 2018. Data-independent acquisition-based SWATH-MS for quantitative proteomics: a tutorial. Mol. Syst. Biol. 14 (8), e8126.
Lund, R., et al., 2009. Efficient isolation and quantitative proteomic analysis of cancer cell plasma membrane proteins for identification of metastasis-associated cell surface markers. J. Proteome Res. 8 (6), 3078–3090.
MacArthur, R.D., Novak, R.M., 2008. Reviews of anti-infective agents: maraviroc: the first of a new class of antiretroviral agents. Clin. Infect. Dis. 47 (2), 236–241.
Maity, P.C., et al., 2015. The nanoscale organization of the B lymphocyte membrane. Biochim. Biophys. Acta 1853 (4), 830–840.
Mansouri, M., et al., 2009. Molecular mechanism of BST2/tetherin downregulation by K5/MIR2 of Kaposi's sarcoma-associated herpesvirus. J. Virol. 83 (19), 9672–9681.
Martell, J.D., et al., 2016. A split horseradish peroxidase for the detection of intercellular protein–protein interactions and sensitive visualization of synapses. Nat. Biotechnol. 34 (7), 774–780.
Martinez-Martin, N., 2017. Technologies for proteome-wide discovery of extracellular host-pathogen interactions. J. Immunol. Res. 2017, 2197615.
Martinez-Martin, N., et al., 2016. The extracellular interactome of the human adenovirus family reveals diverse strategies for immunomodulation. Nat. Commun. 7, 11473.
Matheson, N.J., et al., 2015. Cell surface proteomic map of HIV infection reveals antagonism of amino acid metabolism by Vpu and Nef. Cell Host Microbe 18 (4), 409–423.
Mattila, P.K., Batista, F.D., Treanor, B., 2016. Dynamics of the actin cytoskeleton mediates receptor cross talk: an emerging concept in tuning receptor signaling. J. Cell Biol. 212 (3), 267–280.
Maverakis, E., et al., 2015. Glycans in the immune system and The Altered Glycan Theory of Autoimmunity: a critical review. J. Autoimmun. 57, 1–13.
Meier, F., et al., 2020. diaPASEF: parallel accumulation-serial fragmentation combined with data-independent acquisition. Nat. Methods 17 (12), 1229–1236.
Müller, M., et al., 2020. Light-Mediated Discovery of Surfaceome Nanoscale Organization and Intercellular Receptor Interaction Networks. Cold Spring Harbor Laboratory. https://doi.org/10.1101/2020.08.11.246652.
Muntel, J., et al., 2019. Surpassing 10 000 identified and quantified proteins in a single run by optimizing current LC-MS instrumentation and data analysis strategy. Mol. Omics 15 (5), 348–360.
Neil, S.J.D., Zang, T., Bieniasz, P.D., 2008. Tetherin inhibits retrovirus release and is antagonized by HIV-1 Vpu. Nature 451 (7177), 425–430.
Nejmeddine, M., Bangham, C.R.M., 2010. The HTLV-1 virological synapse. Viruses 2 (7), 1427–1447.
Nejmeddine, M., et al., 2005. Human T-lymphotropic virus, type 1, tax protein triggers microtubule reorientation in the virological synapse. J. Biol. Chem. 280 (33), 29653–29660.

Novy, K., et al., 2018. Proteotype profiling unmasks a viral signalling network essential for poxvirus assembly and transcriptional competence. Nat. Microbiol. 3 (5), 588–599.

Nusinow, D.P., et al., 2020. Quantitative proteomics of the cancer cell line encyclopedia. Cell 180 (2), 387–402.e16.

Olaya-Abril, A., et al., 2014. Surfomics: shaving live organisms for a fast proteomic identification of surface proteins. J. Proteomics 97, 164–176.

Orange, J.S., et al., 2002. Viral evasion of natural killer cells. Nat. Immunol. 3 (11), 1006–1012.

Ou, X., et al., 2020. Characterization of spike glycoprotein of SARS-CoV-2 on virus entry and its immune cross-reactivity with SARS-CoV. Nat. Commun. 11 (1), 1620.

Overington, J.P., Al-Lazikani, B., Hopkins, A.L., 2006. How many drug targets are there? Nat. Rev. Drug Discov. 5 (12), 993–996.

Pegram, H.J., et al., 2011. Activating and inhibitory receptors of natural killer cells. Immunol. Cell Biol. 89 (2), 216–224.

Petersen, J.L., Morris, C.R., Solheim, J.C., 2003. Virus evasion of MHC class I molecule presentation. J. Immunol. 171 (9), 4473–4478.

Peterson, V.M., et al., 2017. Multiplexed quantification of proteins and transcripts in single cells. Nat. Biotechnol. 35 (10), 936–939.

Pham, T.N.Q., et al., 2016. Enhancing virion tethering by BST2 sensitizes productively and latently HIV-infected T cells to ADCC mediated by broadly neutralizing antibodies. Sci. Rep. 6, 37225. https://doi.org/10.1038/srep37225.

Prod'homme, V., et al., 2007. The human cytomegalovirus MHC class I homolog UL18 inhibits LIR-1+ but activates LIR-1-NK cells. J. Immunol. 178 (7), 4473–4481.

Ravenhill, B.J., et al., 2020. Comprehensive cell surface proteomics defines markers of classical, intermediate and non-classical monocytes. Sci. Rep. 10 (1), 4560.

Rees, J.S., et al., 2015. Selective proteomic proximity labeling assay using tyramide (SPPLAT): a quantitative method for the proteomic analysis of localized membrane-bound protein clusters. Curr. Protoc. Protein Sci. 80, 19.27.1–19.27.18.

Reyburn, H.T., et al., 1997. The class I MHC homologue of human cytomegalovirus inhibits attack by natural killer cells. Nature 386 (6624), 514–517.

Rodriguez Boulan, E., Pendergast, M., 1980. Polarized distribution of viral envelope proteins in the plasma membrane of infected epithelial cells. Cell 20 (1), 45–54.

Röst, H.L., Malmström, L., Aebersold, R., 2015. Reproducible quantitative proteotype data matrices for systems biology. Mol. Biol. Cell 26 (22), 3926–3931.

Sakuma, T., et al., 2009. Inhibition of Lassa and Marburg virus production by tetherin. J. Virol. 83 (5), 2382–2385.

Sattentau, Q., 2008. Avoiding the void: cell-to-cell spread of human viruses. Nat. Rev. Microbiol. 6 (11), 815–826.

Schust, D.J., et al., 1998. Trophoblast class I major histocompatibility complex (MHC) products are resistant to rapid degradation imposed by the human cytomegalovirus (HCMV) gene products US2 and US11. J. Exp. Med. 188 (3), 497–503.

Smith, H.R.C., et al., 2002. Recognition of a virus-encoded ligand by a natural killer cell activation receptor. Proc. Natl. Acad. Sci. U. S. A. 99 (13), 8826–8831.

Sobotzki, N., et al., 2018. HATRIC-based identification of receptors for orphan ligands. Nat. Commun. 9 (1), 1–8.

Soh, T.K., et al., 2020. Temporal proteomic analysis of herpes simplex virus 1 infection reveals cell-surface remodeling via pUL56-mediated GOPC degradation. Cell Rep. 33 (1), 108235.

Speth, C., Dierich, M.P., 1999. Modulation of cell surface protein expression by infection with HIV-1. Leukemia 13 (Suppl. 1), S99–105.

Srivastava, M., et al., 2020. Chemical proteomics tracks virus entry and uncovers NCAM1 as Zika virus receptor. Nat. Commun. 11 (1), 3896.

Stergiou, L., et al., 2013. Integrin-mediated signaling induced by simian virus 40 leads to transient uncoupling of cortical actin and the plasma membrane. PLoS One 8 (2), e55799.

Stoeckius, M., et al., 2017. Simultaneous epitope and transcriptome measurement in single cells. Nat. Methods 14 (9), 865–868.

Sugden, S., Cohen, É.A., 2015. Attacking the supply lines: HIV-1 restricts alanine uptake to prevent T cell activation. Cell Host Microbe 18, 514–517.

Tyanova, S., et al., 2016. The Perseus computational platform for comprehensive analysis of (prote)omics data. Nat. Methods 13 (9), 731–740.

Van Damme, N., et al., 2008. The interferon-induced protein BST-2 restricts HIV-1 release and is downregulated from the cell surface by the viral Vpu protein. Cell Host Microbe 3 (4), 245–252.

van Deventer, S., Arp, A.B., van Spriel, A.B., 2021. Dynamic plasma membrane organization: a complex symphony. Trends Cell Biol. 31 (2), 119–129.

van Oostrum, M., et al., 2019. Classification of mouse B cell types using surfaceome proteotype maps. Nat. Commun. 10 (1), 5734.

van Oostrum, M., et al., 2020. Surfaceome dynamics reveal proteostasis-independent reorganization of neuronal surface proteins during development and synaptic plasticity. Nat. Commun. 11 (1), 4990.

Vasiliver-Shamis, G., et al., 2009. Human immunodeficiency virus type 1 envelope gp120-induced partial T-cell receptor signaling creates an F-actin-depleted zone in the virological synapse. J. Virol. 83 (21), 11341–11355.

Vermeire, K., et al., 2002. CADA inhibits human immunodeficiency virus and human herpesvirus 7 replication by down-modulation of the cellular CD4 receptor. Virology 302 (2), 342–353.

Viswanathan, K., et al., 2017. Quantitative membrane proteomics reveals a role for tetraspanin enriched microdomains during entry of human cytomegalovirus. PLoS One 12 (11), e0187899.

Vivier, E., et al., 2008. Functions of natural killer cells. Nat. Immunol. 9 (5), 503–510.

Volz, T., et al., 2013. The entry inhibitor Myrcludex-B efficiently blocks intrahepatic virus spreading in humanized mice previously infected with hepatitis B virus. J. Hepatol. 58 (5), 861–867.

Washburn, M.P., 2016. There is no human interactome. Genome Biol. 17, 48.

Weekes, M.P., et al., 2010. Comparative analysis of techniques to purify plasma membrane proteins. J. Biomol. Tech. 21 (3), 108–115.

Weekes, M.P., et al., 2014. Quantitative temporal viromics: an approach to investigate host-pathogen interaction. Cell 157 (6), 1460–1472.

Wollscheid, B., et al., 2009. Mass-spectrometric identification and relative quantification of N-linked cell surface glycoproteins. Nat. Biotechnol. 27 (4), 378–386.

Xie, J., et al., 2017. Immunochemical engineering of cell surfaces to generate virus resistance. Proc. Natl. Acad. Sci. U. S. A. 114 (18), 4655–4660.

Yamauchi, Y., Helenius, A., 2013. Virus entry at a glance. J. Cell Sci. 126 (Pt. 6), 1289–1295.

Yan, H., et al., 2012. Sodium taurocholate cotransporting polypeptide is a functional receptor for human hepatitis B and D virus. Elife 1, e00049.

Yin, H., Flynn, A.D., 2016. Drugging membrane protein interactions. Annu. Rev. Biomed. Eng. 18, 51–76.

Zhang, Y., et al., 2019. Plasma membrane proteomic profile discovers macrophage-capping protein related to latent HIV-1. Curr. HIV Res. 17 (1), 42–52.

Zhen, Y., et al., 2018. Proximity labeling by a recombinant APEX2-FGF1 fusion protein reveals interaction of FGF1 with the proteoglycans CD44 and CSPG4. Biochemistry 57 (26), 3807–3816.

Zhou, Z., et al., 2020. Surface protein imputation from single cell transcriptomes by deep neural networks. Nat. Commun. 11 (1), 651.

CHAPTER FIVE

Proteomics approaches for the identification of protease substrates during virus infection

Xavier Martiáñez-Vendrell and Marjolein Kikkert*
Molecular Virology Laboratory, Department of Medical Microbiology, LUMC Center for Infectious Diseases (LU-CID), Leiden University Medical Center, Leiden, The Netherlands
*Corresponding author: e-mail address: m.kikkert@lumc.nl

Contents

Abstract

Proteases precisely and irreversibly catalyze the hydrolysis of peptide bonds, regulating the fate, localization, and activity of many proteins. Consequently, proteolytic activity plays an important role in fundamental cellular processes such as differentiation and migration, immunological and inflammatory reactions, apoptosis and survival. During virus infection, host proteases are involved in several processes, from cell entry to initiation, progression and resolution of inflammation. On the other hand, many viruses encode their own highly specific proteases, responsible for the proteolytic processing of viral proteins, but, at the same time, to cleave host proteins to corrupt antiviral host responses and adjust protein activity to favor viral replication. Traditionally, protease substrate identification has been addressed by means of hypothesis-driven approaches, but recent advances in proteomics have made a toolkit available to uncover the extensive repertoire of host proteins cleaved during infection, either by viral or host proteases. Here, we review the currently available proteomics-based methods that can and have contributed to the systematic and unbiased identification of new protease substrates in the context of virus-host interactions. The role of specific proteases during the course of virus infections will also be highlighted.

Advances in Virus Research, Volume 109
ISSN 0065-3527
https://doi.org/10.1016/bs.aivir.2021.03.003

Abbreviations

1D	one-dimensional
2D	two-dimensional
2D-DIGE	two-dimensional difference gel electrophoresis
3Cpro	3C protease
COFRADIC	Combined Fractionation Diagonal Chromatography
CstF-64	cleavage stimulating factor 64
CV	coxsackievirus
EV	enterovirus
GPx8	glutathione peroxidase 8
HCV	hepatitis C virus
HIV	human immunodeficiency virus
HPG-ALD	hyperbranched polyglycerol-aldehyde
HRV	human rhinovirus
HUNTER	High-efficiency Undecanal-based N-Termini Enrichment
HYTANE	hydrophobic tagging-assisted N-termini enrichment
IFN-α	interferon-α
iTRAQ	isobaric tags for relative and absolute quantitation
LC	liquid chromatography
MALDI-TOF	matrix-assisted laser desorption/ionization time-of-flight
MMP	matrix metalloprotease
Mpro	main protease
MS/MS	tandem mass spectrometry
OCIAD1	antigen domain containing protein 1
PLpro	papain-like protease
PPI	protein-protein interaction
Pre-mRNA	precursor messenger ribonucleic acid
PROTOMAP	Protein Topography and Migration Analysis Platform
PV	poliovirus
RP-HPLC	reverse-phase high-performance liquid chromatography
SARS-CoV-2	severe acute respiratory syndrome coronavirus-2
SCX	Strong Cation Exchange Chromatography
SDS-PAGE	sodium dodecyl sulfhate–polyacrylamide gel electrophoresis
SILAC	stable isotope labeling with amino acids in cell culture
TAILS	Terminal Amine Isotopic Labeling of Substrates
TEV	tobacco etch virus
TMT	tandem mass tag
TNBS	2,4,6-trinitrobenzene sulfonic acid
ZVP	Zika virus protease

1. Introduction

Proteases are enzymes with diverse structures and functions that efficiently and irreversibly catalyze the hydrolysis of peptide bonds (Barrett et al., 1998). Initially, proteases were identified as gastric enzymes participating in the non-specific proteolytic breakdown of proteins in food. Over time, the development of new principles, such as limited proteolysis (Linderstrom-Lang, 1952), as well as newly developed methods, unfolded the knowledge on proteolytic processing. They also ascertained the fact that many proteases are exceptionally selective and participate in specific proteolytic cleavage reactions, which are indispensable in all living organisms. The human degradome, defined as the full repertoire of proteases encoded by the human genome, accounts for 588 proteases coded by approximately 3% of the human genome, representing one of the largest enzyme families in vertebrates (Bond, 2019; Pérez-Silva et al., 2016). According to their mechanisms of catalysis, mammalian proteases are classified into five different classes: aspartyl, cysteine, metallo, serine and threonine proteases. Based on amino acid sequence comparison, these classes can be further down-branched into families, and these can be grouped into clans on the basis of structural homology (Puente et al., 2003; Rawlings et al., 2012). Consistently, the proteolytic activity of such a vast and diverse repertoire of proteases is involved in several key processes that regulate the fate, localization and activity of most proteins. This will ultimately contribute to controlling crucial biological and physiological processes such as DNA replication, cell cycle regulation, cell differentiation and migration, immunological and inflammatory reactions, apoptosis and others. Host proteases can also participate in virus entry and egress. In the 70s, it was observed that the proteolytic cleavage and activation of a viral glycoprotein (fusion protein) of paramyxoviruses Sendai virus, simian virus 5, and Newcastle disease virus by trypsin-like proteases are necessary for the fusion of the viral envelope with the target cell plasma membrane to initiate infection (Nagai et al., 1976a, b; Scheid and Choppin, 1974, 1977), and that cleavage activation can determine virus tropism and pathogenicity (Nagai, 1993; Nagai et al., 1976a, 1979).

Viruses are infectious units with diameters ranging from 16 nm (circoviruses) to about 500 nm (mimiviruses) that can only replicate within the cells of a host organism. Virtually, all living organisms can be infected by viruses. Not all viruses encode one or more proteases, but many have evolved their own highly specific proteases that engage in the processing of viral proteins. The first viral proteins with proteolytic activity were identified in the retrovirus Rous sarcoma virus (Von Der Helm, 1977) and the picornavirus encephalomyocarditis virus (EMCV) (Gorbalenya et al., 1979; Palmenberg et al., 1979). More recently, it has been established that several viral proteases engage accessory functions directed to hijack a number of cellular processes. A well-studied example is that of viral proteases that function as deubiquitinating enzymes (DUBs), which have been identified in diverse virus families, including herpesviruses, coronaviruses, arteriviruses and others (reviewed in Bailey-Elkin et al., 2017, Van Huizen and Kikkert, 2020). Next to their role in polyprotein cleavage, viral DUBs can impair cellular ubiquitin-dependent processes to, for example, dampen innate antiviral responses and endorse viral replication.

Although several proteolytic events and their effects have been characterized in a wide range of viral infections so far, the complete repertoire of host proteins cleaved during infection, either by viral or host proteases, has not been entirely delineated. Further identification of host proteins targeted for cleavage will augment our understanding of virus-host interactions and disease mechanisms. Typically, protease substrate identification has been addressed by means of hypothesis-driven approaches. Researchers would identify a specific pathway affected during infection and infer a putative role for a particular protease in cleaving a pathway component, and study this using a variety of *in vitro* assays. However, this is time-consuming and restricted to the identification of a very limited number of protein substrates, and it widely depends on the availability of antibodies able to recognize the potential target proteins as well as the resulting cleavage fragments. Identification of protein substrates can also be achieved using *in silico* approaches. Known protease substrate specificity determined by comparison and alignment of known protease substrates (e.g., viral polyprotein cleavage sites for viral proteases) or by the screening of peptide libraries (Boulware and Daugherty, 2006; Deng et al., 2000; Harris et al., 2000; Nazif and Bogyo, 2001; Schilling and Overall, 2008; Turk et al., 2001) can be used for database searching or to feed bioinformatic algorithms to predict and identify unknown protein substrates (Blom et al., 1996; Cai et al., 1998; López-Otín and Overall, 2002; Thomson et al., 2003). However, these techniques do not consider important determinants of proteolytic processing such as protein expression levels, putative presence of co-factors, accessibility of cleavage sites within the infected cell, and exosites (Turk et al., 2012). Additionally, proteases can cleave sites that are different from the consensus cleavage site, which limits the spectrum of *in silico* predicted protein substrates (Kleifeld et al., 2010). Lastly, over the last two decades significant progress has been accomplished in the development of proteomics-based approaches that allow the identification of proteases and their full repertoire of substrates at the proteome wide level, a discipline more specifically referred as degradomics (López-Otín and Overall, 2002). Here, we review the currently available proteomics-based methods used in the field of degradomics that can benefit the systematic and unbiased identification of new protease substrates in the context of virus-host interactions, and highlight the role of specific proteases in disease mechanisms during virus infections.pt

2. Gel-based proteomics methods

Multiple in-gel proteomics approaches exist that allow for the identification of differentially expressed proteins in proteome samples. Upon incubation of protein mixtures with or without catalytically active protease, samples can be separated by one or two-dimensional electrophoresis. In 1D electrophoresis, if a protein substrate is cleaved by the protease of interest, its corresponding band will either be absent or show decreased intensity compared to the untreated sample on-gel, and fragments of different molecular weight resulting from cleavage will appear. The bands can then be excised from the gel, subjected to in-gel digestion with trypsin and analyzed by mass spectrometry to study the protein identity corresponding to the differentially represented bands between treated and untreated samples. A modified 1D approach in combination with SILAC labeling to allow for protein quantification has been carried out to identify host cellular protein substrates of hepatitis C virus (HCV) NS3-4A protease (Morikawa et al., 2014). In this study, a human osteosarcoma-derived cell line inducibly expressing NS3-4A was cultured in the presence of heavy isotope amino acids $^{13}C_6$-L-lysine and $^{13}C_6$ $^{15}N_4$-L-arginine for isotopic labeling, while control cells not expressing the protease were cultured in medium with standard unlabeled (light) amino acids. NS3-4A protease expression was induced for 48 h by tetracycline withdrawal, and cell extracts were collected and separated by SDS-PAGE. Each lane was then partitioned in several slices, which were in turn digested with trypsin, and resulting peptide mixes from each slice were analyzed by mass spectrometry (MS). The incorporation of heavy isotope labeled amino acids into newly synthesized proteins only in NS3-4A expressing cells results in doublet peptides with different mass spectra between protease-treated and control samples, allowing to determine the relative protein abundance in each slice. The identification of the same protein in two different slices with opposite heavy to light ratios indicates a shift in-gel mobility dependent on the protease, suggesting such a protein as a possible substrate of the protease. Besides MAVS, a previously known substrate of the NS3-4A protease (Rassmann et al., 2006), the authors identified glutathione peroxidase 8

(GPx8) as an NS3-4A substrate and validated its cleavage in models of HCV infection and in liver biopsies from HCV chronic patients. The same screening further identified antigen domain containing protein 1 (OCIAD1) as a novel cellular substrate of HCV NS3-4A protease (Tran et al., 2020). Interestingly, co-expression of tagged OCIAD1 with NS3-4A protease of different hepacivirus in 293 T cells showed all hepaciviral NS3-4A proteases tested to cleave OCIAD1, suggesting this proteolytic event to be a relevant evolutionary conserved aspect of hepaciviruses infections in their respective hosts.

In the 2D approach, protease-treated and untreated samples are each run on two consecutive gels whereby proteins are separated both by molecular weight and isoelectric point. The two protein spot patterns can then be aligned and analyzed to identify spots corresponding to protease substrates. The 2D electrophoresis methods have been used to identify proteome alterations and protease substrates during human rhinovirus (HRV) serotype-2 and coxsackievirus B3 (CVB3) infections (Chan et al., 2015; Meylan et al., 2005; Seipelt et al., 2000). Similarly, 2D electrophoresis followed by MALDI-TOF analysis identified eight cellular nuclear proteins as substrates of enterovirus 71 (EV71) 3C protease (3Cpro) (Weng et al., 2009). This later study further validated and mapped the cleavage of cleavage stimulating factor 64 (CstF-64), and suggested that the cleavage of CstF-64 by EV71 3Cpro inhibits the 3′-end processing and polyadenylation of cellular pre-mRNA.

In addition to conventional 1D and 2D gel electrophoresis methods, other gel-based approaches can be employed in the discovery of protease substrates. By pre-labeling two protein samples (e.g., pre and post-protease digested) using fluorescent dyes with different absorption-emission spectra, 2D difference gel electrophoresis (2D-DIGE) allows to co-separate and visualize the two samples on a single 2D-PAGE gel, improving the reliability of qualitative and quantitative results compared to common 2D gels (Alban et al., 2003; Bredemeyer et al., 2004; Kondo and Hirohashi, 2009; Unlü et al., 1997). Protein Topography and Migration Analysis Platform (PROTOMAP) resolves protease-treated and -untreated samples by 1D electrophoresis. Each lane is cut into bands of fixed length intervals, and proteins in each band are digested into peptides and then sequenced using liquid chromatography with tandem mass spectrometry (LC-MS/MS). Molecular weight changes in SDS-PAGE (Y-axis) and the sequence coverage from the N- to the C-termini (X-axis) for each molecular weight change or band are plotted together and combined with LC-MS/MS analysis spectral counting data, which allows for semiquantitative analysis of the different proteins detected in a sample (Dix et al., 2008; Simon et al., 2009). Finally, although gel-based methods allow identifying protease substrates, they do not allow for the identification of proteolytic cleavage events taking place close to the termini or for low abundant protein products and make it difficult to exactly determine cleavage sites. Furthermore, technical limitations such as limited lane or spot resolution and poor sample recovery restrict their sensitivity.

3. Gel-free proteomics methods

Proteolytic events result in peptides with newly exposed amine (at the N-termini) or carboxyl (at the C-termini) groups (Rogers and Overall, 2013). Enrichment of nascent peptides in a protease-treated sample by positive or negative selection and subsequent MS/MS analysis is used by so-called terminomics approaches, and allows to identify novel protease substrates. Because of the low chemical reactivity of carboxyl groups, most terminomics strategies rely on the selection of newly generated peptide N-termini (neo-N-termini).

3.1 Methods for negative selection of proteolytic fragments

Combined Fractionation Diagonal Chromatography (COFRADIC) was the first N-terminal peptide isolation and enrichment procedure based on diagonal electrophoresis and diagonal chromatography (Gevaert et al., 2002, 2003). Although initially designed to be used for the identification of peptides holding the rare amino acid methionine (Gevaert et al., 2002), COFRADIC can be used for the identification of novel protease substrates. Proteome samples are either treated with the protease of interest or with an inactive catalytic version of the protease (or alternatively without protease), which serves as a control. Upon sample collection, cysteine residues are reduced and alkylated to impede the formation of disulfide bridges, and all N-terminal primary α-amines (N-termini and neo-N-termini), as well as free ε-amines in lysine side chains, are acetylated (Fig. 1A). Subsequently, protein extracts are digested with trypsin. This yields N-terminally acetylated peptides and internal peptides presenting free α-amines at the newly generated N-termini by trypsin cleavage. Internal peptides are then separated from peptides exhibiting acetylated N-termini by Strong Cation Exchange Chromatography (SCX). Because of their additional positive charge conferred by the free amine group, internal peptides will interact more strongly with the SCX resin and elute later, while N-terminal and neo-N-terminal acetylated

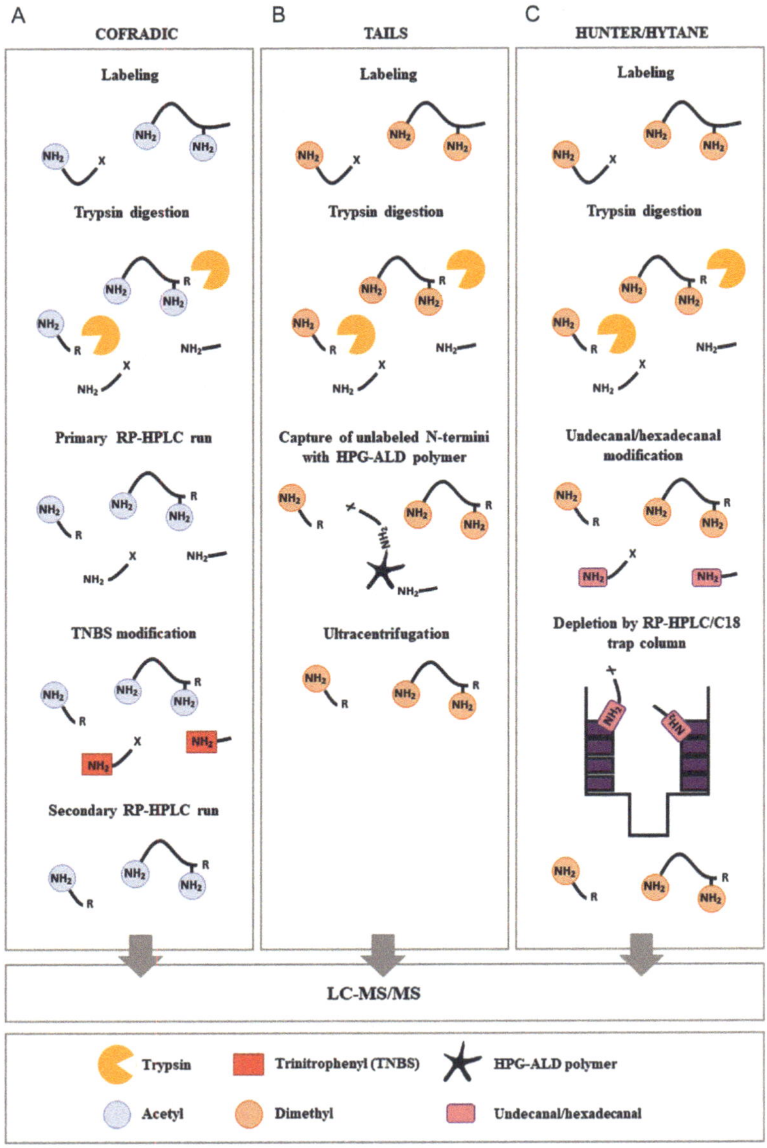

Fig. 1 Schematic workflow of labeled peptide N-termini negative selection methods. (A) COFRADIC, (B) TAILS, and (C) HUNTER. After the sample is exposed to the protease of interest, primary amines are flagged at the protein level using distinct labeling strategies (e.g., dimethylation) and are subsequently digested with, in general, trypsin. Labeled peptide N-termini are then separated from C-termini and internal peptides by negative selection and analyzed by LC-MS/MS. *NH_2*, amine group; *X*, P1 residue for protease of interest; *R*, arginine, P1 residue for trypsin. *Figure adapted from Eckhard, U., Marino, G., Butler, G. S., Overall, C. M., 2016. Positional proteomics in the era of the human proteome project on the doorstep of precision medicine. Biochimie, 122, 110–118.*

peptides will elute in the flow-through. The flow-through is then subjected to reverse-phase high-performance liquid chromatography (RP-HPLC), and each resulting fraction is treated with 2,4,6-trinitrobenzenesulfonic acid (TNBS), which modifies only internal peptides and changes their HPLC retention time compared to all other N-terminal peptides. Following, TNBS treated peptides undergo a second RP-HPLC round to reduce sample complexity further and segregate internal peptides from unaffected original N-terminal peptides, which can then be collected for LC-MS/MS analysis. COFRADIC has been used to identify cleavage events during Fas-induced apoptosis, Taxol-mediated cell death and in anthracycline-treated acute myelogenous leukemia cells, as well as to characterize cleavage specificity of caspases 3 and 7 (Demon et al., 2009; Gausdal et al., 2008; Impens et al., 2008; Van Damme et al., 2005). To our knowledge, this method has only been used once for the systematic identification of viral protease substrates so far. Impens and colleagues used N-terminal COFRADIC to screen for protein cleavages in lysates of human Jurkat T-cells incubated *in vitro* with human immunodeficiency virus-1 (HIV-1) protease (Impens et al., 2012). Their work identified 148 cleavage sites in 123 proteins, only two of which, eIF3D and actin, had been previously identified as substrates of HIV-1 protease (Jäger et al., 2012; Tomasselli et al., 1991). Interestingly, seven other host substrate proteins identified using COFRADIC were found to interact with HIV-1 protease in an interactomics screen (Jäger et al., 2012). Even by the time the study was published, the number of host cell proteins known to be cleaved by HIV-1 PR was approximately 80 (Fu et al., 2009), the small overlap of HIV-1 protease targets validated the approach, but at the same time, it acknowledged its limitations. The multiple chromatographic steps involved in the procedure can result in sample loss, hampering the detection of less abundant peptides, a problem that can be amplified by the fact that only a fraction of the injected peptides into LC-MS/MS instruments is fragmented for identification (Liu et al., 2004). Additionally, tryptic digestion can yield neo-N-terminal peptides that are too short to be confidentially identified by LC-MS/MS.

Chris Overall's group developed another N-terminomics approach named Terminal Amine Isotopic Labeling of Substrates (TAILS) (Kleifeld et al., 2010). Differently from COFRADIC, TAILS was purposely developed for the discovery of protease substrates (Kleifeld et al., 2010; Prudova et al., 2010). In TAILS, the protein extract is incubated with the protease of interest, whereas in the control sample, either the protease is excluded or the protein extract is exposed to a catalytic mutant protease (Prudova et al., 2010). Next, primary amines and lysine side chains are isotopically labeled using dimethyl labeling (Fig. 1B). Amine blocking can also be achieved by metabolic (SILAC) or isobaric (iTRAQ or TMT) labeling. Depending on the labeling strategy, TAILS can be used to compare up to 10 protease-treated and control proteome samples (Kleifeld et al., 2011; Klein et al., 2015), but the recent development of an isobaric 16-plex TMT reagent set could allow for multiplexing up to 16 samples (Li et al., 2020a; Thompson et al., 2019). Treated protein extracts are then mixed in equal amounts and digested with trypsin or GluC, generating new and unlabeled N-terminal peptides. The resulting peptide mixture is incubated with a water-soluble, dendritic hyperbranched polyglycerol-aldehyde (HPG-ALD) polymer that will covalently bind tryptic neo-N-terminal peptides but not previously blocked N-termini. Before LC-MS/MS analysis, the experimentally blocked and naturally blocked N-terminal peptide fraction is then separated from the polymer-bound internal peptide fraction by ultracentrifugation, which enriches for blocked N-terminal peptides by negative selection. TAILS has been utilized to discover host cellular substrates of picornavirus proteases *in vitro*. Jagdeo and colleagues identified 72 and 34 new host protein targets of 3Cpro of poliovirus (PV) and coxsackievirus, respectively (Jagdeo et al., 2018). They further validated several of the TAILS-identified substrates and confirmed their cleavage in infected cells by immunoblotting. The knockdown of some of these protein targets had either a positive or a negative effect on infection, raising the hypothesis that cleavage by 3Cpro promotes virus replication. One of such targets was the Golgi p115 protein, which was previously shown to be degraded during poliovirus infection (Belov et al., 2010). Paradoxically, validation experiments for p115 cleavage showed that both knockdown of p115 gene during infection and overexpression of a non-cleavable mutant variant decreased poliovirus titers in HeLa cells. The authors then hypothesized that p115 has a key transient role in viral replication, but needs to be cleaved at some point during infection for replication to be completed successfully. Using TAILS, Jagdeo and colleagues also identified 63 high confidence potential substrates of CV 2A protease (Jagdeo, 2017). Taken together, their results identified four high confidence common substrates of poliovirus 3Cpro, CV 3Cpro and CV 2A protease, suggesting these proteins to be decisive targets for successful enterovirus infection. Finally, TAILS can also identify substrates of host proteases during virus infection. Marchant and colleagues showed that mice unable to produce matrix metalloprotease-12 (MMP-12) presented with increased morbidity and mortality upon coxsackievirus infection compared to wild type mice and attributed such effects to a defective interferon-α (IFN-α) secretion into plasma (Marchant et al., 2014). Following, they demonstrated that MMP-12 traffics to the nucleus of virus-infected cells where it binds the *NFKBIA* promoter, leading to IFN-α expression, and used TAILS to identify new substrates of MMP-12. Interestingly, TAILS results revealed IFN-α itself to be an MMP-12 substrate, leading the authors to propose a negative-feedback mechanism by which MMP-12 triggers IFN-α export during viral infection and later cleaves systemic IFN-α to dampen the systemic innate immune response. Such examples show that terminomics techniques in combination with validation and mechanistic studies can be very useful to contextualize proteolytic events both in time, location and biological relevance during virus infection.

Over the last few years, the development of new methods and the introduction of modifications to previously existing methods has significantly widened the availability of terminomics approaches. New N-terminomics protocols based on charge reduction or charge reversal of internal peptides combined with SCX have been published (Lai et al., 2015; Shema et al., 2018; Venne et al., 2013), and modified TAILS and COFRADIC protocols can now support the analysis of C-termini as well (Schilling et al., 2010; Van Damme et al., 2010). More recently, Lange's group has developed the High-efficiency Undecanal-based N-Termini Enrichment (HUNTER) method, an automatable workflow similar to TAILS for the enrichment of N-terminal peptides from as little as 2 μg of protein, which makes it the most sensitive terminomics approach currently available (Weng et al., 2019). In HUNTER, protein extracts are lysed and purified using SP3 magnetic beads, and purified proteins are then subjected to N-terminal α- and lysine ε-amine labeling with dimethylation (Fig. 1C). Samples are digested with trypsin, and digestion-generated peptide α-amines are modified with undecanal, an amine-reactive reagent that replaces HPG-ALD polymer in TAILS. Finally, samples are passed through reverse-phase columns, where undecanal-modified peptides will be retained, resulting in an enriched N-terminal peptide pool in the flow-through. An adaptation of this latter method has recently been applied to identify new cleavage events during SARS-CoV-2 infection in VERO E6 and A549-ACE2 cells (Meyer et al., 2020). This work identified novel cleavage sites within viral proteins, including the polyprotein, the nucleocapsid, ORF3a accessory protein and Spike, as well as 14 putative cellular substrates of SARS-CoV-2 encoded papain-like protease (PLpro) and 3C-like protease or main protease (Mpro). The latter were hypothesized as substrates of either viral protease based on the cleavage sites identified. Interestingly, SARS-CoV-2 infections in A549-ACE2 resulted in a reduction in viral titers upon individual knock-down of the 14 identified hits using siRNAs, suggesting the cleaved proteins to be necessary for efficient virus replication. If that is the case, this could indicate viral proteases do not only act on host proteins to corrupt antiviral host responses but to also adjust protein activity to favor viral replication. Similarly, a variation of TAILS, termed as hydrophobic tagging-assisted N-termini enrichment (HYTANE) (Chen et al., 2016), has also been applied to uncover host substrates of the main protease of three human coronaviruses (hCoV-NL63, SARS-CoV and SARS-CoV-2) upon incubation of recombinant Mpro with epithelial and endothelial lung cells (Koudelka et al., 2020). In this case, depletion of internal peptides is achieved by modifying trypsin-generated peptides with hexadecanal, which allows the retention of modified peptides in C18 trap columns, similarly to HUNTER (Fig. 1C). This study revealed 640 high-confidence cleavage sites within 434 unique proteins, some of them involved in antiviral innate immunity (e.g., NEMO and optineurin), suggesting a putative role for Mpro in hijacking host antiviral responses. Additionally, five of the identified substrates overlapped with previously identified hits in SARS-CoV-2 infected A549-ACE2 cells, confirming these to be relevant Mpro substrates during virus infection.

3.2 Methods for positive selection of proteolytic fragments

The first N-terminomics method based on positive enrichment of N-terminal peptides was the subtiligase-mediated N-terminomics approach (Mahrus et al., 2008). In this method, proteins in biological samples are treated with an engineered bacterial subtiligase enzyme and a synthetic peptide coupled to a biotin molecule that will be linked to N-terminal α-amines by the subtiligase enzyme (Fig. 2A). Next, samples are subjected to tryptic digestion and peptides are incubated with avidin affinity media, which will bind biotinylated peptides only. Because the synthetic peptide harbors a cleavage site for tobacco etch virus (TEV) protease (TENLYFQ↓SY), incubation with TEV protease will allow to easily recover avidin-captured

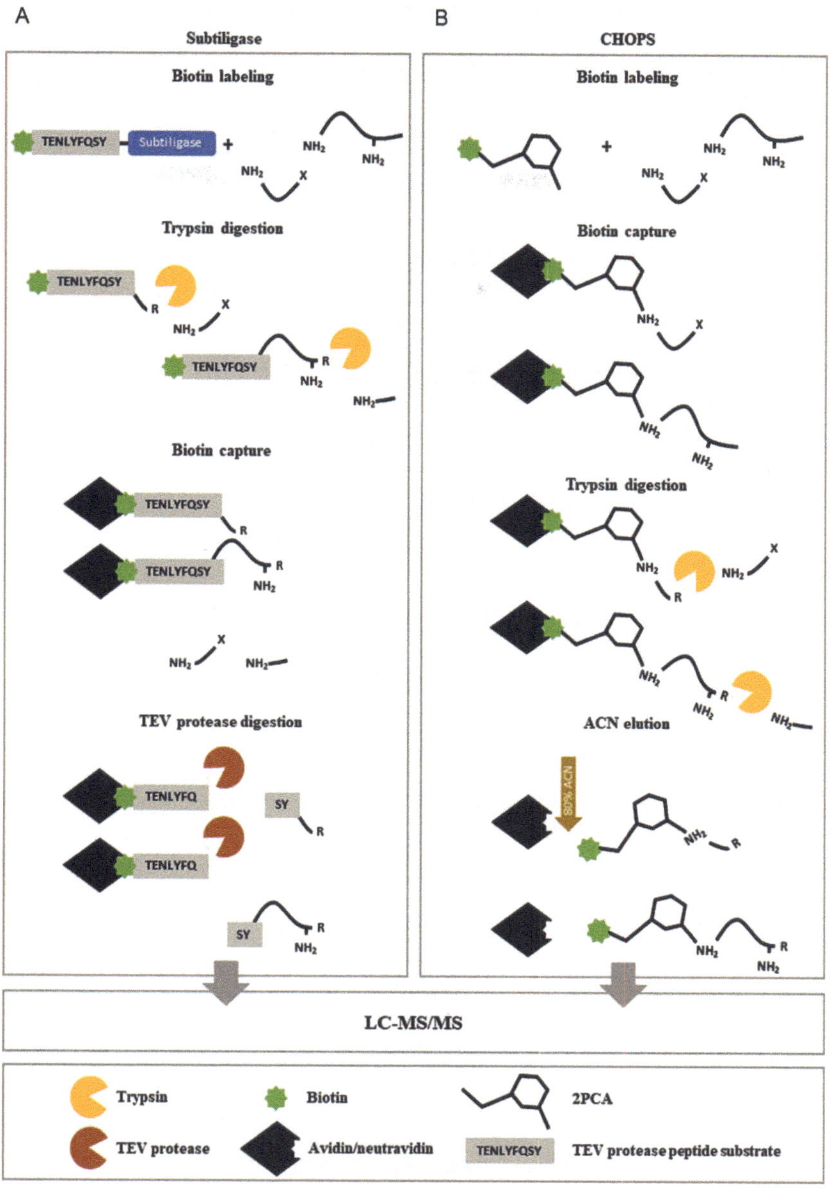

Fig. 2 Schematic workflow of labeled peptide N-termini positive selection methods. (A) Subtiligase method. Proteome samples previously digested with the protease of interest are incubated with a biotinylated peptide ester containing a TEV protease cleavage site that is linked to α-amines by the subtiligase enzyme. Samples are then trypsinized and biotinylated N-terminal peptides are captured with immobilized avidin. Finally, SY-dipeptide modified N-terminal peptides are recovered using TEV protease, and analyzed by LC-MS/MS. (B) CHOPS. Upon treatment of the sample with the protease of interest, biotins molecules are linked to N-termini using 2PCA, and N-terminal biotinylated peptides are captured with neutravidin beads. Samples are then subjected to on-bead trypsin digestion, and neutravidin-bound peptides are finally eluted using 80% acetonitrile 0.1% formic acid, and analyzed by LC-MS/MS. *NH_2*, amine group; *X*, P1 residue for protease of interest; *R*, arginine, P1 residue for trypsin.

peptides. The recovered peptides preserve an N-terminal SY-dipeptide modification resulting from TEV protease cleavage that will allow differentiating between labeled peptides from possible contaminating unlabeled peptides during LC-MS/MS analysis. Subtiligase-based N-terminomics can also incorporate isobaric or metabolic labeling for relative and absolute quantification (Hill et al., 2018; Wildes and Wells, 2010). The subtiligase method, in combination with SILAC labeling, has been used to identify 31 host protein targets of Zika virus NS2B-NS3 protease (ZVP) (Hill et al., 2018). Among these, autophagy-related protein 16-1 (ATG16L1) and eukaryotic translation initiation factor 4 gamma 1 (eIF4G1), which participate in type-II interferon production and autophagy, and host-cell translation, respectively, were found to be cleaved in Zika virus-infected cells, suggesting a role for ZVP in suppressing innate antiviral responses.

Saeed and colleagues have recently used an adaptation of the subtiligase method, in which trypsin digestion is performed on-bead rather than prior to avidin capture (Wiita et al., 2014b), to label newly generated N-termini of peptides in enterovirus-infected HeLa cells (Saeed et al., 2020). Proteomics analysis of subtiligase-labeled neo-N-termini identified 46 proteins being cleaved during the course of infection of all viruses tested, including CVB3, HRV A16, PV, enterovirus D70 (EV70) and EV71. In further validation experiments, a selected set of target proteins were found to also be cleaved in various cell types for which enteroviruses have tropism, suggesting enteroviruses to largely cleave a common set of host proteins. Moreover, proteolytic processing was vastly uniform across cell types. In addition, for most proteins, the detected neo-N-termini were the same, indicating specific proteolytic cleavage rather than non-specific degradation. Experimental designs such as the previously described, in which a process is induced in a certain biological system (e.g., virus infection, induction of apoptosis, etc.) are referred as forward experiments, in contrast to reverse experiments, where an exogenous protease is incubated *in vitro* with a proteome of interest in which endogenous proteases have been inactivated. Thus, forward experiments will yield peptides cleaved by the different proteases present in the studied biological system and make it difficult to pinpoint the proteases responsible for each cleavage. Nonetheless, as terminomics approaches provide information on the identity of N-terminal residues of the cleavage products, it is possible to narrow down on the proteases that generate those. The enteroviruses studied by Saeed and colleagues encode for two proteases, 2Apro and 3Cpro, each with specific cleavage site specificities at the P1 and P1' positions (the two amino acids in between which cleavage occurs). Interestingly, the number of peptides derived from proteolytic events taking place in cleavage sites for both 2Apro and 3Cpro increased in enterovirus-infected cells as compared to the non-infected cells, most probably suggesting these cleavages were mediated by viral proteases (Saeed et al., 2020). In addition, virus infection can lead to the activation of other proteases, such as caspases (Lai et al., 2020; Li et al., 2020b), which can also contribute to a differential identification of terminal peptides in between infected and non-infected cells. Therefore, when conducting forward experiments, prior knowledge on cleavage site specificities for all proteases expressed in the studied system is important to relate the identified cleavage events to specific proteases.

A more recent approach to positively enrich for N-terminal peptides in proteome samples is known as CHOPS (Chemical enrichment of Protease Substrates) (Griswold et al., 2019). CHOPS uses pyridine-2-carboxyaldehyde (2PCA) (Macdonald et al., 2015) to link biotin to protein N-termini (Fig. 2B). 2PCA condenses with N-terminal amines to form imines, which in turn react with the next nitrogen in the peptide to form a stable 5-membered ring. Peptides with an N-terminus followed by a proline as well as lysine side chains will be missed as the cyclization reaction cannot take place in these residues. Biotinylated peptides are then selected using neutravidin beads, and on-bead trypsin digestion is performed. Resulting tryptic peptides are washed, and biotinylated peptides are eluted with 80% acetonitrile (ACN) containing 0.1% formic acid, and finally analyzed either by SDS-PAGE or LC-MS/MS. CHOPS has been shown to perform similarly to the subtiligase method and represents another interesting N-terminomics strategy, but it has not yet been used in the discovery of protease substrates in virus-host interactions.

4. Virus-host protein-protein interactions

The coordinated dynamic interactions between the millions of individual components in a cell confer it with its living properties, and an important number of molecular interactions occurs between proteins. To study the mechanisms that dominate a particular system, it is necessary to delineate and understand the protein-protein interactions (PPIs) in it. During virus infection, cellular and viral proteins compete for interacting partners, rewiring the normal map of cellular PPIs and inducing profound changes in the infected cell. Viruses exploit host protein complexes to adjust host biological processes and thus support virus survival and replication. One example is the hijacking of the interferon response by positive double-stranded RNA viruses directed to suppress the antiviral innate immune system and create a window of opportunity for successful viral replication (reviewed in Beachboard and Horner, 2016, Chen et al., 2017, García-Sastre, 2017, Gokhale et al., 2014, Kindler et al., 2016, Nelemans and Kikkert, 2019, Pardy et al., 2019, Uno and Ross, 2018), a strategy in which viral proteases may play an important role (Lei and Hilgenfeld, 2017).

Several technologies are in place to explore PPIs, including yeast-2-hybrid (Fields and Song, 1989), protein microarray-based technologies (Glick et al., 2016), and mass spectrometry-based approaches such as affinity- and immuno-purification (Gingras et al., 2007), proximity labeling (Roux et al., 2012) and chemical crosslinking (Leitner et al., 2016), which can be complemented with literature mining and bioinformatics approaches. These methods can provide valuable information on host proteins, complexes and pathways that are altered during infection, and bring us close to protease binding partners which could, at the same time, be potential protease substrates. However, the proteolytic activity when using a protease catalytic domain as bait in a PPI study extensively limits the detection of putative protease substrates as many will be unbound upon cleavage. To overcome this constraint, yeast two-hybrid analyses were customized for protease-substrate discovery. To do so, exosites can be exploited to screen for protease interactors (Mcquibban et al., 2000). The binding of protease domains with functional and structural differences from the catalytic domain to a substrate is often a requirement for substrate cleavage. For example, collagen cleavage by certain MMP necessitates the binding of collagen to the hemopexin COOH-C-terminal domain of MMP for proteolytic processing to occur (Clark and Cawston, 1989; Knäuper et al., 1993, 1997; Murphy et al., 1992). Alternatively, catalytic inactive mutant proteases that cannot cut and release their substrates once bound can also be used as baits in PPI studies. Inactive-catalytic-domain capture (ICDC) approaches circumvent the limitation of protease exosite scanning, which cannot identify substrates for which exosite interactions are not a prerequisite for subsequent cleavage (Overall et al., 2004).

Over the last decade, efforts have been made to map and investigate protein interaction networks between viruses and their host cells.

However, high-throughput screens of PPIs have been performed for a relatively small number of human viruses, including HIV (Jäger et al., 2012), human papillomavirus (Eckhardt et al., 2018), HCV (De Chassey et al., 2008), herpesviruses (Fossum et al., 2009; Stellberger et al., 2010; Uetz et al., 2006), Influenza A virus (Shapira et al., 2009), Dengue virus (Khadka et al., 2011), Zika virus (Coyaud et al., 2018), and more recently, SARS-CoV-2 (Gordon et al., 2020; Samavarchi-Tehrani et al., 2020), among others. Today, comprehensive virus-human PPI data is collected in several publically available databases such as HPIDB (Ammari et al., 2016), P-HIPSTer (Lasso et al., 2019) and VirHostNet (Navratil et al., 2009). Interestingly, Jäger and colleagues carried out a systematic affinity purification mass spectrometry (AP-MS) to uncover HIV-human protein complexes that identified several interacting partners of HIV catalytic mutant protease, including 12 subunits of the 13-subunit translation initiation complex eIF3, which has an inhibitory effect on HIV replication (Jäger et al., 2012). Further co-transfection of individual subunits of eIF3 with active HIV protease in HEK293T cells showed subunit eIF3d to be cleaved by HIV protease. Similarly, co-affinity purification followed by mass spectrometry was used to identify substrates of the Zika virus NS2B3 protease (Xing et al., 2020). Following the reasoning behind inactive-catalytic-domain capture, the authors used the proteolytic mutant protease to identify its protein binding partners in human neural precursor cells. Further experiments showed that NS2B3 protease is responsible for the cleavage of Septin-2, leading to modifications in septin cytoskeletal components that result in failed cytokinesis and delayed cell division. PPI studies can also complement and add confidence to results from degradomics studies. It has been reported that viral proteins are highly interconnected (Bailer and Haas, 2009; Shapira et al., 2009), probably to form functional viral (enzymatic) complexes. These complexes can interact with a larger number of host proteins in comparison to the normal level of connectivity in the human interactome (Shapira et al., 2009). In this regard, the identification of substrates of HIV protease (Jurczyszak et al., 2020) and SARS-CoV-2 Mpro (Koudelka et al., 2020; Meyer et al., 2020) that, at the same time, have been described to interact with viral proteins in PPI analysis (Gordon et al., 2020; Jäger et al., 2012; Samavarchi-Tehrani et al., 2020) could suggest that the substrates are put in close contact with viral proteases by other viral proteins.

5. Conclusion and future perspectives

The recent development of proteomics approaches that allow for large-scale identification and characterization of protease substrates has greatly contributed not only to catalog extensive substrate repertoires of several proteases and to profile their substrate specificity, but also to better understand the roles that proteases play in health and disease. The identification of peptide fragments resulting from proteolytic cleavage provides opportunities to develop new biomarkers of disease, as well as targets for drug design (Butler and Overall, 2007; Huesgen et al., 2014; Marino et al., 2015; Vizovišek et al., 2018). For example, N-terminomics was shown to identify proteolytic signatures of apoptosis in plasma samples from hematologic cancer patients undergoing chemotherapy, which correlated with proteolytic fragments released in media by chemotherapeutics-treated hematologic cancer cell lines, suggesting terminomics approaches as a useful tool to monitor treatment success (Wiita et al., 2014a). As disease-specific proteolysis profiles are better characterized for a wider range of diseases, our understanding of both key proteases and substrates underlying disease could further guide the design of treatment strategies (Butler and Overall, 2007).

To date, a small number of studies have used proteomics methods to characterize proteolytic events in the context of virus-host interactions. So far, most large-scale degradomics studies have identified host substrates of viral proteases *in vitro* (Table 1). The identified substrates in *in vitro* models are not necessarily biologically relevant *in vivo*, as substrate recognition and cleavage by a protease *in vivo* are dependent on spatial and temporal colocalization of the protease and its substrate. However, *in vitro* experiments are still highly valuable as they provide a prospect of putatively relevant substrates that can be further studied and validated in virus-infected cells and organisms. In addition, few studies have put the focus on characterizing the proteolytic activity of host proteases during virus infection (Marchant et al., 2014; Meyer et al., 2020; Saeed et al., 2020). The consolidation and improvement of existing techniques will most likely augment their accessibility and broaden their use to more relevant experimental designs (such as their application in *in vivo* studies) and to study an increasing number of protease degradomes. Together with information provided by other methods, degradomics will contribute to a more detailed picture of how proteases modulate cellular processes during virus infection, providing useful information for the development of prevention and treatment strategies.

Table 1 Summary of gel-free terminomics approaches used to identify protease substrates in the context of virus-host interactions.

Method	Virus	Protease	Sample	Host protein substrates identified	References
COFRADIC	HIV	HIV PR	*In vitro* cleavage of cell lysates	123	Impens et al. (2012)
TAILS	CVB3	3Cpro	*In vitro* cleavage of cell lysates	34	Jagdeo et al. (2018)
	CVB3	2A	*In vitro* cleavage of cell lysates	63	Jagdeo (2017)
	PV	3Cpro	*In vitro* cleavage of cell lysates	72	Jagdeo et al. (2018)
	SARS-CoV	Mpro (3Cpro)	*In vitro* cleavage of cell lysates	124	Koudelka et al. (2020)
	SARS-CoV-2	Mpro (3Cpro)	*In vitro* cleavage of cell lysates	328	
	hCoV-NL63	Mpro (3Cpro)	*In vitro* cleavage of cell lysates	247	
HUNTER	SARS-CoV-2	Mpro (3Cpro)[a]	Virus-infected cells	11	Meyer et al. (2020)
		PLpro[a]	Virus-infected cells	3	
Subtiligase	Zika virus	NS2B-NS3	*In vitro* cleavage of cell lysates	31	Hill et al. (2018)
	CVB3	No specific protease	Virus-infected cells	211	Saeed et al. (2020)
	PV	No specific protease	Virus-infected cells	218	
	HRV A16	No specific protease	Virus-infected cells	278	
	EV70	No specific protease	Virus-infected cells	274	
	EV71	No specific protease	Virus-infected cells	206	

[a]Substrates are identified as possible PLpro and Mpro based on neo-N-termini peptide matching with consensus cleavage sequences for either Mpro or PLpro.

Acknowledgments

We thank Mariska van Huizen for carefully and critically reading the manuscript. X.M.V. is supported by the European Union's Horizon 2020 research and innovation program named INITIATE under the Marie Skłodowska-Curie grant agreement No 813343.

References

Alban, A., David, S.O., Bjorkesten, L., Andersson, C., Sloge, E., Lewis, S., Currie, I., 2003. A novel experimental design for comparative two-dimensional gel analysis: two-dimensional difference gel electrophoresis incorporating a pooled internal standard. Proteomics 3, 36–44.

Ammari, M.G., Gresham, C.R., Mccarthy, F.M., Nanduri, B., 2016. HPIDB 2.0: a curated database for host-pathogen interactions. Database (Oxford) 2016, baw103.

Bailer, S.M., Haas, J., 2009. Connecting viral with cellular interactomes. Curr. Opin. Microbiol. 12, 453–459.

Bailey-Elkin, B.A., Knaap, R.C.M., Kikkert, M., Mark, B.L., 2017. Structure and function of viral deubiquitinating enzymes. J. Mol. Biol. 429, 3441–3470.

Barrett, A.J., Rawlings, N.D., Woessner, J.F., 1998. Handbook of Proteolytic Enzymes. Academic Press, San Diego, CA.

Beachboard, D.C., Horner, S.M., 2016. Innate immune evasion strategies ofDNA and RNA viruses. Curr. Opin. Microbiol. 32, 113–119.

Belov, G.A., Kovtunovych, G., Jackson, C.L., Ehrenfeld, E., 2010. Poliovirus replication requires the N-terminus but not the catalytic Sec7 domain of ArfGEF GBF1. Cell. Microbiol. 12, 1463–1479.

Blom, N., Hansen, J., Blaas, D., Brunak, S., 1996. Cleavage site analysis in picornaviral polyproteins: discovering cellular targets by neural networks. Protein Sci. 5, 2203–2216.

Bond, J.S., 2019. Proteases: history, discovery, and roles in health and disease. J. Biol. Chem. 294, 1643–1651.

Boulware, K.T., Daugherty, P.S., 2006. Protease specificity determination by using cellular libraries of peptide substrates (CLiPS). Proc. Natl. Acad. Sci. U. S. A. 103, 7583.

Bredemeyer, A.J., Lewis, R.M., Malone, J.P., Davis, A.E., Gross, J., Townsend, R.R., Ley, T.J., 2004. A proteomic approach for the discovery of protease substrates. Proc. Natl. Acad. Sci. U. S. A. 101, 11785–11790.

Butler, G.S., Overall, C.M., 2007. Proteomic validation of protease drug targets: pharmacoproteomics of matrix metalloproteinase inhibitor drugs using isotope-coded affinity tag labelling and tandem mass spectrometry. Curr. Pharm. Des. 13, 263–270.

Cai, Y.D., Yu, H., Chou, K.C., 1998. Artificial neural network method for predicting HIV protease cleavage sites in protein. J. Protein Chem. 17, 607–615.

Chan, S.Y., Sam, I.C., Lai, J.K., Chan, Y.F., 2015. Cellular proteome alterations in response to enterovirus 71 and coxsackievirus A16 infections in neuronal and intestinal cell lines. J. Proteomics 125, 121–130.

Chen, L., Shan, Y., Weng, Y., Sui, Z., Zhang, X., Liang, Z., Zhang, L., Zhang, Y., 2016. Hydrophobic tagging-assisted N-termini enrichment for in-depth N-terminome analysis. Anal. Chem. 88, 8390–8395.

Chen, S., Wu, Z., Wang, M., Cheng, A., 2017. Innate immune evasion mediated by flaviviridae non-structural proteins. Viruses 9 (10), 291.

Clark, I.M., Cawston, T.E., 1989. Fragments of human fibroblast collagenase. Purification and characterization. Biochem. J. 263, 201–206.

Coyaud, E., Ranadheera, C., Cheng, D., Gonçalves, J., Dyakov, B.J.A., Laurent, E.M.N., St-Germain, J., Pelletier, L., Gingras, A.C., Brumell, J.H., Kim, P.K., Safronetz, D., Raught, B., 2018. Global interactomics uncovers extensive organellar targeting by Zika virus. Mol. Cell. Proteomics 17, 2242–2255.

De Chassey, B., Navratil, V., Tafforeau, L., Hiet, M.S., Aublin-Gex, A., Agaugué, S., Meiffren, G., Pradezynski, F., Faria, B.F., Chantier, T., Le Breton, M., Pellet, J., Davoust, N., Mangeot, P.E., Chaboud, A., Penin, F., Jacob, Y., Vidalain, P.O., Vidal, M., André, P., Rabourdin-Combe, C., Lotteau, V., 2008. Hepatitis C virus infection protein network. Mol. Syst. Biol. 4, 230.

Demon, D., Van Damme, P., Vanden Berghe, T., Deceuninck, A., Van Durme, J., Verspurten, J., Helsens, K., Impens, F., Wejda, M., Schymkowitz, J., Rousseau, F., Madder, A., Vandekerckhove, J., Declercq, W., Gevaert, K., Vandenabeele, P., 2009. Proteome-wide substrate analysis indicates substrate exclusion as a mechanism to generate caspase-7 versus caspase-3 specificity. Mol. Cell. Proteomics 8, 2700–2714.

Deng, S.J., Bickett, D.M., Mitchell, J.L., Lambert, M.H., Blackburn, R.K., Carter 3rd, H.L., Neugebauer, J., Pahel, G., Weiner, M.P., Moss, M.L., 2000. Substrate specificity of human collagenase 3 assessed using a phage-displayed peptide library. J. Biol. Chem. 275, 31422–31427.

Dix, M.M., Simon, G.M., Cravatt, B.F., 2008. Global mapping of the topography and magnitude of proteolytic events in apoptosis. Cell 134, 679–691.

Eckhardt, M., Zhang, W., Gross, A.M., Von Dollen, J., Johnson, J.R., Franks-Skiba, K.E., Swaney, D.L., Johnson, T.L., Jang, G.M., Shah, P.S., Brand, T.M., Archambault, J., Kreisberg, J.F., Grandis, J.R., Ideker, T., Krogan, N.J., 2018. Multiple routes to oncogenesis are promoted by the human papillomavirus-host protein network. Cancer Discov. 8, 1474–1489.

Fields, S., Song, O., 1989. A novel genetic system to detect protein-protein interactions. Nature 340, 245–246.

Fossum, E., Friedel, C.C., Rajagopala, S.V., Titz, B., Baiker, A., Schmidt, T., Kraus, T., Stellberger, T., Rutenberg, C., Suthram, S., Bandyopadhyay, S., Rose, D., Von Brunn, A., Uhlmann, M., Zeretzke, C., Dong, Y.A., Boulet, H., Koegl, M., Bailer, S.M., Koszinowski, U., Ideker, T., Uetz, P., Zimmer, R., Haas, J., 2009. Evolutionarily conserved herpesviral protein interaction networks. PLoS Pathog. 5, e1000570.

Fu, W., Sanders-Beer, B.E., Katz, K.S., Maglott, D.R., Pruitt, K.D., Ptak, R.G., 2009. Human immunodeficiency virus type 1, human protein interaction database at NCBI. Nucleic Acids Res. 37, D417–D422.

García-Sastre, A., 2017. Ten strategies of interferon evasion by viruses. Cell Host Microbe 22, 176–184.

Gausdal, G., Gjertsen, B.T., Mccormack, E., Van Damme, P., Hovland, R., Krakstad, C., Bruserud, Ø., Gevaert, K., Vandekerckhove, J., Døskeland, S.O., 2008. Abolition of stress-induced protein synthesis sensitizes leukemia cells to anthracycline-induced death. Blood 111, 2866–2877.

Gevaert, K., Van Damme, J., Goethals, M., Thomas, G.R., Hoorelbeke, B., Demol, H., Martens, L., Puype, M., Staes, A., Vandekerckhove, J., 2002. Chromatographic isolation of methionine-containing peptides for gel-free proteome analysis: identification of more than 800 *Escherichia coli* proteins. Mol. Cell. Proteomics 1, 896–903.

Gevaert, K., Goethals, M., Martens, L., Van Damme, J., Staes, A., Thomas, G.R., Vandekerckhove, J., 2003. Exploring proteomes and analyzing protein processing by mass spectrometric identification of sorted N-terminal peptides. Nat. Biotechnol. 21, 566–569.

Gingras, A.C., Gstaiger, M., Raught, B., Aebersold, R., 2007. Analysis of protein complexes using mass spectrometry. Nat. Rev. Mol. Cell Biol. 8, 645–654.

Glick, Y., Ben-Ari, Y., Drayman, N., Pellach, M., Neveu, G., Boonyaratanakornkit, J., Avrahami, D., Einav, S., Oppenheim, A., Gerber, D., 2016. Pathogen receptor discovery with a microfluidic human membrane protein array. Proc. Natl. Acad. Sci. U. S. A. 113, 4344–4349.

Gokhale, N.S., Vazquez, C., Horner, S.M., 2014. Hepatitis C virus. Strategies to evade antiviral responses. Futur. Virol. 9, 1061–1075.

Gorbalenya, A.E., Svitkin, Y.V., Kazachkov, Y.A., Agol, V.I., 1979. Encephalomyocarditis virus-specific polypeptide p22 is involved in the processing of the viral precursor polypeptides. FEBS Lett. 108, 1–5.

Gordon, D.E., Jang, G.M., Bouhaddou, M., Xu, J., Obernier, K., White, K.M., O'meara, M.J., Rezelj, V.V., Guo, J.Z., Swaney, D.L., Tummino, T.A., Hüttenhain, R., Kaake, R.M., Richards, A.L., Tutuncuoglu, B., Foussard, H., Batra, J., Haas, K., Modak, M., Kim, M., Haas, P., Polacco, B.J., Braberg, H., Fabius, J.M., Eckhardt, M., Soucheray, M., Bennett, M.J., Cakir, M., Mcgregor, M.J., Li, Q., Meyer, B., Roesch, F., Vallet, T., Mac Kain, A., Miorin, L., Moreno, E., Naing, Z.Z.C., Zhou, Y., Peng, S., Shi, Y., Zhang, Z., Shen, W., Kirby, I.T., Melnyk, J.E., Chorba, J.S., Lou, K., Dai, S.A., Barrio-Hernandez, I., Memon, D., Hernandez-Armenta, C., Lyu, J., Mathy, C.J.P., Perica, T., Pilla, K.B., Ganesan, S.J., Saltzberg, D.J., Rakesh, R., Liu, X., Rosenthal, S.B., Calviello, L., Venkataramanan, S., Liboy-Lugo, J., Lin, Y., Huang, X.-P., Liu, Y., Wankowicz, S.A., Bohn, M., Safari, M., Ugur, F.S., Koh, C., Savar, N.S., Tran, Q.D., Shengjuler, D., Fletcher, S.J., O'neal, M.C., Cai, Y., Chang, J.C.J., Broadhurst, D.J., Klippsten, S., Sharp, P.P., Wenzell, N.A., Kuzuoglu-Ozturk, D., Wang, H.-Y., Trenker, R., Young, J.M., Cavero, D.A., Hiatt, J., Roth, T.L., Rathore, U., Subramanian, A., Noack, J., Hubert, M., Stroud, R.M., Frankel, A.D., Rosenberg, O.S., Verba, K.A., Agard, D.A., Ott, M., Emerman, M., Jura, N., et al., 2020. A SARS-CoV-2 protein interaction map reveals targets for drug repurposing. Nature 583, 459–468.

Griswold, A.R., Cifani, P., Rao, S.D., Axelrod, A.J., Miele, M.M., Hendrickson, R.C., Kentsis, A., Bachovchin, D.A., 2019. A chemical strategy for protease substrate profiling. Cell Chem. Biol. 26, 901–907.e6.

Harris, J.L., Backes, B.J., Leonetti, F., Mahrus, S., Ellman, J.A., Craik, C.S., 2000. Rapid and general profiling of protease specificity by using combinatorial fluorogenic substrate libraries. Proc. Natl. Acad. Sci. U. S. A. 97, 7754–7759.

Hill, M.E., Kumar, A., Wells, J.A., Hobman, T.C., Julien, O., Hardy, J.A., 2018. The unique cofactor region of Zika virus NS2B–NS3 protease facilitates cleavage of key host proteins. ACS Chem. Biol. 13, 2398–2405.

Huesgen, P.F., Lange, P.F., Overall, C.M., 2014. Ensembles of protein termini and specific proteolytic signatures as candidate biomarkers of disease. Proteomics Clin. Appl. 8, 338–350.

Impens, F., Van Damme, P., Demol, H., Van Damme, J., Vandekerckhove, J., Gevaert, K., 2008. Mechanistic insight into taxol-induced cell death. Oncogene 27, 4580–4591.

Impens, F., Timmerman, E., Staes, A., Moens, K., Ariën, K.K., Verhasselt, B., Vandekerckhove, J., Gevaert, K., 2012. A catalogue of putative HIV-1 protease host cell substrates. Biol. Chem. 393, 915–931.

Jagdeo, J.M., 2017. Identification of Novel Picornavirus Proteinase Substrates Using Terminal Amine Isotopic Labeling of Substrates. The University of British Columbia.

Jagdeo, J.M., Dufour, A., Klein, T., Solis, N., Kleifeld, O., Kizhakkedathu, J., Luo, H., Overall, C.M., Jan, E., 2018. N-terminomics TAILS identifies host cell substrates of poliovirus and coxsackievirus B3 3C proteinases that modulate virus infection. J. Virol. 92, e02211–17.

Jäger, S., Cimermancic, P., Gulbahce, N., Johnson, J.R., Mcgovern, K.E., Clarke, S.C., Shales, M., Mercenne, G., Pache, L., Li, K., Hernandez, H., Jang, G.M., Roth, S.L., Akiva, E., Marlett, J., Stephens, M., D'orso, I., Fernandes, J., Fahey, M., Mahon, C., O'donoghue, A.J., Todorovic, A., Morris, J.H., Maltby, D.A., Alber, T., Cagney, G., Bushman, F.D., Young, J.A., Chanda, S.K., Sundquist, W.I., Kortemme, T., Hernandez, R.D., Craik, C.S., Burlingame, A., Sali, A., Frankel, A.D., Krogan, N.J., 2012. Global landscape of HIV–human protein complexes. Nature 481, 365–370.

Jurczyszak, D., Zhang, W., Terry, S.N., Kehrer, T., Bermúdez González, M.C., Mcgregor, E., Mulder, L.C.F., Eckwahl, M.J., Pan, T., Simon, V., 2020. HIV protease cleaves the antiviral m6A reader protein YTHDF3 in the viral particle. PLoS Pathog. 16, e1008305.

Khadka, S., Vangeloff, A.D., Zhang, C., Siddavatam, P., Heaton, N.S., Wang, L., Sengupta, R., Sahasrabudhe, S., Randall, G., Gribskov, M., Kuhn, R.J., Perera, R., Lacount, D.J., 2011. A physical interaction network of dengue virus and human proteins. Mol. Cell. Proteomics 10, M111.012187.

Kindler, E., Thiel, V., Weber, F., 2016. Interaction of SARS and MERS coronaviruses with the antiviral interferon response. Adv. Virus Res. 96, 219–243.

Kleifeld, O., Doucet, A., Auf Dem Keller, U., Prudova, A., Schilling, O., Kainthan, R.K., Starr, A.E., Foster, L.J., Kizhakkedathu, J.N., Overall, C.M., 2010. Isotopic labeling of terminal amines in complex samples identifies protein N-termini and protease cleavage products. Nat. Biotechnol. 28, 281–288.

Kleifeld, O., Doucet, A., Prudova, A., Auf Dem Keller, U., Gioia, M., Kizhakkedathu, J.N., Overall, C.M., 2011. Identifying and quantifying proteolytic events and the natural N terminome by terminal amine isotopic labeling of substrates. Nat. Protoc. 6, 1578–1611.

Klein, T., Fung, S.-Y., Renner, F., Blank, M.A., Dufour, A., Kang, S., Bolger-Munro, M., Scurll, J.M., Priatel, J.J., Schweigler, P., Melkko, S., Gold, M.R., Viner, R.I., Régnier, C.H., Turvey, S.E., Overall, C.M., 2015. The paracaspase MALT1 cleaves HOIL1 reducing linear ubiquitination by LUBAC to dampen lymphocyte NF-κB signalling. Nat. Commun. 6, 8777.

Knäuper, V., Osthues, A., Declerck, Y.A., Langley, K.E., Bläser, J., Tschesche, H., 1993. Fragmentation of human polymorphonuclear-leucocyte collagenase. Biochem. J. 291 (Pt. 3), 847–854.

Knäuper, V., Cowell, S., Smith, B., López-Otin, C., O'shea, M., Morris, H., Zardi, L., Murphy, G., 1997. The role of the C-terminal domain of human collagenase-3 (MMP-13) in the activation of procollagenase-3, substrate specificity, and tissue inhibitor of metalloproteinase interaction. J. Biol. Chem. 272, 7608–7616.

Kondo, T., Hirohashi, S., 2009. Application of 2D-DIGE in cancer proteomics toward personalized medicine. Methods Mol. Biol. 577, 135–154.

Koudelka, T., Boger, J., Henkel, A., Schönherr, R., Krantz, S., Fuchs, S., Rodríguez, E., Redecke, L., Tholey, A., 2020. N-Terminomics for the identification of in vitro substrates and cleavage site specificity of the SARS-CoV-2 main protease. Proteomics 21, e2000246.

Lai, Z.W., Gomez-Auli, A., Keller, E.J., Mayer, B., Biniossek, M.L., Schilling, O., 2015. Enrichment of protein N-termini by charge reversal of internal peptides. Proteomics 15, 2470–2478.

Lai, Y., Wang, M., Cheng, A., Mao, S., Ou, X., Yang, Q., Wu, Y., Jia, R., Liu, M., Zhu, D., Chen, S., Zhang, S., Zhao, X.-X., Huang, J., Gao, Q., Wang, Y., Xu, Z., Chen, Z., Zhu, L., Luo, Q., Liu, Y., Yu, Y., Zhang, L., Tian, B., Pan, L., Rehman, M.U., Chen, X., 2020. Regulation of apoptosis by enteroviruses. Front. Microbiol. 11, 1145.

Lasso, G., Mayer, S.V., Winkelmann, E.R., Chu, T., Elliot, O., Patino-Galindo, J.A., Park, K., Rabadan, R., Honig, B., Shapira, S.D., 2019. A structure-informed atlas of human-virus interactions. Cell 178, 1526–1541.e16.

Lei, J., Hilgenfeld, R., 2017. RNA-virus proteases counteracting host innate immunity. FEBS Lett. 591, 3190–3210.

Leitner, A., Faini, M., Stengel, F., Aebersold, R., 2016. Crosslinking and mass spectrometry: an integrated technology to understand the structure and function of molecular machines. Trends Biochem. Sci. 41, 20–32.

Li, J., Van Vranken, J.G., Pontano Vaites, L., Schweppe, D.K., Huttlin, E.L., Etienne, C., Nandhikonda, P., Viner, R., Robitaille, A.M., Thompson, A.H., Kuhn, K., Pike, I., Bomgarden, R.D., Rogers, J.C., Gygi, S.P., Paulo, J.A., 2020a. TMTpro reagents: a set of isobaric labeling mass tags enables simultaneous proteome-wide measurements across 16 samples. Nat. Methods 17, 399–404.

Li, S., Zhang, Y., Guan, Z., Li, H., Ye, M., Chen, X., Shen, J., Zhou, Y., Shi, Z.-L., Zhou, P., Peng, K., 2020b. SARS-CoV-2 triggers inflammatory responses and cell death through caspase-8 activation. Signal Transduct. Target. Ther. 5, 235.

Linderstrom-Lang, K., 1952. The initial stages in the breakdown of proteins by enzymes. In: Lane Medical Lectures. vol. 6. Lectures.

Liu, H., Sadygov, R.G., Yates 3rd, J.R., 2004. A model for random sampling and estimation of relative protein abundance in shotgun proteomics. Anal. Chem. 76, 4193–4201.

López-Otín, C., Overall, C.M., 2002. Protease degradomics: a new challenge for proteomics. Nat. Rev. Mol. Cell Biol. 3, 509–519.

Macdonald, J.I., Munch, H.K., Moore, T., Francis, M.B., 2015. One-step site-specific modification of native proteins with 2-pyridinecarboxyaldehydes. Nat. Chem. Biol. 11, 326–331.

Mahrus, S., Trinidad, J.C., Barkan, D.T., Sali, A., Burlingame, A.L., Wells, J.A., 2008. Global sequencing of proteolytic cleavage sites in apoptosis by specific labeling of protein N termini. Cell 134, 866–876.

Marchant, D.J., Bellac, C.L., Moraes, T.J., Wadsworth, S.J., Dufour, A., Butler, G.S., Bilawchuk, L.M., Hendry, R.G., Robertson, A.G., Cheung, C.T., Ng, J., Ang, L., Luo, Z., Heilbron, K., Norris, M.J., Duan, W., Bucyk, T., Karpov, A., Devel, L., Georgiadis, D., Hegele, R.G., Luo, H., Granville, D.J., Dive, V., Mcmanus, B.M., Overall, C.M., 2014. A new transcriptional role for matrix metalloproteinase-12 in antiviral immunity. Nat. Med. 20, 493–502.

Marino, G., Eckhard, U., Overall, C.M., 2015. Protein termini and their modifications revealed by positional proteomics. ACS Chem. Biol. 10, 1754–1764.

Mcquibban, G.A., Gong, J.-H., Tam, E.M., Mcculloch, C.A.G., Clark-Lewis, I., Overall, C.M., 2000. Inflammation dampened by gelatinase A cleavage of monocyte chemoattractant protein-3. Science 289, 1202–1206.

Meyer, B., Chiaravalli, J., Brownridge, P., Bryne, D.P., Daly, L.A., Agou, F., Eyers, C.E., Eyers, P.A., Vignuzzi, M., Emmott, E., 2020. Characterisation of Protease Activity During SARS-CoV-2 Infection Identifies Novel Viral Cleavage Sites and Cellular Targets for Drug Repurposing. bioRxiv. 2020.09.16.297945.

Meylan, E., Curran, J., Hofmann, K., Moradpour, D., Binder, M., Bartenschlager, R., Tschopp, J., 2005. Cardif is an adaptor protein in the RIG-I antiviral pathway and is targeted by hepatitis C virus. Nature 437, 1167–1172.

Morikawa, K., Gouttenoire, J., Hernandez, C., Dao Thi, V.L., Tran, H.T., Lange, C.M., Dill, M.T., Heim, M.H., Donzé, O., Penin, F., Quadroni, M., Moradpour, D., 2014. Quantitative proteomics identifies the membrane-associated peroxidase GPx8 as a cellular substrate of the hepatitis C virus NS3-4A protease. Hepatology 59, 423–433.

Murphy, G., Allan, J.A., Willenbrock, F., Cockett, M.I., O'connell, J.P., Docherty, A.J., 1992. The role of the C-terminal domain in collagenase and stromelysin specificity. J. Biol. Chem. 267, 9612–9618.

Nagai, Y., 1993. Protease-dependent virus tropism and pathogenicity. Trends Microbiol. 1, 81–87.

Nagai, Y., Ogura, H., Klenk, H.-D., 1976b. Studies on the assembly of the envelope of Newcastle disease virus. Virology 69, 523–538.

Nagai, Y., Klenk, H.D., Rott, R., 1976a. Proteolytic cleavage of the viral glycoproteins and its significance for the virulence of Newcastle disease virus. Virology 72, 494–508.

Nagai, Y., Shimokata, K., Yoshida, T., Hamaguchi, M., Iinuma, M., Maeno, K., Matsumoto, T., Klenk, H.D., Rott, R., 1979. The spread of a pathogenic and an apathogenic strain of Newcastle disease virus in the chick embryo as depending on the protease sensitivity of the virus glycoproteins. J. Gen. Virol. 45, 263–272.

Navratil, V., De Chassey, B., Meyniel, L., Delmotte, S., Gautier, C., André, P., Lotteau, V., Rabourdin-Combe, C., 2009. VirHostNet: a knowledge base for the management and the analysis of proteome-wide virus-host interaction networks. Nucleic Acids Res. 37, D661–D668.

Nazif, T., Bogyo, M., 2001. Global analysis of proteasomal substrate specificity using positional-scanning libraries of covalent inhibitors. Proc. Natl. Acad. Sci. U. S. A. 98, 2967–2972.

Nelemans, T., Kikkert, M., 2019. Viral innate immune evasion and the pathogenesis of emerging RNA virus infections. Viruses 11, 961.

Overall, C.M., Tam, E.M., Kappelhoff, R., Connor, A., Ewart, T., Morrison, C.J., Puente, X., López-Otín, C., Seth, A., 2004. Protease degradomics: mass spectrometry discovery of protease substrates and the CLIP-CHIP, a dedicated DNA microarray of all human proteases and inhibitors. Biol. Chem. 385, 493–504.

Palmenberg, A.C., Pallansch, M.A., Rueckert, R.R., 1979. Protease required for processing picornaviral coat protein resides in the viral replicase gene. J. Virol. 32, 770–778.

Pardy, R.D., Valbon, S.F., Richer, M.J., 2019. Running interference: interplay between Zika virus and the host interferon response. Cytokine 119, 7–15.

Pérez-Silva, J.G., Español, Y., Velasco, G., Quesada, V., 2016. The Degradome database: expanding roles of mammalian proteases in life and disease. Nucleic Acids Res. 44, D351–D355.

Prudova, A., Auf Dem Keller, U., Butler, G.S., Overall, C.M., 2010. Multiplex N-terminome analysis of MMP-2 and MMP-9 substrate degradomes by iTRAQ-TAILS quantitative proteomics. Mol. Cell. Proteomics 9, 894–911.

Puente, X.S., Sánchez, L.M., Overall, C.M., López-Otín, C., 2003. Human and mouse proteases: a comparative genomic approach. Nat. Rev. Genet. 4, 544–558.

Rassmann, A., Henke, A., Zobawa, M., Carlsohn, M., Saluz, H.P., Grabley, S., Lottspeich, F., Munder, T., 2006. Proteome alterations in human host cells infected with coxsackievirus B3. J. Gen. Virol. 87, 2631–2638.

Rawlings, N.D., Barrett, A.J., Bateman, A., 2012. MEROPS: the database of proteolytic enzymes, their substrates and inhibitors. Nucleic Acids Res. 40, D343–D350.

Rogers, L.D., Overall, C.M., 2013. Proteolytic post-translational modification of proteins: proteomic tools and methodology. Mol. Cell. Proteomics 12, 3532–3542.

Roux, K.J., Kim, D.I., Raida, M., Burke, B., 2012. A promiscuous biotin ligase fusion protein identifies proximal and interacting proteins in mammalian cells. J. Cell Biol. 196, 801–810.

Saeed, M., Kapell, S., Hertz, N.T., Wu, X., Bell, K., Ashbrook, A.W., Mark, M.T., Zebroski, H.A., Neal, M.L., Flodström-Tullberg, M., Macdonald, M.R., Aitchison, J.D., Molina, H., Rice, C.M., 2020. Defining the proteolytic landscape during enterovirus infection. PLoS Pathog. 16, e1008927.

Samavarchi-Tehrani, P., Abdouni, H., Knight, J.D.R., Astori, A., Samson, R., Lin, Z.-Y., Kim, D.-K., Knapp, J.J., St-Germain, J., Go, C.D., Larsen, B., Wong, C.J., Cassonnet, P., Demeret, C., Jacob, Y., Roth, F.P., Raught, B., Gingras, A.-C., 2020. A SARS-CoV-2—Host Proximity Interactome. bioRxiv. 2020.09.03.282103.

Scheid, A., Choppin, P.W., 1974. Identification of biological activities of paramyxovirus glycoproteins. Activation of cell fusion, hemolysis, and infectivity by proteolytic cleavage of an inactive precursor protein of Sendai virus. Virology 57, 475–490.

Scheid, A., Choppin, P.W., 1977. Two disulfide-linked polypeptide chains constitute the active F protein of paramyxoviruses. Virology 80, 54–66.

Schilling, O., Overall, C.M., 2008. Proteome-derived, database-searchable peptide libraries for identifying protease cleavage sites. Nat. Biotechnol. 26, 685–694.
Schilling, O., Barré, O., Huesgen, P.F., Overall, C.M., 2010. Proteome-wide analysis of protein carboxy termini: C terminomics. Nat. Methods 7, 508–511.
Seipelt, J., Liebig, H.-D., Sommergruber, W., Gerner, C., Kuechler, E., 2000. 2A proteinase of human rhinovirus cleaves cytokeratin 8 in infected HeLa cells. J. Biol. Chem. 275, 20084–20089.
Shapira, S.D., Gat-Viks, I., Shum, B.O., Dricot, A., De Grace, M.M., Wu, L., Gupta, P.B., Hao, T., Silver, S.J., Root, D.E., Hill, D.E., Regev, A., Hacohen, N., 2009. A physical and regulatory map of host-influenza interactions reveals pathways in H1N1 infection. Cell 139, 1255–1267.
Shema, G., Nguyen, M.T.N., Solari, F.A., Loroch, S., Venne, A.S., Kollipara, L., Sickmann, A., Verhelst, S.H.L., Zahedi, R.P., 2018. Simple, scalable, and ultrasensitive tip-based identification of protease substrates. Mol. Cell. Proteomics 17, 826–834.
Simon, G.M., Dix, M.M., Cravatt, B.F., 2009. Comparative assessment of large-scale proteomic studies of apoptotic proteolysis. ACS Chem. Biol. 4, 401–408.
Stellberger, T., Häuser, R., Baiker, A., Pothineni, V.R., Haas, J., Uetz, P., 2010. Improving the yeast two-hybrid system with permutated fusions proteins: the varicella zoster virus interactome. Proteome Sci. 8, 8.
Thompson, A., Wölmer, N., Koncarevic, S., Selzer, S., Böhm, G., Legner, H., Schmid, P., Kienle, S., Penning, P., Höhle, C., Berfelde, A., Martinez-Pinna, R., Farztdinov, V., Jung, S., Kuhn, K., Pike, I., 2019. TMTpro: design, synthesis, and initial evaluation of a proline-based isobaric 16-plex tandem mass tag reagent set. Anal. Chem. 91, 15941–15950.
Thomson, R., Hodgman, T.C., Yang, Z.R., Doyle, A.K., 2003. Characterizing proteolytic cleavage site activity using bio-basis function neural networks. Bioinformatics 19, 1741–1747.
Tomasselli, A.G., Hui, J.O., Adams, L., Chosay, J., Lowery, D., Greenberg, B., Yem, A., Deibel, M.R., Zürcher-Neely, H., Heinrikson, R.L., 1991. Actin, troponin C, Alzheimer amyloid precursor protein and pro-interleukin 1 beta as substrates of the protease from human immunodeficiency virus. J. Biol. Chem. 266, 14548–14553.
Tran, H.T.L., Morikawa, K., Anggakusuma, Zibi, R., Thi, V.L.D., Penin, F., Heim, M.H., Quadroni, M., Pietschmann, T., Gouttenoire, J., Moradpour, D., 2020. OCIAD1 is a host mitochondrial substrate of the hepatitis C virus NS3-4A protease. PLoS One 15, e0236447.
Turk, B.E., Huang, L.L., Piro, E.T., Cantley, L.C., 2001. Determination of protease cleavage site motifs using mixture-based oriented peptide libraries. Nat. Biotechnol. 19, 661–667.
Turk, B., Turk, D., Turk, V., 2012. Protease signalling: the cutting edge. EMBO J. 31, 1630–1643.
Uetz, P., Dong, Y.A., Zeretzke, C., Atzler, C., Baiker, A., Berger, B., Rajagopala, S.V., Roupelieva, M., Rose, D., Fossum, E., Haas, J., 2006. Herpesviral protein networks and their interaction with the human proteome. Science 311, 239–242.
Unlü, M., Morgan, M.E., Minden, J.S., 1997. Difference gel electrophoresis: a single gel method for detecting changes in protein extracts. Electrophoresis 18, 2071–2077.
Uno, N., Ross, T.M., 2018. Dengue virus and the host innate immune response. Emerg. Microbes Infect. 7, 167.
Van Damme, P., Martens, L., Van Damme, J., Hugelier, K., Staes, A., Vandekerckhove, J., Gevaert, K., 2005. Caspase-specific and nonspecific in vivo protein processing during Fas-induced apoptosis. Nat. Methods 2, 771–777.

Van Damme, P., Staes, A., Bronsoms, S., Helsens, K., Colaert, N., Timmerman, E., Aviles, F.X., Vandekerckhove, J., Gevaert, K., 2010. Complementary positional proteomics for screening substrates of endo- and exoproteases. Nat. Methods 7, 512–515.

Van Huizen, M., Kikkert, M., 2020. The role of atypical ubiquitin chains in the regulation of the antiviral innate immune response. Front. Cell Dev. Biol. 7, 392.

Venne, A.S., Vögtle, F.N., Meisinger, C., Sickmann, A., Zahedi, R.P., 2013. Novel highly sensitive, specific, and straightforward strategy for comprehensive N-terminal proteomics reveals unknown substrates of the mitochondrial peptidase Icp55. J. Proteome Res. 12, 3823–3830.

Vizovišek, M., Vidmar, R., Drag, M., Fonović, M., Salvesen, G.S., Turk, B., 2018. Protease specificity: towards in vivo imaging applications and biomarker discovery. Trends Biochem. Sci. 43, 829–844.

Von Der Helm, K., 1977. Cleavage of Rous sarcoma viral polypeptide precursor into internal structural proteins in vitro involves viral protein p15. Proc. Natl. Acad. Sci. U. S. A. 74, 911–915.

Weng, K.F., Li, M.L., Hung, C.T., Shih, S.R., 2009. Enterovirus 71 3C protease cleaves a novel target CstF-64 and inhibits cellular polyadenylation. PLoS Pathog. 5, e1000593.

Weng, S.S.H., Demir, F., Ergin, E.K., Dirnberger, S., Uzozie, A., Tuscher, D., Nierves, L., Tsui, J., Huesgen, P.F., Lange, P.F., 2019. Sensitive determination of proteolytic proteoforms in limited microscale proteome samples. Mol. Cell. Proteomics 18, 2335–2347.

Wiita, A.P., Seaman, J.E., Wells, J.A., 2014b. Global analysis of cellular proteolysis by selective enzymatic labeling of protein N-termini. Methods Enzymol. 544, 327–358.

Wiita, A.P., Hsu, G.W., Lu, C.M., Esensten, J.H., Wells, J.A., 2014a. Circulating proteolytic signatures of chemotherapy-induced cell death in humans discovered by N-terminal labeling. Proc. Natl. Acad. Sci. U. S. A. 111, 7594–7599.

Wildes, D., Wells, J.A., 2010. Sampling the N-terminal proteome of human blood. Proc. Natl. Acad. Sci. U. S. A. 107, 4561.

Xing, H., Xu, S., Jia, F., Yang, Y., Xu, C., Qin, C., Shi, L., 2020. Zika NS2B is a crucial factor recruiting NS3 to the ER and activating its protease activity. Virus Res. 275, 197793.

CHAPTER SIX

Post-translational modification control of viral DNA sensors and innate immune signaling

Bokai Song[†], **Dawei Liu**[†], **Todd M. Greco, and Ileana M. Cristea***
Department of Molecular Biology, Princeton University, Princeton, NJ, United States
*Corresponding author: e-mail address: icristea@princeton.edu

Contents

Abstract

The vertebrate innate immune system confers host cells with mechanisms to protect against both evolutionarily ancient pathogens and newly emerging pathogenic strains. Innate immunity relies on the host cell's ability to distinguish between self and pathogen-derived molecules. To achieve this, the innate immune system uses germline encoded receptors called pattern recognition receptors (PRRs), which recognize various molecular signatures, including nucleic acids, proteins, lipids, glycans and glycolipids. Among these molecules, the recognition of pathogenic, mislocalized, or damaged DNA by cellular protein receptors, commonly called DNA sensors, represents a major surveillance pathway for initiating immune signaling. The ability of cells to temporally regulate DNA sensor activation and subsequent signal termination is critical for effective immune signaling. These same mechanisms are also co-opted by pathogens to promote their replication. Therefore, there is significant interest in understanding DNA

[†] Equal contribution as co-first authors.

Advances in Virus Research, Volume 109
ISSN 0065-3527
https://doi.org/10.1016/bs.aivir.2021.03.001

sensor regulatory networks during microbial infections and autoimmune disease. One emerging aspect of DNA sensor regulation is through post-translational modifications (PTMs), including phosphorylation, acetylation, ubiquitination, ADP-ribosylation, SUMOylation, methylation, deamidation, glutamylation. In this chapter, we discuss how PTMs have been shown to positively or negatively impact DNA sensor functions via diverse mechanisms, including direct regulation of enzymatic activity, protein-protein and protein-DNA interactions, protein translocations and protein turnover. In addition, we highlight the ability of virus-induced PTMs to promote immune evasion. We also discuss the recent evidence linking PTMs on DNA sensors with human diseases and more broadly, highlight promising directions for future research on PTM-mediated regulation of DNA sensor-dependent immune signaling.

Abbreviations

AdV	adenovirus
AGS	Aicardi-Goutières syndrome
AIM2	absent in melanoma 2
BLK	B-lymphoid tyrosine kinase
cGAMP	cyclic GMP-AMP
cGAS	cyclic-GMP-AMP synthase
CpG	cytosine-phosphate-guanine
DAI	DNA-dependent activator of IFN-regulatory factor
DDR	DNA damage response
DDX41	DEAD box polypeptide 41
DHX36	DEAH box helicase 36
DHX9	DEAH box helicase 9
DNA-PK	DNA-dependent protein kinase
DNA-PKcs/PRKDC	DNA-dependent protein kinase catalytic subunit
D-NHEJ	deletional nonhomologous end joining
ER	endoplasmic reticulum
GAR	glycine-arginine-rich
HCMV	human cytomegalovirus
HDAC	histone deacetylase
HIV-1	human immunodeficiency virus type 1
hnRNPA2B1	heterogeneous nuclear ribonucleoproteins A2/B1
HPV	human papilloma virus
HSV-1	herpes simplex virus type 1
IFI16	interferon-inducible protein 16
IFIX	interferon-inducible protein X
IFN	interferon
IMAC	immobilized metal affinity chromatography
IP	immunoaffinity purification
IRF3	interferon regulatory factor 3
ISG	interferon-stimulated gene
Jak	Janus kinase
JMJD6	histone arginine demethylase
KAT5	lysine acetyltransferase 5
KSHV	Kaposi's sarcoma-associated herpesvirus

LRRFIP1	leucine-rich repeat flightless interacting protein-1
m^6A	N^6-methyladenosine
MRE11	meiotic recombination 11 homolog A
MS	mass spectrometry
MyD88	myeloid differentiation primary response protein
nLC	nano liquid chromatography
OASL	2′-5′-oligoadenylate synthase-like protein
PAMP	pathogen-associated molecular pattern
PARP	poly ADP-ribose polymerase
pDC	plasmacytoid dendritic cell
PKR	protein kinase RNA-activated
POI	protein of interest
PRM	parallel reaction monitoring
PRMT1	protein arginine *N*-methyltransferase 1
PRR	pattern recognition receptor
PTM	post-translational modifications
PyV	polyomavirus
RGG	arginine-glycine-glycine–rich
RNA Pol III	RNA polymerase III
SDS-PAGE	sodium dodecyl sulfate–polyacrylamide gel electrophoresis
SENP2	sentrin-specific protease 2
STAT	signal transducer and activator of transcription
STING	endoplasmic reticulum-associated adaptor
TBK1	TANK-binding kinase 1
TLR	Toll-like receptor
TREX1	DNA 3′ repair exonuclease 1
Triad3A	triad domain-containing protein 3
TRIM14	tripartite motif-containing protein 14
TRIM21	tripartite motif-containing protein 21
TRIM56	tripartite motif-containing protein 56
TTLL4	tubulin polyglutamylase 4
TTLL6	tubulin polyglutamylase 6
USP14	ubiquitin-specific-processing protease 14
VSV	vesicular stomatitis virus
XRCC5	X-ray repair cross-complementing protein 6
XRCC6	X-ray repair cross-complementing protein 5

1. Introduction

The vertebrate innate immune system is at the front line of defense against pathogens. This ancient defense mechanism confers host cells protection not only against pathogens that have existed for much of the vertebrate evolution, but also against novel pathogenic strains. A critical function of innate immunity is to distinguish between self and pathogen- or damage-derived molecules. To recognize pathogen-associated molecular patterns (PAMPs), the innate immune system employs a series of germline encoded receptors, known as pattern recognition receptors (PRRs). PRRs recognize a vast range of molecular signatures, including nucleic acids, proteins, lipids, glycans and glycolipids, and among them, the sensing of DNA has been studied for decades (Chen et al., 2016b; Gordon, 2002; Tan et al., 2018; Thompson et al., 2011).

DNA (both double- and single-stranded) is a common genetic material for many pathogens (DNA viruses, bacteria, etc.), and thus detecting pathogen-associated DNA provides a unique signature for initiating innate immune programs. In addition, host DNA that is either damaged or mislocalized due to pathological conditions can also be sensed and elicit an immune response. This gives rise to autoimmunity, where the host immune system initiates responses against healthy tissues (Chen et al., 2016b; Tan et al., 2018). To date, researchers have identified and characterized the function of many host receptors for pathogenic DNA, which collectively are classified as DNA sensors. At minimum, a protein that functions as a DNA sensor should directly bind to pathogenic DNA, and then trigger innate immune signaling through the expression of antiviral or inflammatory cytokines and chemokines (Crow et al., 2015; Liu et al., 2016). Additionally, DNA sensors can protect host cells by inducing apoptosis (Zierhut et al., 2019), autophagy (Lei et al., 2018), or repressing virus gene expression upon binding to viral DNA (Diner et al., 2016; Johnson et al., 2014).

The first protein to be characterized as a DNA sensor was Toll-like receptor 9 (TLR9). TLR9 recognizes unmethylated cytosine-phosphate-guanine (CpG) nucleotide sequences, which is rare in vertebrate genomes compared to DNA viruses and bacteria (Krieg et al., 1995; Takeshita et al., 2001). After unmethylated CpG stimulation, TLR9 moves from the endoplasmic reticulum (ER) to the Golgi apparatus and then to endolysosomes, where it interacts with MyD88 to trigger downstream proinflammatory cytokine responses (Chockalingam et al., 2009; Leifer et al., 2004) (Fig. 1).

Unlike TLR9, many other DNA sensors do not use unique molecule features to distinguish pathogenic vs host DNA. Rather, their DNA sensing ability is intimately tied to their subcellular localization. For example, DNA sensors are often localized to the cytosol and are positioned to detect mislocalized (cytosolic) DNA, a marker for viral infection, cellular DNA damage, and mitochondrial DNA leakage (Chen et al., 2016b; Riley and

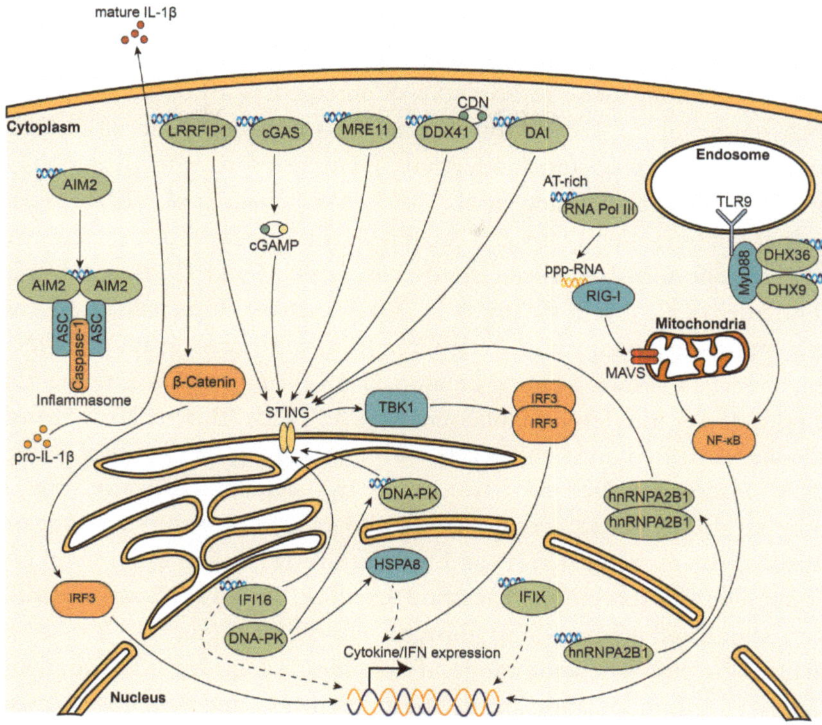

Fig. 1 DNA sensors and their downstream signaling pathways. Known cytoplasmic (AIM2, cGAS, DAI, DDX41, DHX36, DHX9, DNAPK, LRRFIP1, MRE11, RNA Pol III) and nuclear (IFI16, IFIX, hnRNPA2B1) DNA sensors are indicated. Most DNA sensors converge on STING-TBK1-IRF3 axis, where the activated DNA sensor eventually leads to the dimerization of STING. Subsequently, this results in the phosphorylation of TBK1 and IRF3. Phosphorylated IRF3 dimerizes and translocates into the nucleus, leading to gene transcription of cytokines and interferons. Other pathways include the NF-κB pathway and the β-catenin pathway.

Tait, 2020; Tan et al., 2018). The first cytosolic DNA sensor discovered was the cytosolic protein, DNA-dependent activator of IFN-regulatory factors (DAI), also known as Z-DNA binding protein 1 (Takaoka et al., 2007). Subsequent to the identification of DAI, many other "cytosolic" DNA sensors have been identified, including absent in melanoma 2 (AIM2) (Burckstummer et al., 2009; Hornung et al., 2009), DNA-protein kinase (DNA-PK) (Burleigh et al., 2020; Ferguson et al., 2012; Zhang et al., 2011a), DEAD box polypeptide 41 (DDX41) (Miyashita et al., 2011; Zhang et al., 2011b), DEAH box helicase 9 and 36 (DHX9 and DHX36) (Kim et al., 2010), meiotic recombination 11 homolog A (MRE11) (Kondo et al., 2013), leucine-rich repeat flightless interacting protein-1 (LRRFIP1) (Yang et al., 2010), RNApolymerase III

(RNA Pol III) (Ablasser et al., 2009; Chiu et al., 2009), and cyclic-GMP-AMP synthase (cGAS) (Sun et al., 2013; Wu et al., 2013) (Table 1). Among these sensors, cGAS emerged as a dominant cytosolic DNA sensor during DNA virus or retrovirus infection (Gao et al., 2013; Sun et al., 2013), bacterial infections (Hansen et al., 2014), as well as self-DNA leakage (Harding et al., 2017; Mackenzie et al., 2017). cGAS binds double stranded DNA and catalyzes the production of a second messenger, cyclic GMP-AMP (cGAMP), using ATP and GTP as substrates (Lohofener et al., 2015; Zhang et al., 2013b, 2014). Intracellular cGAMP then binds the endoplasmic reticulum (ER)-associated adaptor STING, which recruits TANK-binding kinase 1 (TBK1) to phosphorylate the transcription factor interferon regulatory factor 3 (IRF3) (Bowie, 2012; Tanaka and Chen, 2012). The STING-TBK1-IRF3 signaling axis is critical, as most DNA sensors identified to date activate this signaling pathway (Fig. 1). Phosphorylated IRF3 then dimerizes and translocates into the nucleus, activating the expression of Type I interferons (IFNs) and other cytokines. Through both autocrine and paracrine signaling, Type I IFNs activate interferon stimulated genes (ISGs) through Jak-STAT signaling (Yan et al., 2016). ISGs possess a wide range of activities, which include stimulating cell-to-cell communication and cell death pathways, and targeting pathways required for viral entry, translation, replication, and egress. Overall, these activities convey a concerted effort to control and/or restrict the pathogenic infection process (Schneider et al., 2014).

Although cytosolic DNA sensors are deployed to detect aberrant cytosolic DNA, many DNA viruses are known to replicate their genomes in the nucleus, including the herpesviruses, herpes simplex virus type 1 (HSV-1) and human cytomegalovirus (HCMV), as well as adenovirus (AdV), human papilloma virus (HPV) and polyomavirus (PyV) (Schmid et al., 2014). Researchers have identified several DNA sensors that reside in the host cell nucleus, including interferon-inducible protein 16 (IFI16) (Kerur et al., 2011; Unterholzner et al., 2010) and interferon-inducible protein X (IFIX) (Diner et al., 2015a). Notably, even the cytosolic DNA sensor cGAS has been detected in the nucleus (Liu et al., 2018). Like cytosolic sensors, IFN-mediated signaling of many nuclear DNA sensors is also STING-dependent (Unterholzner et al., 2010; Wang et al., 2019), although STING-independent mechanisms have also been proposed (Diner et al., 2016). As an additional layer of versatility, some DNA sensors may act synergically and thus may often compensate for each other. For example, the well-known ribonucleoprotein, hnRNPA2B1, moonlights as a nuclear DNA sensor activating

Table 1 Identified PTMs on DNA sensors.

DNA sensor PTM: amino acid(s)	Sensing location/pathway PTM characterization	References
AIM2	**Cytoplasm/ Inflammasome**	Burckstummer et al. (2009) and Hornung et al. (2009)
–	–	
cGAS	**Cytoplasm/STING**	Sun et al. (2013) and Wu et al. (2013)
Acetylation: K7[a], K21[c], K47[c], K50[abc], K62[c], K63[c], K82[c], K83[c], K198[b], K285[b], K292[b], K355[b], K384[ab], K392[a], K394[a], K414[a], K432[b], and K479[b]	Ab-based detection	Dai et al. (2019)[a]
Deamidation: N210[d], N389[d], Q451[d] and Q454[d]	Mutagenesis	Song et al. (2020a)[b]
Glutamylation: E272[e], E302[e]	In-vitro assay	Song et al. (2020b)[c]
Phosphorylation: S13[l], S37[bl], S64[l], T69[l], T91[l], S116[bl], S129[l], S143[lm], S201[b], Y215[g], S221[b], S263[b], and S305[bfl]	IP-MS identification	Zhang et al. (2018)[d]
SUMOylation: K217[h], K464[h]	Targeted MS (PRM)	Xia et al. (2016)[e]
Ubiquitination[k]: K271[i], K335[j], K414[i]		Seo et al. (2015)[f] and Liu et al. (2018)[g]
		Hu et al. (2016)[h]
		Chen et al. (2016a)[i], Seo et al. (2018)[j] and Chen and Chen (2019)[k]
		Li et al. (2021)[l] and Olsen et al. (2010)[m]
DAI	**Cytoplasm/STING**	Takaoka et al. (2007)
–	–	
DDX41	**Cytoplasm/STING**	Zhang et al. (2011b)
Ubiquitination[a]	Ab-based detection	Zhang et al. (2013a)[a]
Phosphorylation: S4[b], S21[b], and S23[b]	Mutagenesis	Dephoure et al. (2008)[b], Mayya et al. (2009)[b], Olsen et al. (2010)[b] and Zhou et al. (2013a)[b]
SUMOylation: K416[c] and K442[c]	In-vitro assay	Hendriks et al. (2017)[c]

Continued

Table 1 Identified PTMs on DNA sensors.—cont'd

DNA sensor PTM: amino acid(s)	Sensing location/pathway PTM characterization	References
DHX9 **Methylation**[a]: K146[c] and R1175[c] **Phosphorylation**[b]: S87[d], S125[d], S321[d], S449[d], and S506[d] Acetylation: K191[e], K199[e], and K1024[e] SUMOylation: K697[f]	**Cytoplasm/MyD88** Ab-based detection Mutagenesis In-vitro assay	Kim et al. (2010) Smith et al. (2004)[a] and Sadler et al. (2009)[b] Guo et al. (2014)[c] Bian et al. (2014)[d], Dephoure et al. (2008)[d], Olsen et al. (2010)[d] and Zhou et al. (2013a)[d] Choudhary et al. (2009)[e] Hendriks et al. (2017)[f]
DHX36 Acetylation: K947[a] Phosphorylation: S161[b] and S963[b]	**Cytoplasm/MyD88** –	Kim et al. (2010) Choudhary et al. (2009)[a] Zhou et al. (2013a)[b]
DNA-PK: **PRKDC** **Phosphorylation: S2056**[c]**, T2609**[abc]**, S2612**[bf]**, T2638**[bf]**, T2647**[bf], S511[g], S687[g], S841[g], S1065[g], T2535[g], S2789[g], S3205[g], S3731[g], S3821[g], and S4026[g] Acetylation: K117[h], K828[h], K1209[h], K1970[h], K2259[h], K3241[h], K3260[h], K3621[h], K3638[h], and K3642[h] **Ku70** **ADP-ribosylation**[e] **Phosphorylation: S6**[a]**, S51**[d], S2[g], S27[g], S306[g], T455[g], S477[g], S520[g], S550[g], S560[g] SUMOylation: K287[i], K317[i], and K556[i] **Ku80** **ADP-ribosylation**[e]	**Cytoplasm/HSPA8** Ab-based detection Mutagenesis In-vitro assay MS identification	Burleigh et al. (2020), Ferguson et al. (2012) and Zhang et al. (2011a) Chan et al. (2002)[a] Douglas et al. (2002)[b] Wechsler et al. (2004)[c] Jin and Weaver (1997)[d] Beck et al. (2014)[e] Ismail et al. (2015)[f] Olsen et al. (2010)[g], Rigbolt et al. (2011)[g] and Zhou et al. (2013a)[g] Choudhary et al. (2009)[h] Hendriks et al. (2014)[i] and Hendriks et al. (2017)[i]

Phosphorylation: S577[a]**, S579**[a]**, S580**[a]**, T715**[a], S255[g], S258[g], S318[g], and T535[g] **Ubiquitination**[f] Acetylation: K144[h], K265[h], K332[h], K660[h], and K665[h] SUMOylation: K195[i], K532[i], K534[i], K566[i], K568[i], K669[i], and K688[i]		
hnRNPA2B1 **Methylation/Demethylation: Arg226**[a] Acetylation: K168[b] and K173[b] Methylation: R203[c], R213[c], R228[c], R238[c], R266[c], R325[c], and R350[c] Phosphorylation: T4[d], S29[d], S85[d], T140[d], T159[d], T176[d], S189[d], S201[d], S212[d], S225[d], S231[d], S236[d], S259[d], S324[d], Y331[d], S341[d], S344[d], and Y347[d] SUMOylation: K22[e], K104[e], K112[e], K120[e], K137[e], K152[e], K168[e] and K173[e]	**Nucleus/STING** Ab-based detection Mutagenesis	Wang et al. (2019) Wang et al. (2019)[a] Choudhary et al. (2009)[b] Guo et al. (2014)[c] Bian et al. (2014)[d], Mayya et al. (2009)[d], Olsen et al. (2010)[d], Rigbolt et al. (2011)[d] and Zhou et al. (2013a)[d] Hendriks et al. (2017)[e]
IFI16 **Acetylation**: K45[a], **K99**[a], **K128**[a], K214[a], K444[a], K451[a], K561[a], K598[a], and K614[a] **Phosphorylation**[b]**: S95**[a]**, S106**[a]**, S153**[a], S168[a], S174[a], S575[a], S780[a] **Ubiquitination**[c] SUMOylation: K116[d], K128[d], K561[d], and K683[d]	**Nucleus/STING** Ab-based detection Mutagenesis In-vitro assay IP-MS identification	Kerur et al. (2011) and Unterholzner et al. (2010) Li et al. (2012)[a] Dell'Oste et al. (2014)[b] Li et al. (2019)[c] Hendriks et al. (2014)[d], Hendriks et al. (2017)[d] and Hendriks et al. (2015)[d]
IFIX –	**Nucleus/STING** –	Diner et al. (2015a)

Continued

Table 1 Identified PTMs on DNA sensors.—cont'd

DNA sensor PTM: amino acid(s)	Sensing location/pathway PTM characterization	References
LRRFIP1	**Cytoplasm/β-catenin**	Yang et al. (2010)
Phosphorylation: S16[a], S115[a], S116[a], S120[a], S555[a], S581[a], S618[a], T676[a], S714[a], S733[a], S766[a], S768[a]	–	Bian et al. (2014)[a], Dephoure et al. (2008)[a], Mayya et al. (2009)[a], Olsen et al. (2010)[a] and Zhou et al. (2013a)[a]
SUMOylation: K305[b], K606[b]		Hendriks et al. (2017)[b] and Impens et al. (2014)[b]
MRE11	**Cytoplasm/STING**	Kondo et al. (2013)
Methylation in gly-arg-rich region[a]	Ab-based detection	Zhuang et al. (2009)[a]
Phosphorylation: S275[b], S619[b], S649[b], S678[b], S688[b], S689[b], and S701[b]	Mutagenesis In-vitro assay IP-MS validation	Bian et al. (2014)[b], Dephoure et al. (2008)[b], Mayya et al. (2009)[b], Olsen et al. (2010)[b] and Zhou et al. (2013a)[b]
SUMOylation: K255[c], K416[c], and K625[c]		Hendriks et al. (2017)[c]
RNA Pol III	**Cytoplasm/RIG-I**	Ablasser et al. (2009) and Chiu et al. (2009)
Acetylation: K445	–	Choudhary et al. (2009)[a]
TLR9	**Endosome/MyD88**	Krieg et al. (1995) and Takeshita et al. (2001)
Ubiquitination	Ab-based detection	Chuang and Ulevitch (2004)

Listed in the table are the PTM types and sites, PTM characterization approaches, DNA sensing location and pathway involved, and the papers that reported the DNA sensor identification (first line) or the identification of the PTMs. The functionally characterized sites are labeled in bold. Superscript letters denote the papers that reported the identification of specific PTM sites.

cGAS-dependent and -independent pathways to stimulate type I interferon signaling (Wang et al., 2019). During HSV-1 infection, hnRNPA2B1 binds viral DNA in the nucleus, dimerizes and translocates to the cytoplasm to activate the STING-TBK1-IRF3 axis and thus initiate antiviral signaling. hnRNPA2B1 also potentiates cGAS-STING-dependent cytokine expression by facilitating m^6A modification and nucleocytoplasmic trafficking of CGAS, IFI16, and STING mRNAs (Wang et al., 2019). Overall, these findings (and those from many other laboratories) highlight the importance of both cytosolic and nuclear DNA sensors and demonstrate the convergence many sensors have on the TBK1-IRF3 signaling axis. Additional studies are required to help define how these DNA sensors are regulated to coordinate a unified immune response.

The accelerated study of DNA sensors in recent years has highlighted their important roles in microbial infections and autoimmune diseases (Gao et al., 2015; Zhou et al., 2020). By understanding the regulation of these cellular sensors, there is the potential to develop more effective treatments for these conditions. However, despite a rapid expansion in our knowledge of DNA sensing mechanisms, several key questions remain unanswered. For example, it remains to be fully elucidated how host cells selectively recognize the DNA genome of the nuclear-replicating DNA viruses in the context of the host genome, and how cells maintain the DNA sensors in the inactive state during cellular homeostasis. These open questions emphasize the need to continue investigating the molecular mechanisms of DNA sensor regulation.

Given the immune activating nature of DNA sensing signal transduction, the ability of cells to temporally regulate DNA sensor activation is critical. From a fundamental perspective, many aspects of DNA sensor regulation invoke basic biochemical principles. For example, signal termination can occur via degradation of the immunostimulatory DNA, the second messenger, or the DNA sensor protein itself (Paludan, 2015). These mechanisms often involve protein-protein interactions, which may exert either negative or positive regulation depending on the context. For example, cGAS interacts with the DNA sensor IFI16 to promote interferon signaling (Orzalli et al., 2015), while its interaction with OASL, a protein that classically functions in the RNA sensing pathway, inhibits cGAS activity (Lum et al., 2018). Viral proteins have also been found to interact with DNA sensors to suppress host immune signaling, for example, Kaposi's sarcoma-associated herpesvirus (KSHV) ORF52 (Wu et al., 2015), HSV-1 pUL37 (Zhang et al., 2018) and VP22 (Huang et al., 2018a), and HCMV pUL31 (Huang et al., 2018b) and pUL83 (Biolatti et al., 2018) interact with and inhibit cGAS.

More recently, post-translational modifications (PTMs) of DNA sensor proteins have emerged as an additional layer of regulation. The rapid identification of potential functional PTMs has been aided by the development of high-resolution tandem mass spectrometry (MS), which confidently identifies the specific modified residue(s). Like most proteins, DNA sensors also have been found to be decorated by a diverse range of PTMs, including phosphorylation, acetylation, methylation, glutamylation, deamidation, ADP-ribosylation, SUMOylation, and ubiquitination. While only a fraction of the identified PTMs have been functionally assessed, for the sites that have been characterized, they provide key toggles of sensor functions. In this chapter, we will focus on painting the current PTM landscape of DNA sensors. Specifically, we will highlight how PTM identification using MS-based proteomics has accelerated the discovery of regulatory hubs of DNA sensor bioactivity. We will provide key examples of PTMs that impact the activity of DNA sensors. This regulation can occur directly through enzymatic activation/inhibition or protein degradation, as well as indirectly by altering the subcellular localization of DNA sensors. We will also highlight specific PTMs that are pathogen- or disease-associated. Finally, we will discuss future implications of PTM characterization studies in contributing to the development of selective therapeutics under pathogenic or autoimmune states.

2. Mass spectrometry-based proteomic characterization of PTMs

Without the aid of mass spectrometry, PTM research is largely based on prior knowledge or predictive analyses, and frequently begins with identification of an interaction between a protein of interest (POI) and a specific PTM regulatory enzyme (kinase, phosphatase, acetyltransferase, etc.) (Fig. 2, *bottom*). For example, cGAS was discovered to interact with tubulin polyglutamylase TTLL4 and TTLL6, two enzymes involved in protein glutamylation (Xia et al., 2016). Subsequent biochemical and molecular experiments confirmed that these enzymes were responsible for glutamylation of cGAS at two sites. Practically, PTM identification by these hypothesis-driven approaches often employs in vitro assays with purified enzymes and recombinant proteins or synthetic peptides from the POI. This approach has been invaluable for demonstrating direct modification of DNA sensors (Chan et al., 2002; Sadler et al., 2009; Zhang et al., 2018). Historically, PTM detection and visualization has employed antibody-based approaches with (2D)-SDS-PAGE and western blotting techniques. The antibodies may

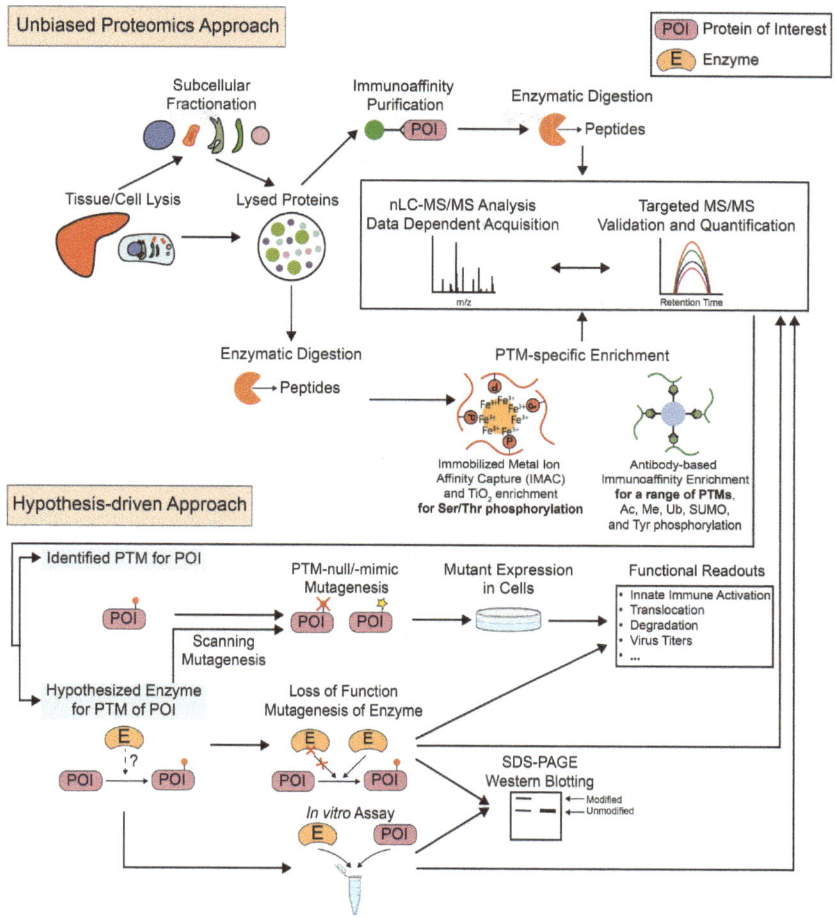

Fig. 2 Representative biochemical and proteomic methods for identifying PTMs and modifying enzymes for DNA sensors. (*Top*) In unbiased proteomics approaches, following tissue/cell lysis and optional subcellular fractionation, immunoaffinity purification can be performed using antibodies that target the DNA sensor of interest. The isolated proteins will be subjected to enzymatic digestion and nano liquid chromatography-tandem mass spectrometry (nLC-MS/MS) to identify modified peptides on the DNA sensor itself, as well as any interacting protein which may serve as a potential modifying enzyme. Alternatively, one can carry out PTM-specific enrichment after enzymatic digestion of the lysed proteins. This allows for systematic detection of specific modified peptides in the proteome. Ac, acetylation; Me, methylation; SUMO, SUMOylation; Ub, ubiquitination. (*Bottom*) In contrast, hypothesis-driven approaches usually begin with prior knowledge of candidate PTM sites or modifying enzymes, which can be followed up by in-depth characterization using mutagenesis, in vitro enzymatic assays, and in vivo cellular assays. The impact of PTM-null or loss of function enzymes in these assays can be monitored by SDS-PAGE/western blot, functional readouts, or mass spectrometry.

be PTM-specific for direct detection, or raised against the POI, which can indirectly suggest the presence of modifications through molecular weight and/or charge shifts (Fig. 2, *bottom*). These biochemical approaches are often followed with site-directed mutagenesis for downstream functional characterization assays (Fig. 1, *bottom*). One example is the IFI16 acetylation at K99 within its nuclear localization signal, which mutagenesis studies indicated plays a role in the subcellular localization of this DNA sensor (Li et al., 2012).

More recently, with the rapid development of analytical and proteomic technologies, researchers can detect and analyze PTMs from a wide range of tissues and cells with increased specificity and sensitivity (Fig. 1, *top*) (Gillen et al., 2020; Greco et al., 2014). Nonetheless, as PTMs are often present at low stoichiometry, adequate depth of analysis is often achieved with biochemical fractionation or protein and peptide (immuno)affinity enrichment techniques (Fig. 2, *top*). For example, if the goal is to identify different types of PTMs present on a POI, the enrichment of the protein by immunoaffinity purification coupled with mass spectrometry (IP-MS) is commonly performed. IP-MS approaches benefit from high sequence coverage of the POI, and therefore, have the potential to identify the greatest number of unique PTM sites. For example, IP-MS analyses of IFI16 and cGAS have contributed the majority of the known phosphorylation and acetylation sites, accelerating the identification of PTM sites that regulate innate immune signaling (Dai et al., 2019; Li et al., 2012; Song et al., 2020a,b). In addition to studying multiple PTMs of one protein of interest, mature proteomic technologies are available to broadly detect and quantify certain types of PTMs on different proteins across diverse cellular and tissue proteome landscapes (Ke et al., 2016). Enrichment strategies have been developed for different PTM types, such as the enrichment of phosphorylated peptides by affinity-based approaches using metal oxides, e.g., titanium dioxide (TiO_2) (Larsen et al., 2005; Pinkse et al., 2004) or immobilized metal ions (IMAC), e.g., Fe^{3+}, Ti^{4+}, or Zn^{4+} (Beausoleil et al., 2004; Feng et al., 2007; Nuhse et al., 2003; Zhou et al., 2013b) (Fig. 2, *top*). These core enrichment techniques have provided the foundation on which phosphorylation signaling networks have been studied and cataloged at the proteome scale (Hornbeck et al., 2015; Olsen et al., 2006; White and Wolf-Yadlin, 2016; Yu et al., 2019). As PTMs are exquisitely regulated in cellular space and time, proteome scale characterizations of these PTM dynamics present unique challenges. Yet, recent work shows promise in addressing some of these issues (Lun and Bodenmiller, 2020; Martinez-Val et al., 2021). For example, recently a high-throughput approach was developed to decipher spatio-temporal phospho-proteome regulation using sequential cell fractionation and titanium IMAC (Bekker-Jensen et al., 2020), coupled with rapid LC-MS/MS and data independent acquisition (Martinez-Val et al., 2021).

Although TiO_2- and IMAC-based methods have been shown to provide effective enrichment of peptides containing Ser and Thr phosphorylation sites, they are known to be less suitable for enriching for Tyr phosphorylated peptides. Therefore, immunoaffinity purifications using anti-phosphotyrosine antibodies are often used as a complementary enrichment method (Rush et al., 2005). Antibody-based immunoaffinity approaches have been proven valuable for investigating a range of different PTMs, including acetylation, methylation, ubiquitination, and SUMOylation (Choudhary et al., 2009; Guo et al., 2014; Hendriks et al., 2018; Kim et al., 2011; Murray et al., 2018; Zhang et al., 2009). The reliance on antibodies comes with their own inherent drawbacks, including specificity and affinity issues. For example, motif bias has been observed with anti-acetyllysine antibodies, motivating the use of a cocktail of monoclonal antibodies to enrich acetylated lysines under different sequence contexts (Shaw et al., 2011; Svinkina et al., 2015). Specific antibodies against mono- and dimethyl arginine, and mono-, di-, and trimethyl lysine have been developed for methylation enrichment (Guo et al., 2014; Rothbart et al., 2015), while ubiquitination can be enriched by antibodies against the diglycyl lysine (K-GG) ubiquitin (Ub) remnant after trypsinization (Bustos et al., 2012). In addition, a linear-linkage-specific antibody can be applied for IP-MS analysis of polyubiquitinations (Matsumoto et al., 2012). Overall, both affinity and immunoaffinity-based approaches can introduce biases for specific motifs. Therefore, a combination of chemistries (Herring et al., 2015; Yue et al., 2015) and antibodies (Carabetta et al., 2016; Svinkina et al., 2015; Zhang et al., 2009) provide complementary methods to eliminate or reduce these biases.

In addition to identification of PTM site-localization, mass spectrometry enables quantitative analysis of PTMs. The advent of accurate mass and high-resolution mass spectrometers coupled to nano liquid chromatography separations (nLC) have facilitated global quantification of intact modified peptides by label-free methods using MS^1 peak intensity or area under the curve metrics (Montoya et al., 2011). Additionally, targeted mass spectrometry approaches, such as parallel reaction monitoring (PRM) (Peterson et al., 2012), have further increased the sensitivity and specificity of detection (Manes and Nita-Lazar, 2018). These targeted approaches are designed to use signature features (nLC retention time, peptide mass) and fragmentation information to uniquely detect and quantify a peptide of interest. In contrast to global quantification, targeted-MS/MS quantification is performed at the MS/MS level, i.e., using fragment ions. These targeted approaches can be designed for the detection of hundreds of proteins, comprised of either unmodified or modified peptides. Once the signature parameters are determined, the assays are transferable between different laboratories. This method improves quantification sensitivity and reproducibility, which are particularly welcome benefits when analyzing low abundance PTMs that are modulated under different biological conditions (Song et al., 2020a). In addition, targeted-MS/MS is uniquely suited to resolve ambiguities in the sites of modification when multiple potential residues are present in the modified peptide sequence. Overall, this method provides an effective mean for detecting PTMs for which site-specific antibodies are not available and for accurately quantifying changes in a certain modification during a biological process.

3. DNA sensors are regulated by a combination of PTMs

One way that PTMs affect protein activity is by causing conformational changes to the protein structure. For DNA sensor proteins, these conformation changes may impact their DNA binding ability, enzymatic activity, and/or protein-protein interactions. PTMs may also influence the subcellular localization or turnover of DNA sensor proteins. This section highlights three prominent mechanisms by which PTMs regulate DNA sensors: enzymatic activity (Section 3.1), subcellular localization (Section 3.2), and half-life (Section 3.3). The subset of PTMs discussed below and that provide these functional molecular toggles are enumerated in Table 1 and graphically summarized in Fig. 3.

3.1 PTM-mediated activation or inhibition of DNA sensor activity

3.1.1 cGAS

Perhaps not surprisingly, the most well-studied cytosolic DNA sensor, cGAS, also has the distinction of possessing the most extensive compilation and functional characterization of PTMs, with modifications on at least 42 different amino acid residues. Interestingly, opposing regulation of cGAS functions is achieved not only through its variety of PTMs, but also for the same type of PTM, when present at different sites. The first functionally

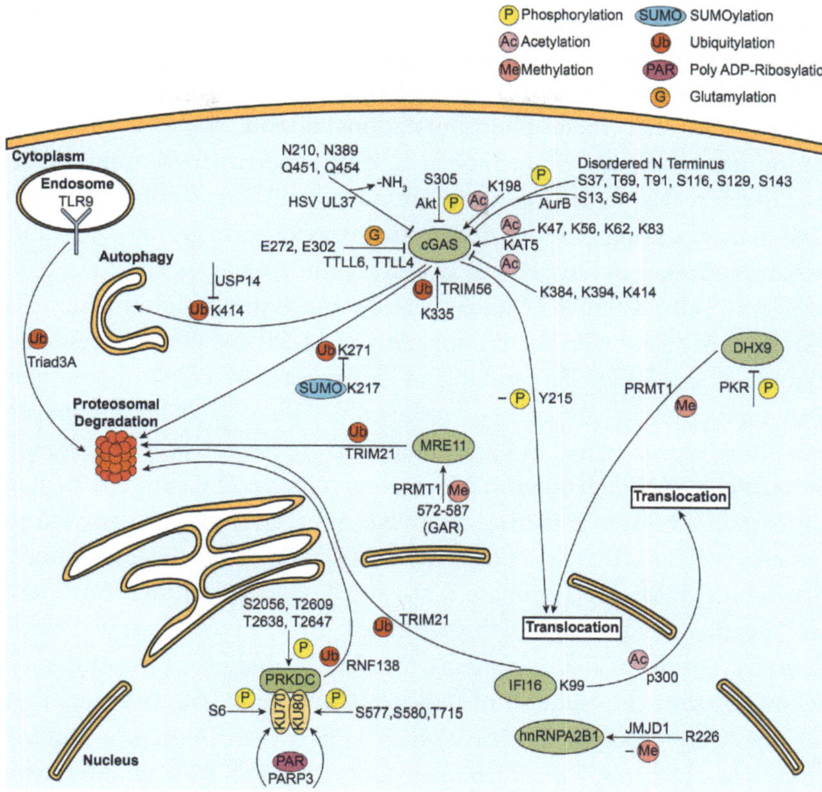

Fig. 3 Characterized PTMs on DNA sensors and their reported functions. The corresponding modifying enzymes (if any) are presented beside each arrow. Illustrated are PTMs reported to impact DNA sensor functions by leading to activation/inhibition of sensor-induced immune signaling, intracellular shuttling of the sensor, or proteasome- or autophagy-mediated sensor degradation.

characterized PTM on cGAS was phosphorylation at Ser305 (Seo et al., 2015). During vesicular stomatitis virus (VSV) infection in L929 cells, Akt kinase mediated Ser305 phosphorylation was shown to down-regulate cGAS-dependent immune signaling (Seo et al., 2015). The inhibition of enzymatic activity is consistent with the location of Ser305 within the catalytic core of cGAS. In addition to Ser305, multiple phosphorylation sites have been identified (Ser37, Ser116, Ser201, Ser221, Ser263), however, mutational analyses at these residues indicated that none of these phosphorylations significantly impact cGAS-mediated immune signaling or apoptosis induction (Song et al., 2020a). Most recently, cGAS has been demonstrated to be hyperphosphorylated at its N-terminus during mitosis, with Ser13,

Ser37, Ser64, Thr69, Thr91, Ser116, Ser129, and Ser143 found as the phosphorylated residues (Li et al., 2021). Aurora B was identified as the kinase that phosphorylates the cGAS N-terminus, and mutagenesis of these residues showed that these phosphorylations inhibit cGAS activity. The N-terminus was further characterized as being necessary for sensing chromatin. Therefore, the hyperphosphorylation likely inhibits chromatin-binding cGAS activity through the negatively charged phosphate groups, providing a regulatory mechanism to prevent an autoimmune response (Li et al., 2021).

cGAS is also a target of glutamylation and lysine ubiquitination. The interaction of cGAS with the E3 ubiquitin ligase TRIM56, promotes monoubiquitination of Lys335, resulting in an increase in cGAS dimerization, DNA-binding ability, and cGAMP production (Seo et al., 2018). E3 ubiquitin ligase TRAF6 promotes cGAS polyubiquitination, and this polyubiquitination upregulates cGAS-dependent IFN signaling (Chen and Chen, 2019). cGAS is also modified at two glutamate residues by glutamylation. Glutamylation is a reversible PTM that has been well-characterized in the regulation of microtubule dynamics (Garnham et al., 2015). Glutamylation forms a covalent, "peptide bond" between the γ-carboxy group of the target protein and the amino group of free glutamate. Similar to ubiquitination, this modification supports the addition of multiple PTM. In cGAS, two glutamylation sites were reported, both of which inhibit cGAS function, but with different mechanisms. The polyglutamylation of Glu272 by tubulin tyrosine ligase-like protein 6 (TTLL6) weakens its DNA binding ability, while monoglutamylation at Glu302 by TTLL4 dampens its synthase activity (Xia et al., 2016). Additionally, deamidation of Asn and Gln residues occurs on cGAS during HSV-1 infection. Specifically, mouse cGAS is deamidated by HSV-1 pUL37 both in cells and in vitro at Asn196, Asn377, Gln436 and Gln439 (corresponding to Asn210, Asn389, Gln451 and Gln454 of human cGAS). While single deamidations do not abolish cGAS function, three or four co-occurring deamidations led to a reduction in cGAS-dependent cytokine expression due to impaired enzymatic activity (Zhang et al., 2018).

Recently, many cGAS acetylation sites (Lys7, Lys21, Lys47, Lys50, Lys62, Lys63, Lys82, Lys83, Lys198, Lys285, Lys292, Lys355, Lys384, Lys392, Lys394, Lys414, Lys432, and Lys479) have been discovered using mass spectrometry (MS) (Dai et al., 2019; Song et al., 2020a,b). Among these sites, acetylation of Lys47, Lys56, Lys62, Lys83, and Lys198 appear to enhance cGAS immune signaling function (Dai et al., 2019; Song et al., 2020a,b). In contrast, Lys384, Lys394, and Lys414 acetylations impair cGAS enzymatic activity by inhibiting cGAMP production, while cGAS-DNA binding remains largely preserved and its dimerization is unaffected (Dai et al., 2019). In addition to inhibiting cytokine induction, Lys384 and Lys414 acetylations inhibit cGAS-dependent apoptosis (Song et al., 2020a). Interestingly, the regulation of Lys384 and Lys414 acetylation status is mediated in part through opposing acetylation and deacetylation by aspirin and histone deacetylase 3 (HDAC3), respectively (Dai et al., 2019). While the roles of lysine acetylation during active pathogen infections remains to be fully elucidated, several recent studies point to their regulatory roles. For instance, acetylation of Lys198 has been shown to be down regulated during herpesvirus infections with either HSV-1 or HCMV. As Lys198 acetylation promotes cGAS activity, the reduced PTM level during infection might reflect a viral mechanism to inhibit cGAS-dependent signaling (Song et al., 2020a). Most recently, Lys47, Lys56, Lys62, and Lys83 were shown to be acetylated by KAT5, increasing the DNA-binding ability of cGAS (Song et al., 2020b). Upon HSV-1 infection, mice with inactivated KAT5 had lower cytokine responses and survival rates, indicating these cGAS acetylations positively regulate cGAS function in vivo (Song et al., 2020b).

3.1.2 DNA-PK

The DNA-dependent protein kinase (DNA-PK) is a serine/threonine protein kinase consisting of a catalytic subunit (DNA-PKcs, also known as PRKDC), as well as a Ku heterodimer composed of Ku70 (XRCC6) and Ku80 (XRCC5) subunits (Chan et al., 1999; Jin and Weaver, 1997). While DNA-PK has been generally thought of as playing a critical role in the DNA damage response (DDR) pathways, more recent evidence suggests that it can also function as a DNA sensor during virus infection. Interestingly, it can activate IRF3-dependent innate immunity independently of its DNA-PKcs kinase activity, suggesting that its function in DNA sensing and DDR may be decoupled (Ferguson et al., 2012). DNA-PK has also been found to activate a STING-independent DNA sensing pathway (Burleigh et al., 2020). Although the exact molecular mechanisms underlying DNA-PK's DNA sensing abilities remain to be elucidated, it is evident that this protein contributes to the regulation of innate immunity in response to foreign DNA.

The catalytic subunit of DNA-PK, PRKDC, is regulated by multiple phosphorylations. PRKDC is autophosphorylated on Ser2056, Ser2612, Thr2609, Thr2638 and Thr2647. Autophosphorylation induces a conformational change that leads to remodeling of the DNA-PK complex, a requisite for efficient end processing during DNA repair (Chan et al., 2002; Douglas et al., 2002). Protein phosphatase 5 catalyzes the dephosphorylation of PRKDC. The phosphorylation status of these residues is tightly regulated as hypo- or hyperphosphorylation of these sites leads to increased sensitivity to DNA damage induction (Wechsler et al., 2004).

The Ku70 and Ku80 subunits of DNA-PK are also regulated by PTMs. Both Ku70 and Ku80 are physiological targets of PRKDC. Researchers found Ser51 in Ku70 could be phosphorylated by PRKDC in vitro, however, this phosphorylation did not affect DNA repair (Jin and Weaver, 1997). Subsequently, it was found that Ku70 is phosphorylated at Ser6, and Ku80 is phosphorylated at Ser577, Ser579, Ser580 and Thr715, and phosphorylation can enhance the helicase activity of their respective Ku subunits (Chan et al., 1999). In addition, Ku70/Ku80 can be ADP-ribosylated by Poly(ADP-ribose) polymerase 3 (PARP3). This is functionally relevant as the depletion of PARP3 inhibits recruitment of Ku80 to sites of DNA damage and reduces double strand break repair (Beck et al., 2014). Finally, Ku80 can be ubiquitinated by RNF138, leading to the removal of Ku complex from the DNA breaks (Ismail et al., 2015).

3.1.3 DHX9

Nuclear DNA helicase II (NDH II), also known as ATP-dependent RNA helicase A (RHA), is encoded by the DHX9 gene and functions as a sensor for both viral DNA and RNA (Kim et al., 2010). Although we know DHX9 can be regulated by phosphorylations, the impact of the modifications on its DNA or RNA sensing ability is not clear. DHX9 is phosphorylated by PRKDC, the catalytic subunit of DNA-PK, and this phosphorylation can be stimulated by both viral DNA and RNA, consistent with its functions in DNA and RNA metabolism (Yang et al., 2012). In addition, DHX9 is phosphorylated by protein kinase R (PKR), which has an overall inhibitory effect on DHX9 function (Sadler et al., 2009). DHX9 is known to regulate expression of the HIV-1 transactivation response via direct binding with viral RNA. Of note, the regions of DHX9 that bind viral RNA are also targets of PKR phosphorylation, and thus present a plausible therapeutic target (Sadler et al., 2009).

3.1.4 MRE11

MRE11 has been shown to be recruited to exogenous dsDNA in the cytosol, and can induce STING-dependent Type I interferon expression (Kondo et al., 2013). In addition to its DNA sensing function, MRE11 is a DNA double-strand break repair protein, containing a DNA binding domain and a phosphodiester domain, hence functioning also as an ssDNA endonuclease and dsDNA exonuclease. The exonuclease function is involved in deletional nonhomologous end joining (D-NHEJ), contributing to DNA damage repair (Boisvert et al., 2005; Zhuang et al., 2009). The exonuclease function of MRE11 is positively regulated by arginine methylation in its glycine-arginine-rich (GAR) region, between amino acids 572 and 587 (Boisvert et al., 2005; Zhuang et al., 2009). Although methylation is required for MRE11 functions during DNA damage, it remains unknown whether the DNA sensor activity is also affected by methylation (Zhuang et al., 2009).

3.1.5 *hnRNPA2B1*

Prior to the characterization of hnRNPA2B1 as a DNA sensor, methylation was identified within its arginine-glycine-glycine–rich (RGG) domain (Gary and Clarke, 1998). Recent work characterizing hnRNPA2B1 functions during HSV-1 infection found that it senses nuclear DNA (Wang et al., 2019). Specifically, nuclear hnRNPA2B1 binds viral DNA, dimerizes, then becomes demethylated within the RGG domain at Arg226, upon which the homodimer translocates to the cytoplasm to initiate downstream type I interferon signaling (Wang et al., 2019). This dimerization-dependent demethylation is mediated by the arginine demethylase JMJD6, as an JMJD6 inhibitor prevented demethylation and also inhibited downstream signaling. Given that the JMJD6 inhibitor did not affect hnRNPA2B1 translocation, it is thought that demethylation at Arg226 is selectively required for hnRNPA2B1 activation (Wang et al., 2019).

3.2 PTM-controlled translocation of DNA sensors

In addition to the direct regulation of the activity of DNA sensors, PTMs can also indirectly regulate their functions by controlling their subcellular localization. This allows for a spatially restricted activation of DNA sensors. For example, IFI16, commonly regarded as a nuclear DNA sensor, has been shown to be acetylated at K99 within one of its nuclear localization sequence motifs by the acetyltransferase p300, leading to its localization to the cytoplasm (Li et al., 2012). This allows for its DNA sensing function to occur in the cytoplasm, as well as for the activation of inflammasome, leading to the expression of antiviral cytokines (Ansari et al., 2015; Li et al., 2012). In addition, the DEAH-box family protein DHX9, which was found to recognize cytosolic microbial DNA in plasmacytoid dendritic cells (pDCs) (Kim et al., 2010), can interact with and be methylated by PRMT1, which mediates its nuclear import (Smith et al., 2004). Although direct evidence is lacking, it is possible that PRMT1-mediated DHX9 methylation negatively regulates the nucleic acid sensing of DHX9 to prevent overactivation. This is supported by a previous study that reported a higher proinflammatory response in myeloid-specific PRMT1 knock-out mice (Tikhanovich et al., 2017).

Changes in DNA sensor subcellular localization can also be regulated by PTMs outside the context of pathogen invasion. cGAS is phosphorylated at Y215 by B-lymphoid tyrosine kinase (BLK), facilitating its cytoplasmic retention. Upon dephosphorylation, cGAS was shown to shuttle into the nucleus, where it inhibits PARP1-mediated homologous recombination pathways upon DNA damage (Liu et al., 2018). These initial reports of localization-dependent activities of DNA sensors provide a promising avenue for future mechanistic investigations of PTMs with currently unknown functions.

3.3 PTM-modulated protein degradation of DNA sensors

It is critical for cells to either maintain or return to a state of immune homeostasis in the absence of or after being challenged by pathogens, respectively, in order to prevent overactivation of immune responses that are linked to different human diseases. One of the major negative feedback mechanisms is the degradation of the DNA sensors that initiated innate immune signaling. Frequently, this is achieved through ubiquitination of the proteins to facilitate degradation through the proteasome. Both IFI16 and DDX41 have been shown to be ubiquitinated by the E3 ubiquitin ligase TRIM21 for subsequent degradation (Li et al., 2019; Zhang et al., 2013a). TRIM21-mediated degradation of IFI16 is dependent on STING activation, a downstream signaling protein of IFI16. This highlights the importance of negative feedback regulation in the context of DNA sensing. Similarly, the Toll-like receptor proteins TLR4 and TLR9 were found to be ubiquitinated by Triad3A and targeted for degradation to inhibit their signaling functions (Chuang and Ulevitch, 2004).

cGAS was reported to be ubiquitinated at K271 and K414 which facilitate degradation through the proteasome and p62-mediated autophagy, respectively (Chen et al., 2016a; Hu et al., 2016). In addition, both sites have mechanisms in place to inhibit their polyubiquitination. Hu et al. reported that in uninfected cells, cGAS can be SUMOylated at K217, which resides in relative proximity to the ubiquitination site K271, thereby being proposed to serve as a steric hindrance to the E3 ligase (Hu et al., 2016). Chen et al. found that TRIM14, which can be upregulated by interferons, recruits the ubiquitin carboxyl-terminal hydrolase (USP14) to K414 to cleave the polyubiquitination chains, thereby preventing the interaction of cGAS with p62 (Chen et al., 2016a). cGAS was also found to be SUMOylated at K464 early during HSV-1 infection, but later de-SUMOylated by sentrin-specific protease 2 (SENP2) to allow for polyubiquitination on the same site, promoting its degradation (Hu et al., 2016). Given the proximity of K414 and K464, it is also possible that SUMOylation on K464 could sterically inhibit the interaction between cGAS and p62 mediated by K414 polyubiquitination. Altogether, these modifications, their crosstalk, and their functional significance highlight the intricate balance cells have to maintain between activation and inhibition of DNA sensors.

4. Virus-induced PTMs for dampening or evading the host innate immune response

The abilities of PTMs to regulate DNA sensor functions are not exclusive to the host cell. Viruses often hijack host PTM modifying enzymes or alternatively encode viral enzymes that directly modify DNA sensors to promote immune evasion and facilitate viral replication. For instance, cGAS was reported to be deamidated in its Mab21 enzymatic domain by the viral deamidase pUL37 from HSV-1 (Zhang et al., 2018). Deamidation of human cGAS at one single site (N210) was shown to be sufficient to decrease the cGAMP-synthesizing activity of cGAS and inhibit downstream interferon and ISG production. Furthermore, the N210 site was found to be under strong positive selection through analysis of the protein sequences in non-human primates. This further supports the notion that the deamidation site in cGAS is an active target for viral evasion (Zhang et al., 2018).

IFI16 is also known to be a primary target of viral proteins. Orzalli et al. reported that, during HSV-1 infection, the RING finger domain of the viral protein ICP0 promotes the degradation of IFI16 by acting as an E3 ubiquitin ligase (Orzalli et al., 2012). Although others have reported that ICP0 is neither necessary nor sufficient for IFI16 degradation (Cuchet-Lourenco et al., 2013), the HSV-1-induced degradation of IFI16 is well recognized and shown to be mediated via the targeting of the IFI16 pyrin domain (Diner et al., 2015b). The β-herpesvirus HCMV also acquired several mechanisms to suppress IFI16's DNA sensing ability, one of which is driven by regulatory PTM event. The viral kinase pUL97 phosphorylates IFI16, causing its mislocalization to the cytoplasm early in infection, while later in infection, trapping IFI16 into egressing virions, likely by promoting the interaction between IFI16 and the multivesicular body protein Vps4A (Dell'Oste et al., 2014). Although the specific site of pUL97-mediated IFI16 phosphorylation is currently unknown, many of the previously reported but as yet uncharacterized phosphorylation sites reside near the nuclear localization signal and are therefore candidates for regulating subcellular distribution during infection. Although not directly added to the DNA sensor, but rather impacting its function, PTM-driven regulation was also shown to counteract one of the mechanisms of HCMV inhibition of IFI16. HCMV was reported to use its tegument protein pUL83 to clamp the IFI16 pyrin domain, preventing its oligomerization and the ability of IFI16 to induce immune signaling. Host cell induced phosphorylation at Ser364 within the pyrin-associating domain of pUL83 was shown to restore the ability of IFI16 to oligomerize, likely by interfering with the ability of pUL83 to fully clamp the IFI16 pyrin domain (Li et al., 2013). To date, no systematic comparison of the PTM landscape during different types of virus infections has been performed. It is expected that continued global host PTM characterization, as well studies focused on DNA sensors, under various infection conditions will help point to unique or shared PTM signatures that underlie mechanisms of host defense or virus immune evasion.

5. PTMs on DNA sensors affect disease progression and treatment

Innate immunity is a double-edged sword: the activation of innate immune system protects us from pathogen infection and cellular damages; however, its over-activation drives autoimmune diseases and compromises human health. As self-DNA sensing is linked to autoimmune and autoinflammatory diseases, cells employ multiple DNases to cleave aberrantly located and accumulated DNA, and thus avoid self-immune activation (Motwani et al., 2019). Therefore, mutations on DNases could cause autoimmune diseases in a DNA-sensor dependent manner. For example, the loss of DNase II endonuclease activity would lead to excessive production of type I interferons, and further lead to neonatal anemia, membranoproliferative glomerulonephritis, liver fibrosis, and deforming arthropathy (Rodero et al., 2017). Another clinically relevant example is the loss-of-function of three-prime repair exonuclease 1 (TREX1; also known as DNase III), which can lead to systemic lupus erythematosus and Aicardi–

Goutieres Syndrome (AGS) (Crow and Manel, 2015; Stetson et al., 2008). Aicardi-Goutières syndrome (AGS) is a genetic disorder that affects the white matter in the brain and spinal cord. AGS patients with DNA 3′ repair exonuclease 1 (TREX1) loss-of-function mutations cannot degrade mis-localized cytosolic DNA (Crow and Manel, 2015). This causes elevated levels of cGAS-dependent IFN production, contributing to autoimmune signaling that destructively targets the white matter (Gao et al., 2015). A promising treatment for AGS patients was suggested to be the inhibition of cGAS using the common anti-inflammatory drug aspirin, which directly acetylates cGAS at Lys384 and Lys414, leading to reduced interferon signaling. In patient cells and an AGS mouse model, aspirin effectively represses AGS symptoms (Dai et al., 2019). This demonstrates the potential of aspirin to broadly mitigate the impact of cGAS-dependent autoimmunity on disease progression.

6. Future perspectives

The combination of hypothesis-driven molecular approaches and discovery-based global proteomics methods has facilitated the functional characterization of many PTMs on DNA sensors. Yet, owing to the systems-wide advantages of proteomic approaches to simultaneously survey the PTM landscape, novel PTMs are being identified faster than they can be functionally characterized. For example, IFI16 has been found to be SUMOylated and to also possess many phosphorylation sites that remain to be linked to specific protein functions (Impens et al., 2014; Li et al., 2012). Moreover, we have recently reported that cGAS is decorated with numerous acetylation and phosphorylation sites that could differentially impact immune signaling and apoptosis (Song et al., 2020a), but additional characterization of these sites is needed to determine whether they reflect bona fide regulatory elements during pathogen infections.

The ability of PTMs to govern a protein's localization, for example, by inducing shuttling events, is a fascinating post-translational regulatory mechanism, particularly in the context of DNA sensors. The core functionality of DNA sensors to surveil the intracellular environment requires precise spatial and temporal regulation of DNA recognition. Given the different cellular routes for the trafficking of pathogenic DNA, sensors must have the ability to respond to these diverse stimuli. Mechanistically, much remains to be understood about how PTMs govern protein localization and how they participate in host immune defense and viral immune evasion. However, these early reports of PTM-dependent regulation of sensor localization have shown the remarkable ability of a single modification to establish the DNA sensor's localization. Moreover, the fact that localization-defining PTM sites use both addition and removal mechanisms suggests a mechanistic flexibility that can leverage opposing PTM modifying enzymes, e.g., kinases and phosphatases. As DNA sensor PTMs continue to be characterized, it is likely that a greater appreciation will be realized for the ability of PTMs to act as dynamic, stimulus-dependent molecular switches to control protein translocations.

Recent work suggests that DNA sensor PTMs do not exclusively regulate protein activities in isolation, but rather through a combination of co-occurring modifications that give rise to different proteoforms. For example, modifications that occur on lysine residues (acetylation, ubiquitination, and SUMOylation) are often found to target the same residue, and thus the existence of one PTM precludes the others. A more intricate mechanism of regulation occurs when proximal modified residues either always co-occur, or are present only in isolation on the same protein molecule. Given the large number of potential residues and types of modifications, characterizing the proteoforms of a single gene product is no trivial task (Schaffer et al., 2019). Toward this goal, intact, often lower molecular weight proteins are directly analyzed by mass spectrometry. This approach is commonly referred to as top-down MS. In contrast to bottom-up proteomics, top-down analysis is able to better preserve unique PTM combinations (Schaffer et al., 2019). Depending on the proteoform complexity and molecular weight of a particular gene product, a variation on top-down MS, called middle-down MS, can be used. Middle-down approaches use a selective enzymatic proteolysis that generates larger polypeptides than traditional digestion methods (e.g., using trypsin), thereby retaining regions of the protein intact for mass spectrometry analysis (Wu et al., 2012). As these approaches continue to be refined (Cristobal et al., 2017; Cupp-Sutton and Wu, 2020), they have the potential to provide a unique perspective to proteoform-specific functions of DNA sensors.

In the interim, clues to the inter-relationships between DNA sensor PTMs are emerging. For example, a mapping study of the SUMO proteome, found a significant degree of co-modification with phosphorylation (Hendriks et al., 2017), and crosstalks were also investigated for ADP-ribosylation and phosphorylation (Daniels et al., 2020), to name just a few. So far, a limited number of examples of PTM co-regulation are known for DNA sensors. For example, cGAS ubiquitination of cGAS (K271) is inhibited by its SUMOylation (K217) (Hu et al., 2016). Thus,

SUMOylation provides a "brake" on ubiquitin-dependent proteasomal degradation of cGAS. The existence of histone "PTM code" is well known as a primary determinant of their epigenetic regulation (Onder et al., 2015). While a PTM code for DNA sensors is still being explored, it is likely that as their PTM landscapes become richer, a systems-level synthesis of PTM interdependence will emerge. Also, comparison of PTM codes between DNA sensors may provide novel mechanisms for how specific immune responses achieve specificity despite redundancies and convergent signaling pathways.

Globally, the link between disease states and the PTM profiles of individual patients is of great clinical interest (Thygesen et al., 2018). Analogous to patient-specific proteome profiling, the PTM status of patients may also serve to establish disease biomarkers or contribute to understanding disease progression, and therefore, assist more in the development of precise therapeutics. From the perspective of DNA sensors, their cooperation and convergence to stimulate interferon signaling suggests that future therapies may benefit from targeting multiple DNA sensors, and present selective therapeutic targets. This underscores the need to continue characterizing functional PTMs on DNA sensor proteins, and specifically, PTMs that are associated with overactivation of the immune response and self-DNA sensing states. Profiling the PTM landscape of DNA sensors in inflammatory states may be informative, for example, during virus-associated inflammation or in lung inflammatory diseases where the STING-mediated sensing pathway plays a prominent role (Benmerzoug et al., 2019; Heil and Brockmeyer, 2019). Ultimately, by increasing PTM coverage across DNA sensors, we will gain mechanistic insights into key post-translational regulatory events that fine-tune sensor activities across a wide range of physiological time scales.

Acknowledgments

Funding was provided by NIH NIGMS (GM114141) and by the Edward Mallinckrodt Foundation to I.M.C., and China Scholarship Council (CSC) scholarship (201506210052) to B.S.

References

Ablasser, A., Bauernfeind, F., Hartmann, G., Latz, E., Fitzgerald, K.A., Hornung, V., 2009. RIG-I-dependent sensing of poly(dA:dT) through the induction of an RNA polymerase III-transcribed RNA intermediate. Nat. Immunol. 10 (10), 1065–1072. https://doi.org/10.1038/ni.1779.

Ansari, M.A., Dutta, S., Veettil, M.V., Dutta, D., Iqbal, J., Kumar, B., et al., 2015. Herpesvirus genome recognition induced acetylation of nuclear IFI16 is essential for its cytoplasmic translocation, inflammasome and IFN-beta responses. PLoS Pathog. 11 (7), e1005019. https://doi.org/10.1371/journal.ppat.1005019.

Beausoleil, S.A., Jedrychowski, M., Schwartz, D., Elias, J.E., Villen, J., Li, J., et al., 2004. Large-scale characterization of HeLa cell nuclear phosphoproteins. Proc. Natl. Acad. Sci. U. S. A. 101 (33), 12130–12135. https://doi.org/10.1073/pnas.0404720101.

Beck, C., Boehler, C., Guirouilh Barbat, J., Bonnet, M.E., Illuzzi, G., Ronde, P., et al., 2014. PARP3 affects the relative contribution of homologous recombination and non-homologous end-joining pathways. Nucleic Acids Res. 42 (9), 5616–5632. https://doi.org/10.1093/nar/gku174.

Bekker-Jensen, D.B., Bernhardt, O.M., Hogrebe, A., Martinez-Val, A., Verbeke, L., Gandhi, T., et al., 2020. Rapid and site-specific deep phosphoproteome profiling by data-independent acquisition without the need for spectral libraries. Nat. Commun. 11 (1), 787. https://doi.org/10.1038/s41467-020-14609-1.

Benmerzoug, S., Ryffel, B., Togbe, D., Quesniaux, V.F.J., 2019. Self-DNA sensing in lung inflammatory diseases. Trends Immunol. 40 (8), 719–734. https://doi.org/10.1016/j.it.2019.06.001.

Bian, Y., Song, C., Cheng, K., Dong, M., Wang, F., Huang, J., et al., 2014. An enzyme assisted RP-RPLC approach for in-depth analysis of human liver phosphoproteome. J. Proteomics 96, 253–262. https://doi.org/10.1016/j.jprot.2013.11.014.

Biolatti, M., Dell'Oste, V., Pautasso, S., Gugliesi, F., von Einem, J., Krapp, C., et al., 2018. Human cytomegalovirus tegument protein pp65 (pUL83) dampens type I interferon production by inactivating the DNA sensor cGAS without affecting sting. J. Virol. 92 (6). https://doi.org/10.1128/JVI.01774-17, e01774-17.

Boisvert, F.M., Dery, U., Masson, J.Y., Richard, S., 2005. Arginine methylation of MRE11 by PRMT1 is required for DNA damage checkpoint control. Genes Dev. 19 (6), 671–676. https://doi.org/10.1101/gad.1279805.

Bowie, A., 2012. The STING in the tail for cytosolic DNA-dependent activation of IRF3. Sci. Signal. 5 (214), pe9. https://doi.org/10.1126/scisignal.2002919.

Burckstummer, T., Baumann, C., Bluml, S., Dixit, E., Durnberger, G., Jahn, H., et al., 2009. An orthogonal proteomic-genomic screen identifies AIM2 as a cytoplasmic DNA sensor for the inflammasome. Nat. Immunol. 10 (3), 266–272. https://doi.org/10.1038/ni.1702.

Burleigh, K., Maltbaek, J.H., Cambier, S., Green, R., Gale Jr., M., James, R.C., Stetson, D.B., 2020. Human DNA-PK activates a STING-independent DNA sensing pathway. Sci. Immunol. 5 (43), eaba4219. https://doi.org/10.1126/sciimmunol.aba4219.

Bustos, D., Bakalarski, C.E., Yang, Y., Peng, J., Kirkpatrick, D.S., 2012. Characterizing ubiquitination sites by peptide-based immunoaffinity enrichment. Mol. Cell. Proteomics 11 (12), 1529–1540. https://doi.org/10.1074/mcp.R112.019117.

Carabetta, V.J., Greco, T.M., Tanner, A.W., Cristea, I.M., Dubnau, D., 2016. Temporal regulation of the Bacillus subtilis acetylome and evidence for a role of MreB acetylation in cell wall growth. mSystems 1 (3). https://doi.org/10.1128/mSystems.00005-16, e00005-16.

Chan, D.W., Ye, R., Veillette, C.J., Lees-Miller, S.P., 1999. DNA-dependent protein kinase phosphorylation sites in Ku 70/80 heterodimer. Biochemistry 38 (6), 1819–1828. https://doi.org/10.1021/bi982584b.

Chan, D.W., Chen, B.P., Prithivirajsingh, S., Kurimasa, A., Story, M.D., Qin, J., Chen, D.J., 2002. Autophosphorylation of the DNA-dependent protein kinase catalytic subunit is required for rejoining of DNA double-strand breaks. Genes Dev. 16 (18), 2333–2338. https://doi.org/10.1101/gad.1015202.

Chen, X., Chen, Y., 2019. Ubiquitination of cGAS by TRAF6 regulates anti-DNA viral innate immune responses. Biochem. Biophys. Res. Commun. 514 (3), 659–664. https://doi.org/10.1016/j.bbrc.2019.05.022.

Chen, M., Meng, Q., Qin, Y., Liang, P., Tan, P., He, L., et al., 2016a. TRIM14 inhibits cGAS degradation mediated by selective autophagy receptor p62 to promote innate immune responses. Mol. Cell 64 (1), 105–119. https://doi.org/10.1016/j.molcel.2016.08.025.

Chen, Q., Sun, L., Chen, Z.J., 2016b. Regulation and function of the cGAS-STING pathway of cytosolic DNA sensing. Nat. Immunol. 17 (10), 1142–1149. https://doi.org/10.1038/ni.3558.

Chiu, Y.H., Macmillan, J.B., Chen, Z.J., 2009. RNA polymerase III detects cytosolic DNA and induces type I interferons through the RIG-I pathway. Cell 138 (3), 576–591. https://doi.org/10.1016/j.cell.2009.06.015.

Chockalingam, A., Brooks, J.C., Cameron, J.L., Blum, L.K., Leifer, C.A., 2009. TLR9 traffics through the Golgi complex to localize to endolysosomes and respond to CpG DNA. Immunol. Cell Biol. 87 (3), 209–217. https://doi.org/10.1038/icb.2008.101.

Choudhary, C., Kumar, C., Gnad, F., Nielsen, M.L., Rehman, M., Walther, T.C., et al., 2009. Lysine acetylation targets protein complexes and co-regulates major cellular functions. Science 325 (5942), 834–840. https://doi.org/10.1126/science.1175371.

Chuang, T.H., Ulevitch, R.J., 2004. Triad3A, an E3 ubiquitin-protein ligase regulating Toll-like receptors. Nat. Immunol. 5 (5), 495–502. https://doi.org/10.1038/ni1066.

Cristobal, A., Marino, F., Post, H., van den Toorn, H.W., Mohammed, S., Heck, A.J., 2017. Toward an optimized workflow for middle-down proteomics. Anal. Chem. 89 (6), 3318–3325. https://doi.org/10.1021/acs.analchem.6b03756.

Crow, Y.J., Manel, N., 2015. Aicardi-Goutieres syndrome and the type I interferonopathies. Nat. Rev. Immunol. 15 (7), 429–440. https://doi.org/10.1038/nri3850.

Crow, M.S., Javitt, A., Cristea, I.M., 2015. A proteomics perspective on viral DNA sensors in host defense and viral immune evasion mechanisms. J. Mol. Biol. 427 (11), 1995–2012. https://doi.org/10.1016/j.jmb.2015.02.016.

Cuchet-Lourenco, D., Anderson, G., Sloan, E., Orr, A., Everett, R.D., 2013. The viral ubiquitin ligase ICP0 is neither sufficient nor necessary for degradation of the cellular DNA sensor IFI16 during herpes simplex virus 1 infection. J. Virol. 87 (24), 13422–13432. https://doi.org/10.1128/JVI.02474-13.

Cupp-Sutton, K.A., Wu, S., 2020. High-throughput quantitative top-down proteomics. Mol. Omics 16 (2), 91–99. https://doi.org/10.1039/c9mo00154a.

Dai, J., Huang, Y.J., He, X., Zhao, M., Wang, X., Liu, Z.S., et al., 2019. Acetylation blocks cGAS activity and inhibits self-DNA-induced autoimmunity. Cell 176 (6), 1447–1460. e1414. https://doi.org/10.1016/j.cell.2019.01.016.

Daniels, C.M., Kaplan, P.R., Bishof, I., Bradfield, C., Tucholski, T., Nuccio, A.G., et al., 2020. Dynamic ADP-ribosylome, phosphoproteome, and interactome in LPS-activated macrophages. J. Proteome Res. 19 (9), 3716–3731. https://doi.org/10.1021/acs.jproteome.0c00261.

Dell'Oste, V., Gatti, D., Gugliesi, F., De Andrea, M., Bawadekar, M., Lo Cigno, I., et al., 2014. Innate nuclear sensor IFI16 translocates into the cytoplasm during the early stage of in vitro human cytomegalovirus infection and is entrapped in the egressing virions during the late stage. J. Virol. 88 (12), 6970–6982. https://doi.org/10.1128/JVI.00384-14.

Dephoure, N., Zhou, C., Villen, J., Beausoleil, S.A., Bakalarski, C.E., Elledge, S.J., Gygi, S.P., 2008. A quantitative atlas of mitotic phosphorylation. Proc. Natl. Acad. Sci. U. S. A. 105 (31), 10762–10767. https://doi.org/10.1073/pnas.0805139105.

Diner, B.A., Li, T., Greco, T.M., Crow, M.S., Fuesler, J.A., Wang, J., Cristea, I.M., 2015a. The functional interactome of PYHIN immune regulators reveals IFIX is a sensor of viral DNA. Mol. Syst. Biol. 11 (1), 787. https://doi.org/10.15252/msb.20145808.

Diner, B.A., Lum, K.K., Javitt, A., Cristea, I.M., 2015b. Interactions of the antiviral factor interferon gamma-inducible protein 16 (IFI16) mediate immune signaling and herpes simplex Virus-1 immunosuppression. Mol. Cell. Proteomics 14 (9), 2341–2356. https://doi.org/10.1074/mcp.M114.047068.

Diner, B.A., Lum, K.K., Toettcher, J.E., Cristea, I.M., 2016. Viral DNA sensors IFI16 and cyclic GMP-AMP synthase possess distinct functions in regulating viral gene expression, immune defenses, and apoptotic responses during herpesvirus infection. MBio 7 (6). https://doi.org/10.1128/mBio.01553-16, e01553-16.

Douglas, P., Sapkota, G.P., Morrice, N., Yu, Y., Goodarzi, A.A., Merkle, D., et al., 2002. Identification of in vitro and in vivo phosphorylation sites in the catalytic subunit of the DNA-dependent protein kinase. Biochem. J. 368 (Pt. 1), 243–251. https://doi.org/10.1042/BJ20020973.

Feng, S., Ye, M., Zhou, H., Jiang, X., Jiang, X., Zou, H., Gong, B., 2007. Immobilized zirconium ion affinity chromatography for specific enrichment of phosphopeptides in phosphoproteome analysis. Mol. Cell. Proteomics 6 (9), 1656–1665. https://doi.org/10.1074/mcp.T600071-MCP200.

Ferguson, B.J., Mansur, D.S., Peters, N.E., Ren, H., Smith, G.L., 2012. DNA-PK is a DNA sensor for IRF-3-dependent innate immunity. Elife 1, e00047. https://doi.org/10.7554/eLife.00047.

Gao, D., Wu, J., Wu, Y.T., Du, F., Aroh, C., Yan, N., et al., 2013. Cyclic GMP-AMP synthase is an innate immune sensor of HIV and other retroviruses. Science 341 (6148), 903–906. https://doi.org/10.1126/science.1240933.

Gao, D., Li, T., Li, X.D., Chen, X., Li, Q.Z., Wight-Carter, M., Chen, Z.J., 2015. Activation of cyclic GMP-AMP synthase by self-DNA causes autoimmune diseases. Proc. Natl. Acad. Sci. U. S. A. 112 (42), E5699–E5705. https://doi.org/10.1073/pnas.1516465112.

Garnham, C.P., Vemu, A., Wilson-Kubalek, E.M., Yu, I., Szyk, A., Lander, G.C., et al., 2015. Multivalent microtubule recognition by tubulin tyrosine ligase-like family glutamylases. Cell 161 (5), 1112–1123. https://doi.org/10.1016/j.cell.2015.04.003.

Gary, J.D., Clarke, S., 1998. RNA and protein interactions modulated by protein arginine methylation. Prog. Nucleic Acid Res. Mol. Biol. 61, 65–131. https://doi.org/10.1016/s0079-6603(08)60825-9.

Gillen, J., Bridgwater, C., Nita-Lazar, A., 2020. Approaching complexity: systems biology and ms-based techniques to address immune signaling. Expert Rev. Proteomics 17 (5), 341–354. https://doi.org/10.1080/14789450.2020.1780920.

Gordon, S., 2002. Pattern recognition receptors: doubling up for the innate immune response. Cell 111 (7), 927–930. https://doi.org/10.1016/s0092-8674(02)01201-1.

Greco, T.M., Diner, B.A., Cristea, I.M., 2014. The impact of mass spectrometry-based proteomics on fundamental discoveries in virology. Annu. Rev. Virol. 1 (1), 581–604. https://doi.org/10.1146/annurev-virology-031413-085527.

Guo, A., Gu, H., Zhou, J., Mulhern, D., Wang, Y., Lee, K.A., et al., 2014. Immunoaffinity enrichment and mass spectrometry analysis of protein methylation. Mol. Cell. Proteomics 13 (1), 372–387. https://doi.org/10.1074/mcp.O113.027870.

Hansen, K., Prabakaran, T., Laustsen, A., Jorgensen, S.E., Rahbaek, S.H., Jensen, S.B., et al., 2014. *Listeria monocytogenes* induces IFNbeta expression through an IFI16-, cGAS- and STING-dependent pathway. EMBO J. 33 (15), 1654–1666. https://doi.org/10.15252/embj.201488029.

Harding, S.M., Benci, J.L., Irianto, J., Discher, D.E., Minn, A.J., Greenberg, R.A., 2017. Mitotic progression following DNA damage enables pattern recognition within micronuclei. Nature 548 (7668), 466–470. https://doi.org/10.1038/nature23470.

Heil, M., Brockmeyer, N.H., 2019. Self-DNA sensing fuels HIV-1-associated inflammation. Trends Mol. Med. 25 (11), 941–954. https://doi.org/10.1016/j.molmed.2019.06.004.

Hendriks, I.A., D'Souza, R.C., Yang, B., Verlaan-de Vries, M., Mann, M., Vertegaal, A.C., 2014. Uncovering global SUMOylation signaling networks in a site-specific manner. Nat. Struct. Mol. Biol. 21 (10), 927–936. https://doi.org/10.1038/nsmb.2890.

Hendriks, I.A., Treffers, L.W., Verlaan-de Vries, M., Olsen, J.V., Vertegaal, A.C.O., 2015. SUMO-2 orchestrates chromatin modifiers in response to DNA damage. Cell Rep. 10 (10), 1778–1791. https://doi.org/10.1016/j.celrep.2015.02.033.

Hendriks, I.A., Lyon, D., Young, C., Jensen, L.J., Vertegaal, A.C., Nielsen, M.L., 2017. Site-specific mapping of the human SUMO proteome reveals co-modification with phosphorylation. Nat. Struct. Mol. Biol. 24 (3), 325–336. https://doi.org/10.1038/nsmb.3366.

Hendriks, I.A., Lyon, D., Su, D., Skotte, N.H., Daniel, J.A., Jensen, L.J., Nielsen, M.L., 2018. Site-specific characterization of endogenous SUMOylation across species and organs. Nat. Commun. 9 (1), 2456. https://doi.org/10.1038/s41467-018-04957-4.

Herring, L.E., Grant, K.G., Blackburn, K., Haugh, J.M., Goshe, M.B., 2015. Development of a tandem affinity phosphoproteomic method with motif selectivity and its application in analysis of signal transduction networks. J. Chromatogr. B Analyt. Technol. Biomed. Life Sci. 988, 166–174. https://doi.org/10.1016/j.jchromb.2015.02.017.

Hornbeck, P.V., Zhang, B., Murray, B., Kornhauser, J.M., Latham, V., Skrzypek, E., 2015. PhosphoSitePlus, 2014: mutations, PTMs and recalibrations. Nucleic Acids Res. 43 (Database issue), D512–D520. https://doi.org/10.1093/nar/gku1267.

Hornung, V., Ablasser, A., Charrel-Dennis, M., Bauernfeind, F., Horvath, G., Caffrey, D.R., et al., 2009. AIM2 recognizes cytosolic dsDNA and forms a caspase-1-activating inflammasome with ASC. Nature 458 (7237), 514–518. https://doi.org/10.1038/nature07725.

Hu, M.M., Yang, Q., Xie, X.Q., Liao, C.Y., Lin, H., Liu, T.T., et al., 2016. Sumoylation promotes the stability of the DNA sensor cGAS and the adaptor STING to regulate the kinetics of response to DNA virus. Immunity 45 (3), 555–569. https://doi.org/10.1016/j.immuni.2016.08.014.

Huang, J., You, H., Su, C., Li, Y., Chen, S., Zheng, C., 2018a. Herpes simplex virus 1 tegument protein VP22 abrogates cGAS/STING-mediated antiviral innate immunity. J. Virol. 92 (15), e00841-18. https://doi.org/10.1128/JVI.00841-18.

Huang, Z.F., Zou, H.M., Liao, B.W., Zhang, H.Y., Yang, Y., Fu, Y.Z., et al., 2018b. Human cytomegalovirus protein UL31 inhibits DNA sensing of cGAS to mediate immune evasion. Cell Host Microbe 24 (1), 69–80.e64. https://doi.org/10.1016/j.chom.2018.05.007.

Impens, F., Radoshevich, L., Cossart, P., Ribet, D., 2014. Mapping of SUMO sites and analysis of SUMOylation changes induced by external stimuli. Proc. Natl. Acad. Sci. U. S. A. 111 (34), 12432–12437. https://doi.org/10.1073/pnas.1413825111.

Ismail, I.H., Gagne, J.P., Genois, M.M., Strickfaden, H., McDonald, D., Xu, Z., et al., 2015. The RNF138 E3 ligase displaces Ku to promote DNA end resection and regulate DNA repair pathway choice. Nat. Cell Biol. 17 (11), 1446–1457. https://doi.org/10.1038/ncb3259.

Jin, S., Weaver, D.T., 1997. Double-strand break repair by Ku70 requires heterodimerization with Ku80 and DNA binding functions. EMBO J. 16 (22), 6874–6885. https://doi.org/10.1093/emboj/16.22.6874.

Johnson, K.E., Bottero, V., Flaherty, S., Dutta, S., Singh, V.V., Chandran, B., 2014. IFI16 restricts HSV-1 replication by accumulating on the HSV-1 genome, repressing HSV-1 gene expression, and directly or indirectly modulating histone modifications. PLoS Pathog. 10 (11), e1004503. https://doi.org/10.1371/journal.ppat.1004503.

Ke, M., Shen, H., Wang, L., Luo, S., Lin, L., Yang, J., Tian, R., 2016. Identification, quantification, and site localization of protein posttranslational modifications via mass spectrometry-based proteomics. Adv. Exp. Med. Biol. 919, 345–382. https://doi.org/10.1007/978-3-319-41448-5_17.

Kerur, N., Veettil, M.V., Sharma-Walia, N., Bottero, V., Sadagopan, S., Otageri, P., Chandran, B., 2011. IFI16 acts as a nuclear pathogen sensor to induce the inflammasome in response to Kaposi Sarcoma-associated herpesvirus infection. Cell Host Microbe 9 (5), 363–375. https://doi.org/10.1016/j.chom.2011.04.008.

Kim, T., Pazhoor, S., Bao, M., Zhang, Z., Hanabuchi, S., Facchinetti, V., et al., 2010. Aspartate-glutamate-alanine-histidine box motif (DEAH)/RNA helicase A helicases sense microbial DNA in human plasmacytoid dendritic cells. Proc. Natl. Acad. Sci. U. S. A. 107 (34), 15181–15186. https://doi.org/10.1073/pnas.1006539107.

Kim, W., Bennett, E.J., Huttlin, E.L., Guo, A., Li, J., Possemato, A., et al., 2011. Systematic and quantitative assessment of the ubiquitin-modified proteome. Mol. Cell 44 (2), 325–340. https://doi.org/10.1016/j.molcel.2011.08.025.

Kondo, T., Kobayashi, J., Saitoh, T., Maruyama, K., Ishii, K.J., Barber, G.N., et al., 2013. DNA damage sensor MRE11 recognizes cytosolic double-stranded DNA and induces type I interferon by regulating STING trafficking. Proc. Natl. Acad. Sci. U. S. A. 110 (8), 2969–2974. https://doi.org/10.1073/pnas.1222694110.

Krieg, A.M., Yi, A.K., Matson, S., Waldschmidt, T.J., Bishop, G.A., Teasdale, R., et al., 1995. CpG motifs in bacterial DNA trigger direct B-cell activation. Nature 374 (6522), 546–549. https://doi.org/10.1038/374546a0.

Larsen, M.R., Thingholm, T.E., Jensen, O.N., Roepstorff, P., Jorgensen, T.J., 2005. Highly selective enrichment of phosphorylated peptides from peptide mixtures using titanium dioxide microcolumns. Mol. Cell. Proteomics 4 (7), 873–886. https://doi.org/10.1074/mcp.T500007-MCP200.

Lei, Z., Deng, M., Yi, Z., Sun, Q., Shapiro, R.A., Xu, H., et al., 2018. cGAS-mediated autophagy protects the liver from ischemia-reperfusion injury independently of STING. Am. J. Physiol. Gastrointest. Liver Physiol. 314 (6), G655–G667. https://doi.org/10.1152/ajpgi.00326.2017.

Leifer, C.A., Kennedy, M.N., Mazzoni, A., Lee, C., Kruhlak, M.J., Segal, D.M., 2004. TLR9 is localized in the endoplasmic reticulum prior to stimulation. J. Immunol. 173 (2), 1179–1183. https://doi.org/10.4049/jimmunol.173.2.1179.

Li, T., Diner, B.A., Chen, J., Cristea, I.M., 2012. Acetylation modulates cellular distribution and DNA sensing ability of interferon-inducible protein IFI16. Proc. Natl. Acad. Sci. U. S. A. 109 (26), 10558–10563. https://doi.org/10.1073/pnas.1203447109.

Li, T., Chen, J., Cristea, I.M., 2013. Human cytomegalovirus tegument protein pUL83 inhibits IFI16-mediated DNA sensing for immune evasion. Cell Host Microbe 14 (5), 591–599. https://doi.org/10.1016/j.chom.2013.10.007.

Li, D., Wu, R., Guo, W., Xie, L., Qiao, Z., Chen, S., et al., 2019. STING-mediated IFI16 degradation negatively controls type I interferon production. Cell Rep. 29 (5), 1249–1260.e1244. https://doi.org/10.1016/j.celrep.2019.09.069.

Li, T., Huang, T., Du, M., Chen, X., Du, F., Ren, J., Chen, Z.J., 2021. Phosphorylation and chromatin tethering prevent cGAS activation during mitosis. Science 371, eabc5386. https://doi.org/10.1126/science.abc5386.

Liu, J., Qian, C., Cao, X., 2016. Post-translational modification control of innate immunity. Immunity 45 (1), 15–30. https://doi.org/10.1016/j.immuni.2016.06.020.

Liu, H., Zhang, H., Wu, X., Ma, D., Wu, J., Wang, L., et al., 2018. Nuclear cGAS suppresses DNA repair and promotes tumorigenesis. Nature 563 (7729), 131–136. https://doi.org/10.1038/s41586-018-0629-6.

Lohofener, J., Steinke, N., Kay-Fedorov, P., Baruch, P., Nikulin, A., Tishchenko, S., et al., 2015. The activation mechanism of 2′-5'-oligoadenylate synthetase gives new insights into OAS/cGAS triggers of innate immunity. Structure 23 (5), 851–862. https://doi.org/10.1016/j.str.2015.03.012.

Lum, K.K., Song, B., Federspiel, J.D., Diner, B.A., Howard, T., Cristea, I.M., 2018. Interactome and proteome dynamics uncover immune modulatory associations of the pathogen sensing factor cGAS. Cell Syst. 7 (6), 627–642.e626. https://doi.org/10.1016/j.cels.2018.10.010.

Lun, X.K., Bodenmiller, B., 2020. Profiling cell signaling networks at single-cell resolution. Mol. Cell. Proteomics 19 (5), 744–756. https://doi.org/10.1074/mcp.R119.001790.

Mackenzie, K.J., Carroll, P., Martin, C.A., Murina, O., Fluteau, A., Simpson, D.J., et al., 2017. cGAS surveillance of micronuclei links genome instability to innate immunity. Nature 548 (7668), 461–465. https://doi.org/10.1038/nature23449.

Manes, N.P., Nita-Lazar, A., 2018. Application of targeted mass spectrometry in bottom-up proteomics for systems biology research. J. Proteomics 189, 75–90. https://doi.org/10.1016/j.jprot.2018.02.008.

Martinez-Val, A., Bekker-Jensen, D.B., Steigerwald, S., Mehta, A., Tran, T., Sikorski, K., et al., 2021. Spatial-proteomics reveal in-vivo phospho-signaling dynamics at subcellular resolution. bioRxiv. https://doi.org/10.1101/2021.02.02.425898.

Matsumoto, M.L., Dong, K.C., Yu, C., Phu, L., Gao, X., Hannoush, R.N., et al., 2012. Engineering and structural characterization of a linear polyubiquitin-specific antibody. J. Mol. Biol. 418 (3–4), 134–144. https://doi.org/10.1016/j.jmb.2011.12.053.

Mayya, V., Lundgren, D.H., Hwang, S.I., Rezaul, K., Wu, L., Eng, J.K., et al., 2009. Quantitative phosphoproteomic analysis of T cell receptor signaling reveals system-wide modulation of protein-protein interactions. Sci. Signal. 2 (84), ra46. https://doi.org/10.1126/scisignal.2000007.

Miyashita, M., Oshiumi, H., Matsumoto, M., Seya, T., 2011. DDX60, a DEXD/H box helicase, is a novel antiviral factor promoting RIG-I-like receptor-mediated signaling. Mol. Cell. Biol. 31 (18), 3802–3819. https://doi.org/10.1128/MCB.01368-10.

Montoya, A., Beltran, L., Casado, P., Rodriguez-Prados, J.C., Cutillas, P.R., 2011. Characterization of a TiO(2) enrichment method for label-free quantitative phosphoproteomics. Methods 54 (4), 370–378. https://doi.org/10.1016/j.ymeth.2011.02.004.

Motwani, M., Pesiridis, S., Fitzgerald, K.A., 2019. DNA sensing by the cGAS-STING pathway in health and disease. Nat. Rev. Genet. 20 (11), 657–674. https://doi.org/10.1038/s41576-019-0151-1.

Murray, L.A., Sheng, X., Cristea, I.M., 2018. Orchestration of protein acetylation as a toggle for cellular defense and virus replication. Nat. Commun. 9 (1), 4967. https://doi.org/10.1038/s41467-018-07179-w.

Nuhse, T.S., Stensballe, A., Jensen, O.N., Peck, S.C., 2003. Large-scale analysis of in vivo phosphorylated membrane proteins by immobilized metal ion affinity chromatography and mass spectrometry. Mol. Cell. Proteomics 2 (11), 1234–1243. https://doi.org/10.1074/mcp.T300006-MCP200.

Olsen, J.V., Blagoev, B., Gnad, F., Macek, B., Kumar, C., Mortensen, P., Mann, M., 2006. Global, in vivo, and site-specific phosphorylation dynamics in signaling networks. Cell 127 (3), 635–648. https://doi.org/10.1016/j.cell.2006.09.026.

Olsen, J.V., Vermeulen, M., Santamaria, A., Kumar, C., Miller, M.L., Jensen, L.J., et al., 2010. Quantitative phosphoproteomics reveals widespread full phosphorylation site occupancy during mitosis. Sci. Signal. 3 (104), ra3. https://doi.org/10.1126/scisignal.2000475.

Onder, O., Sidoli, S., Carroll, M., Garcia, B.A., 2015. Progress in epigenetic histone modification analysis by mass spectrometry for clinical investigations. Expert Rev. Proteomics 12 (5), 499–517. https://doi.org/10.1586/14789450.2015.1084231.

Orzalli, M.H., DeLuca, N.A., Knipe, D.M., 2012. Nuclear IFI16 induction of IRF-3 signaling during herpesviral infection and degradation of IFI16 by the viral ICP0 protein. Proc. Natl. Acad. Sci. U. S. A. 109 (44), E3008–E3017. https://doi.org/10.1073/pnas.1211302109.

Orzalli, M.H., Broekema, N.M., Diner, B.A., Hancks, D.C., Elde, N.C., Cristea, I.M., Knipe, D.M., 2015. cGAS-mediated stabilization of IFI16 promotes innate signaling during herpes simplex virus infection. Proc. Natl. Acad. Sci. U. S. A. 112 (14), E1773–E1781. https://doi.org/10.1073/pnas.1424637112.

Paludan, S.R., 2015. Activation and regulation of DNA-driven immune responses. Microbiol. Mol. Biol. Rev. 79 (2), 225–241. https://doi.org/10.1128/MMBR.00061-14.

Peterson, A.C., Russell, J.D., Bailey, D.J., Westphall, M.S., Coon, J.J., 2012. Parallel reaction monitoring for high resolution and high mass accuracy quantitative, targeted proteomics. Mol. Cell. Proteomics 11 (11), 1475–1488. https://doi.org/10.1074/mcp.O112.020131.

Pinkse, M.W., Uitto, P.M., Hilhorst, M.J., Ooms, B., Heck, A.J., 2004. Selective isolation at the femtomole level of phosphopeptides from proteolytic digests using 2D-NanoLC-ESI-MS/MS and titanium oxide precolumns. Anal. Chem. 76 (14), 3935–3943. https://doi.org/10.1021/ac0498617.

Rigbolt, K.T., Prokhorova, T.A., Akimov, V., Henningsen, J., Johansen, P.T., Kratchmarova, I., et al., 2011. System-wide temporal characterization of the proteome and phosphoproteome of human embryonic stem cell differentiation. Sci. Signal. 4 (164), rs3. https://doi.org/10.1126/scisignal.2001570.

Riley, J.S., Tait, S.W., 2020. Mitochondrial DNA in inflammation and immunity. EMBO Rep. 21 (4), e49799. https://doi.org/10.15252/embr.201949799.

Rodero, M.P., Tesser, A., Bartok, E., Rice, G.I., Della Mina, E., Depp, M., et al., 2017. Type I interferon-mediated autoinflammation due to DNase II deficiency. Nat. Commun. 8 (1), 2176. https://doi.org/10.1038/s41467-017-01932-3.

Rothbart, S.B., Dickson, B.M., Raab, J.R., Grzybowski, A.T., Krajewski, K., Guo, A.H., et al., 2015. An interactive database for the assessment of histone antibody specificity. Mol. Cell 59 (3), 502–511. https://doi.org/10.1016/j.molcel.2015.06.022.

Rush, J., Moritz, A., Lee, K.A., Guo, A., Goss, V.L., Spek, E.J., et al., 2005. Immunoaffinity profiling of tyrosine phosphorylation in cancer cells. Nat. Biotechnol. 23 (1), 94–101. https://doi.org/10.1038/nbt1046.

Sadler, A.J., Latchoumanin, O., Hawkes, D., Mak, J., Williams, B.R., 2009. An antiviral response directed by PKR phosphorylation of the RNA helicase A. PLoS Pathog. 5 (2), e1000311. https://doi.org/10.1371/journal.ppat.1000311.

Schaffer, L.V., Millikin, R.J., Miller, R.M., Anderson, L.C., Fellers, R.T., Ge, Y., et al., 2019. Identification and quantification of proteoforms by mass spectrometry. Proteomics 19 (10), e1800361. https://doi.org/10.1002/pmic.201800361.

Schmid, M., Speiseder, T., Dobner, T., Gonzalez, R.A., 2014. DNA virus replication compartments. J. Virol. 88 (3), 1404–1420. https://doi.org/10.1128/JVI.02046-13.

Schneider, W.M., Chevillotte, M.D., Rice, C.M., 2014. Interferon-stimulated genes: a complex web of host defenses. Annu. Rev. Immunol. 32, 513–545. https://doi.org/10.1146/annurev-immunol-032713-120231.

Seo, G.J., Yang, A., Tan, B., Kim, S., Liang, Q., Choi, Y., et al., 2015. Akt kinase-mediated checkpoint of cGAS DNA sensing pathway. Cell Rep. 13 (2), 440–449. https://doi.org/10.1016/j.celrep.2015.09.007.

Seo, G.J., Kim, C., Shin, W.J., Sklan, E.H., Eoh, H., Jung, J.U., 2018. TRIM56-mediated monoubiquitination of cGAS for cytosolic DNA sensing. Nat. Commun. 9 (1), 613. https://doi.org/10.1038/s41467-018-02936-3.

Shaw, P.G., Chaerkady, R., Zhang, Z., Davidson, N.E., Pandey, A., 2011. Monoclonal antibody cocktail as an enrichment tool for acetylome analysis. Anal. Chem. 83 (10), 3623–3626. https://doi.org/10.1021/ac1026176.

Smith, W.A., Schurter, B.T., Wong-Staal, F., David, M., 2004. Arginine methylation of RNA helicase a determines its subcellular localization. J. Biol. Chem. 279 (22), 22795–22798. https://doi.org/10.1074/jbc.C300512200.

Song, B., Greco, T.M., Lum, K.K., Taber, C.E., Cristea, I.M., 2020a. The DNA sensor cGAS is decorated by acetylation and phosphorylation modifications in the context of immune signaling. Mol. Cell. Proteomics 19 (7), 1193–1208. https://doi.org/10.1074/mcp.RA120.001981.

Song, Z.M., Lin, H., Yi, X.M., Guo, W., Hu, M.M., Shu, H.B., 2020b. KAT5 acetylates cGAS to promote innate immune response to DNA virus. Proc. Natl. Acad. Sci. U. S. A. 117 (35), 21568–21575. https://doi.org/10.1073/pnas.1922330117.

Stetson, D.B., Ko, J.S., Heidmann, T., Medzhitov, R., 2008. Trex1 prevents cell-intrinsic initiation of autoimmunity. Cell 134 (4), 587–598. https://doi.org/10.1016/j.cell.2008.06.032.

Sun, L., Wu, J., Du, F., Chen, X., Chen, Z.J., 2013. Cyclic GMP-AMP synthase is a cytosolic DNA sensor that activates the type I interferon pathway. Science 339 (6121), 786–791. https://doi.org/10.1126/science.1232458.

Svinkina, T., Gu, H., Silva, J.C., Mertins, P., Qiao, J., Fereshetian, S., et al., 2015. Deep, quantitative coverage of the lysine acetylome using novel anti-acetyl-lysine antibodies and an optimized proteomic workflow. Mol. Cell. Proteomics 14 (9), 2429–2440. https://doi.org/10.1074/mcp.O114.047555.

Takaoka, A., Wang, Z., Choi, M.K., Yanai, H., Negishi, H., Ban, T., et al., 2007. DAI (DLM-1/ZBP1) is a cytosolic DNA sensor and an activator of innate immune response. Nature 448 (7152), 501–505. https://doi.org/10.1038/nature06013.

Takeshita, F., Leifer, C.A., Gursel, I., Ishii, K.J., Takeshita, S., Gursel, M., Klinman, D.M., 2001. Cutting edge: role of toll-like receptor 9 in CpG DNA-induced activation of human cells. J. Immunol. 167 (7), 3555–3558. https://doi.org/10.4049/jimmunol.167.7.3555.

Tan, X., Sun, L., Chen, J., Chen, Z.J., 2018. Detection of microbial infections through innate immune sensing of nucleic acids. Annu. Rev. Microbiol. 72, 447–478. https://doi.org/10.1146/annurev-micro-102215-095605.

Tanaka, Y., Chen, Z.J., 2012. STING specifies IRF3 phosphorylation by TBK1 in the cytosolic DNA signaling pathway. Sci. Signal. 5 (214), ra20. https://doi.org/10.1126/scisignal.2002521.

Thompson, M.R., Kaminski, J.J., Kurt-Jones, E.A., Fitzgerald, K.A., 2011. Pattern recognition receptors and the innate immune response to viral infection. Viruses 3 (6), 920–940. https://doi.org/10.3390/v3060920.

Thygesen, C., Boll, I., Finsen, B., Modzel, M., Larsen, M.R., 2018. Characterizing disease-associated changes in post-translational modifications by mass spectrometry. Expert Rev. Proteomics 15 (3), 245–258. https://doi.org/10.1080/14789450.2018.1433036.

Tikhanovich, I., Zhao, J., Olson, J., Adams, A., Taylor, R., Bridges, B., et al., 2017. Protein arginine methyltransferase 1 modulates innate immune responses through regulation of peroxisome proliferator-activated receptor gamma-dependent macrophage differentiation. J. Biol. Chem. 292 (17), 6882–6894. https://doi.org/10.1074/jbc.M117.778761.

Unterholzner, L., Keating, S.E., Baran, M., Horan, K.A., Jensen, S.B., Sharma, S., et al., 2010. IFI16 is an innate immune sensor for intracellular DNA. Nat. Immunol. 11 (11), 997–1004. https://doi.org/10.1038/ni.1932.

Wang, L., Wen, M., Cao, X., 2019. Nuclear hnRNPA2B1 initiates and amplifies the innate immune response to DNA viruses. Science 365 (6454), eaav0758. https://doi.org/10.1126/science.aav0758.

Wechsler, T., Chen, B.P., Harper, R., Morotomi-Yano, K., Huang, B.C., Meek, K., et al., 2004. DNA-PKcs function regulated specifically by protein phosphatase 5. Proc. Natl. Acad. Sci. U. S. A. 101 (5), 1247–1252. https://doi.org/10.1073/pnas.0307765100.

White, F.M., Wolf-Yadlin, A., 2016. Methods for the analysis of protein phosphorylation-mediated cellular signaling networks. Annu. Rev. Anal. Chem. (Palo Alto, Calif.) 9 (1), 295–315. https://doi.org/10.1146/annurev-anchem-071015-041542.

Wu, C., Tran, J.C., Zamdborg, L., Durbin, K.R., Li, M., Ahlf, D.R., et al., 2012. A protease for 'middle-down' proteomics. Nat. Methods 9 (8), 822–824. https://doi.org/10.1038/nmeth.2074.

Wu, J., Sun, L., Chen, X., Du, F., Shi, H., Chen, C., Chen, Z.J., 2013. Cyclic GMP-AMP is an endogenous second messenger in innate immune signaling by cytosolic DNA. Science 339 (6121), 826–830. https://doi.org/10.1126/science.1229963.

Wu, J.J., Li, W., Shao, Y., Avey, D., Fu, B., Gillen, J., et al., 2015. Inhibition of cGAS DNA sensing by a herpesvirus virion protein. Cell Host Microbe 18 (3), 333–344. https://doi.org/10.1016/j.chom.2015.07.015.

Xia, P., Ye, B., Wang, S., Zhu, X., Du, Y., Xiong, Z., et al., 2016. Glutamylation of the DNA sensor cGAS regulates its binding and synthase activity in antiviral immunity. Nat. Immunol. 17 (4), 369–378. https://doi.org/10.1038/ni.3356.

Yan, Z., Gibson, S.A., Buckley, J.A., Qin, H., Benveniste, E.N., 2016. Role of the JAK/STAT signaling pathway in regulation of innate immunity in neuroinflammatory diseases. Clin. Immunol. 189, 4–13. https://doi.org/10.1016/j.clim.2016.09.014.

Yang, P., An, H., Liu, X., Wen, M., Zheng, Y., Rui, Y., Cao, X., 2010. The cytosolic nucleic acid sensor LRRFIP1 mediates the production of type I interferon via a beta-catenin-dependent pathway. Nat. Immunol. 11 (6), 487–494. https://doi.org/10.1038/ni.1876.

Yang, H., Lundback, P., Ottosson, L., Erlandsson-Harris, H., Venereau, E., Bianchi, M.E., 2012. Redox modification of cysteine residues regulates the cytokine activity of high mobility group box-1 (HMGB1). Mol. Med. 18, 250–259. https://doi.org/10.2119/molmed.2011.00389.

Yu, K., Zhang, Q., Liu, Z., Zhao, Q., Zhang, X., Wang, Y., et al., 2019. qPhos: a database of protein phosphorylation dynamics in humans. Nucleic Acids Res. 47 (D1), D451–D458. https://doi.org/10.1093/nar/gky1052.

Yue, X., Schunter, A., Hummon, A.B., 2015. Comparing multistep immobilized metal affinity chromatography and multistep TiO2 methods for phosphopeptide enrichment. Anal. Chem. 87 (17), 8837–8844. https://doi.org/10.1021/acs.analchem.5b01833.

Zhang, J., Sprung, R., Pei, J., Tan, X., Kim, S., Zhu, H., et al., 2009. Lysine acetylation is a highly abundant and evolutionarily conserved modification in *Escherichia coli*. Mol. Cell. Proteomics 8 (2), 215–225. https://doi.org/10.1074/mcp.M800187-MCP200.

Zhang, X., Brann, T.W., Zhou, M., Yang, J., Oguariri, R.M., Lidie, K.B., et al., 2011a. Cutting edge: Ku70 is a novel cytosolic DNA sensor that induces type III rather than type I IFN. J. Immunol. 186 (8), 4541–4545. https://doi.org/10.4049/jimmunol.1003389.

Zhang, Z., Yuan, B., Bao, M., Lu, N., Kim, T., Liu, Y.J., 2011b. The helicase DDX41 senses intracellular DNA mediated by the adaptor STING in dendritic cells. Nat. Immunol. 12 (10), 959–965. https://doi.org/10.1038/ni.2091.

Zhang, Z., Bao, M., Lu, N., Weng, L., Yuan, B., Liu, Y.J., 2013a. The E3 ubiquitin ligase TRIM21 negatively regulates the innate immune response to intracellular double-stranded DNA. Nat. Immunol. 14 (2), 172–178. https://doi.org/10.1038/ni.2492.

Zhang, X., Shi, H., Wu, J., Zhang, X., Sun, L., Chen, C., Chen, Z.J., 2013b. Cyclic GMP-AMP containing mixed phosphodiester linkages is an endogenous high-affinity ligand for sting. Mol. Cell 51 (2), 226–235. https://doi.org/10.1016/j.molcel.2013.05.022.

Zhang, X., Wu, J., Du, F., Xu, H., Sun, L., Chen, Z., et al., 2014. The cytosolic DNA sensor cGAS forms an oligomeric complex with DNA and undergoes switch-like conformational changes in the activation loop. Cell Rep. 6 (3), 421–430. https://doi.org/10.1016/j.celrep.2014.01.003.

Zhang, J., Zhao, J., Xu, S., Li, J., He, S., Zeng, Y., et al., 2018. Species-specific deamidation of cGAS by herpes simplex virus UL37 protein facilitates viral replication. Cell Host Microbe 24 (2), 234–248.e235. https://doi.org/10.1016/j.chom.2018.07.004.

Zhou, H., Di Palma, S., Preisinger, C., Peng, M., Polat, A.N., Heck, A.J., Mohammed, S., 2013a. Toward a comprehensive characterization of a human cancer cell phosphoproteome. J. Proteome Res. 12 (1), 260–271. https://doi.org/10.1021/pr300630k.

Zhou, H., Ye, M., Dong, J., Corradini, E., Cristobal, A., Heck, A.J., et al., 2013b. Robust phosphoproteome enrichment using monodisperse microsphere-based immobilized titanium (IV) ion affinity chromatography. Nat. Protoc. 8 (3), 461–480. https://doi.org/10.1038/nprot.2013.010.

Zhou, R., Xie, X., Li, X., Qin, Z., Wei, C., Liu, J., Luo, Y., 2020. The triggers of the cGAS-STING pathway and the connection with inflammatory and autoimmune diseases. Infect. Genet. Evol. 77, 104094. https://doi.org/10.1016/j.meegid.2019.104094.

Zhuang, J., Jiang, G., Willers, H., Xia, F., 2009. Exonuclease function of human Mre11 promotes deletional nonhomologous end joining. J. Biol. Chem. 284 (44), 30565–30573. https://doi.org/10.1074/jbc.M109.059444.

Zierhut, C., Yamaguchi, N., Paredes, M., Luo, J.D., Carroll, T., Funabiki, H., 2019. The cytoplasmic DNA sensor cGAS promotes mitotic cell death. Cell 178 (2), 302–315. e323. https://doi.org/10.1016/j.cell.2019.05.035.

CHAPTER SEVEN

Proteomic approaches to investigate gammaherpesvirus biology and associated tumorigenesis

Danielle L. Chappell[a,b,†], Maria C. White[a,†], and Blossom Damania[a,b,c,*]
[a]Lineberger Comprehensive Cancer Center, The University of North Carolina at Chapel Hill, Chapel Hill, NC, United States
[b]Department of Pharmacology, The University of North Carolina at Chapel Hill, Chapel Hill, NC, United States
[c]Department of Microbiology and Immunology, The University of North Carolina at Chapel Hill, Chapel Hill, NC, United States
*Corresponding author: e-mail address: blossom_damania@med.unc.edu

Contents

[†] Contributed equally.

Advances in Virus Research, Volume 109
ISSN 0065-3527
https://doi.org/10.1016/bs.aivir.2020.10.001

Abstract

The DNA viruses, Kaposi's sarcoma-associated herpesvirus (KSHV) and Epstein-Barr virus (EBV), are members of the gammaherpesvirus subfamily, a group of viruses whose infection is associated with multiple malignancies, including cancer. The primary host for these viruses is humans and, like all herpesviruses, infection with these pathogens is lifelong. Due to the persistence of gammaherpesvirus infection and the potential for cancer formation in infected individuals, there is a driving need to understand not only the biology of these viruses and how they remain undetected in host cells but also the mechanism(s) by which tumorigenesis occurs. One of the methods that has provided much insight into these processes is proteomics. Proteomics is the study of all the proteins that are encoded by a genome and allows for (i) identification of existing and novel proteins derived from a given genome, (ii) interrogation of protein-protein interactions within a system, and (iii) discovery of druggable targets for the treatment of malignancies. In this chapter, we explore how proteomics has contributed to our current understanding of gammaherpesvirus biology and their oncogenic processes, as well as the clinical applications of proteomics for the detection and treatment of gammaherpesvirus-associated cancers.

1. Introduction

Herpesviruses are DNA viruses that replicate their genomes in the nucleus of host cells. These viruses establish lifelong infections and have two phases of their lifecycle, including a lytic phase and a latent phase. The quiescent latent cycle expresses few viral proteins, while the lytic cycle includes the expression of most of the viral genes and results in viral replication. At least eight species of human herpesviruses have been characterized to date, and they are classified into three distinct subfamilies: α, β, and γ. The α-herpesviridae subfamily includes herpes simplex virus type 1 (HSV-1), herpes simplex virus type 2 (HSV-2), and varicella-zoster virus (VZV).

The β-herpesviridae subfamily includes human cytomegalovirus (HCMV), by which congenital infection causes the most considerable burden of disease, as well as human herpesvirus-6 and human herpesvirus-7 (HHV-6 and HHV-7). The γ-herpesviridae subfamily includes Epstein-Barr virus (EBV) and Kaposi's sarcoma-associated herpesvirus (KSHV), both of which are viruses linked to tumorigenesis (Jha et al., 2016). EBV is the etiological agent of Burkitt's lymphoma (BL), nasopharyngeal carcinoma (NPC), post-transplant lymphoma (PTL), and Hodgkin's lymphoma. KSHV infection is associated with its namesake Kaposi's sarcoma (KS) as well as primary effusion lymphoma (PEL) and the plasmablastic variant of multicentric Castleman's disease (MCD).

Currently, clinically available inhibitors for herpesviruses target the viral DNA polymerase, a vital protein necessary for the completion of the herpesvirus lifecycle (Pagano et al., 2018). The most common class of these inhibitors are nucleoside analogs, which resemble nucleosides produced by cells, but their incorporation results in termination of DNA replication. However, acyclic nucleoside phosphonates (ANPs) such as cidofovir can also be used to treat herpesvirus infections (De Clercq and Holy, 2005). Collectively, these anti-herpesvirus drugs target the lytic lifecycle of the virus, as they become active only once the viral thymidine kinase (a lytic protein) has phosphorylated the compound. Presently, for treatment of KS and EBV-associated cancers in HIV-positive individuals, highly active antiretroviral therapy (HAART) is used. Although these drugs do not directly target KSHV or EBV, they help to improve immune function by targeting HIV (Cheung et al., 2005; Manners et al., 2018). Likewise, PTL can be treated by stopping the use of immunosuppressive agents to help restore the patient's immune system and therefore limit tumor progression; however, this approach may result in graft rejection. Thus, the generation of novel therapeutics for gammaherpesvirus infections and their associated cancers is of paramount importance.

Proteomics can offer insights that genomics or traditional laboratory techniques fail to elucidate, such as protein-protein interactions and the presence or modulation of post-translational modifications (PTMs). Through an understanding of (i) the proteins contained within gammaherpesvirus particles, (ii) which proteins these viruses interact with during infection, and (iii) how these viruses modulate the protein content within the host, new therapeutic strategies against KSHV and EBV infections and their associated cancers can be developed. In this chapter, we review proteomic methodologies that have been utilized in understanding gammaherpesvirus biology and their roles in tumorigenesis, and highlight proteomic studies that identified key viral proteins and their interaction networks throughout the viral lifecycle. Additionally, we explore in-depth the cancers associated with gammaherpesvirus infection and how proteomics is beneficial to the diagnosis and treatment of these virally-associated cancers. Finally, we conclude by discussing the future potential of proteomics for cancer research and viral oncology.

2. Proteomics approaches discussed in this chapter

Proteomics is the characterization of all of the proteins encoded by the genome of a given organism, tissue, or cell type (Larance and Lamond, 2015). Since proteins are responsible for cellular phenotypes, it is important to characterize individual protein binding partners as well as their functional roles in disease progression. Proteomics uses several approaches to determine protein-protein interactions (PPIs), whether they are direct or indirect. In this section, we discuss proteomic approaches that have been applied toward further characterizing gammaherpesvirus PPIs, either virus-virus or virus-host, as well as some of the drawbacks to each of these approaches.

2.1 Mass spectrometry

The most used and accepted method of proteomics is mass spectrometry (MS). MS measures the mass-to-charge ratio of a given sample. This measured ratio provides peptide predictions of what the sample identity could be by comparing actual sample spectra against central databases such as SwissProt, NCBI, NIST, and predicted peptide sequences. Major search engines include Mascot-Matrix Science, Sequest-John Yeats, and X-Tandem-GPM. These search engines compare peptides against decoy libraries to calculate false discovery rates, and software programs such as MAXQUANT/Perseus are used for quantitative proteomics analysis. Accordingly, each search engine uses different methods for scoring and removing false positives.

2.2 LC-MS and MALDI-MS

LC-MS and MALDI-MS are the two basic types of MS. In liquid chromatography-tandem mass spectrometry (LC-MS/MS), a protein sample is digested into peptides and then separated by reverse phase chromatography. Following chromatographic separation, an energy source ionizes the molecules. These ionized molecules strike a detector and further separate as they pass through a set of magnets based on their mass-to-charge ratio (Aebersold and Mann, 2016).

In contrast to LC-MS/MS, the matrix-assisted laser desorption ionization (MALDI)-MS approach uses single positively-charged ions generated by a laser. Samples are spotted on a substrate with a matrix that assists ionization where a laser vaporizes and ionizes the sample. Ionized molecules are injected into the mass spectrometer, where determinations of molecular charge and mass are then made (Aebersold and Mann, 2003). Readouts from this assay are given in the composition of the sample determined from the spectra. As MALDI-MS relies on single positively-charged ions, the loss of a charge can interfere with the identification, rendering the prediction less accurate. For a novice to the practice of MS, MALDI-MS may represent a better approach to attempt to answer questions regarding protein interactions. Even though MALDI-MS is more time consuming, the samples are salvageable after processing on the machine, unlike when running LC-MS. MALDI-MS is also ideal when working with a complex mixture containing peptides that are low in abundance. However, a noteworthy caveat of both of these techniques is that accuracy relies on the purity of the protein isolated.

2.3 Proteomic approaches to detect protein complexes

A commonly used proteomic method to confirm the high-throughput data generated from MS is antibody-based immunoaffinity purification (IP) of protein complexes, also known as pull-down assays, followed by immunoblot analysis that detects the antibody used in the pull-down (Larance and Lamond, 2015) (Fig. 1). These IP/pull-down assays utilize the "bait and prey" system. The "bait proteins" are prepared with tags to fish for and detect the "prey proteins." Here, we will highlight three antibody affinity assays commonly used to detect PPIs.

The first assay is affinity purification mass spectrometry (AP-MS), by which PPIs are detected for a protein of interest that was purified from cell lysates using antibodies to precipitate the specific protein and its interacting partners (Fig. 1A). These precipitated samples are separated by the molecular weights of proteins in the immunoprecipitate using sodium dodecyl sulfate-polyacrylamide gel electrophoresis (SDS-PAGE). Protein band(s) are prepared for MS analysis to determine if an interaction occurs and, if so, at which amino acid location. In-gel and in-solution digestion of proteins are commonly used preparation strategies prior to LC-MS/MS analysis (Dunham et al., 2012; Gingras et al., 2007). However, limitations of AP-MS include that it does not indicate whether the PPI is direct or indirect, nor the abundance or stoichiometry of the protein.

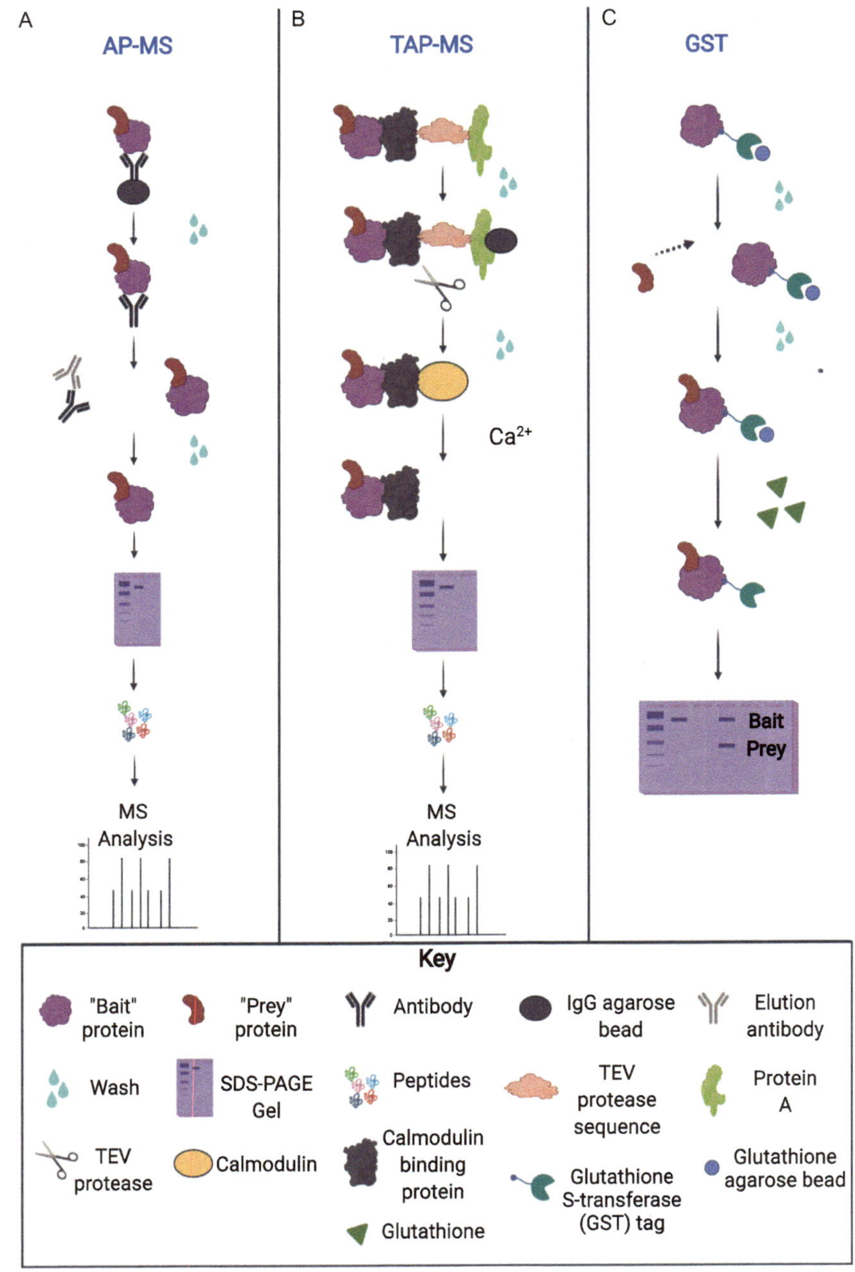

Fig. 1 Overview of the different affinity pull-down approaches. (A) Affinity purification mass spectrometry (AP-MS) approach. AP-MS involves immobilization of "bait" protein on beads/resins, and then unbound proteins are washed away. The protein complexes are then eluted from the beads via antibody incubation. Protein complexes are purified and then separated by SDS-PAGE. Separated proteins are digested into small peptides,

(Continued)

Tandem affinity purification coupled with MS (TAP-MS) can be performed to circumvent the issue of determining whether an interaction is direct or indirect. TAP-MS is similar in concept to epitope tagging except that it uses sequential tags. The first tag is the IgG Sepharose tag that is eventually cleaved off by the tobacco etch virus (TEV) protease (Fig. 1B). The second tag is the calmodulin-binding protein (CBP) that binds to calmodulin. Calcium addition elutes calmodulin off the column, leaving only the CBP, the bait protein, and any other protein(s) that interact with the bait protein in this pull-down assay (Gingras et al., 2007).

Glutathione S-transferase (GST) affinity chromatography is a commonly used research tool over the last several decades that enables the purification of proteins and, ultimately, the determination of the biological function of proteins, by use of recombinant GST-fusion proteins. To produce these recombinant proteins, the functional GST protein of 26 kDa can be tagged at either the N- or C-terminus of the recombinant protein and integrated into an expression vector. Purification of the GST-tagged fusion protein is achieved through the binding of GST to its substrate, glutathione (Dunham et al., 2012; Harper and Speicher, 2011) (Fig. 1C). GST is similar to AP-MS and TAP-MS; however, GST requires an additional step to permit the binding of the protein of interest with the substrate after the extraction of the GST-fusion protein.

Fig. 1—Cont'd and the bait-prey interaction is analyzed by LC-MS/MS. (B) Tandem affinity purification (TAP-MS) approach. TAP-MS employs two different affinity purification steps for a protein complex containing the TAP-tagged protein. The first step immobilizes "bait" protein on an IgG bead/resin for protein A. Any unbound proteins are washed away. The protein complex is then immobilized on calmodulin-containing resins for the second affinity purification for calmodulin-binding protein (CBP). The addition of the tobacco etch virus (TEV) protease induces cleavage at the TEV recognition site. The addition of calcium (Ca^{2+}) elutes the bound protein complex from the CBP. Protein complexes are purified and then separated by SDS-PAGE. Separated proteins are subsequently digested into small peptides that can be identified by LC-MS/MS. (C) Glutathione S-transferase (GST) pull-down assay. The recombinant fusion protein containing both GST and the "bait" is fixed to beads, and unbound proteins are washed away. The "prey" protein is added to the GST-tagged "bait" protein on beads to bind, and unbound "prey" proteins are washed away. Next, the bound protein complex is incubated with free glutathione, allowing for protein elution from the beads. Protein complexes are denatured by heating under reducing conditions. SDS-PAGE is then performed to separate the proteins, and the bait-prey interaction is analyzed by western blot.

2.4 Proteomic approaches to detect direct PPIs

Another commonly used proteomic method to determine PPIs is yeast two-hybrid (Y2H). Unlike AP-MS, Y2H can detect direct PPIs. Y2H involves transfecting yeast cells with two different plasmids, each containing a different reporter protein (such as fluorescent proteins or a LacZ reporter cassette) and the suspected binding protein. The first plasmid is the bait plasmid that contains a DNA binding domain (DBD) as well as coding sequences for protein X. The second plasmid is the prey plasmid that contains a transcription activation domain (AD) as well as coding sequences for protein Y. Upon interaction between proteins X and Y, the DBD and AD domains reconstruct, resulting in the transcription of the reporter gene (Brückner et al., 2009).

2.5 Quantitative techniques

Isotope-coded affinity tag (ICAT) is a method that labels cysteine-rich proteins for quantification via LC-MS/MS. The cysteine residues are recognized by the ICAT's cysteine-specific reactive group that is connected to an isotope-coded linker (heavy or light can be used for comparison) that also contains a biotin affinity tag on the end of the linker (Aslam et al., 2017). It is through the biotin affinity tag that these labeled proteins are isolated (Tao and Aebersold, 2003). ICAT is ideal for complex mixtures of proteins as it is highly automated, and peptides are directly sequenced using MS/MS (Gygi et al., 1999). However, ICAT data can be biased due to its reliance on cysteine-rich proteins.

Unlike ICAT, stable isotopic labeling with amino acids in cell culture (SILAC) labels the whole cellular proteome, not just cysteine-rich proteins, by using non-radioactive isotopic labeling. Amino acid isotopes based on molecular weight are labeled as either "light" or "heavy," added to cell culture media, and then inserted into new polypeptide chains within the cultured cells. This allows for MS-based detection of differences in protein abundance as well as regulation of gene expression, cell signaling, and PTMs. A limitation of using SILAC for proteomic experimentation is that it restricts the user to the examination of biological systems under a very controlled condition (Chahrour et al., 2015). Therefore, these labeling methods are not suitable for use with primary tissue, such as patient samples, but would be suitable for endogenous labeling of cell lines.

2.6 Protein arrays/chips

Protein microarrays, also known as protein chips, detect protein interactions from a small amount of sample and permit the investigation of multiple interactions on a single chip, saving both time and sample material (Aslam et al., 2017). Protein microarrays are similar in concept to ELISAs in that both utilize antibodies to probe against the desired protein of interest. There are three types of protein microarrays: functional microarrays, analytical microarrays, and reverse-phase microarrays. Functional protein microarrays utilize antibodies to detect isolated proteins. Analytical microarrays use a chip that is pre-coated with a capture antibody prior to the addition of the protein and fluorophore detection, similar to a sandwich ELISA. In reverse-phase microarrays, a whole-cell lysate can be added to the chip (Sutandy et al., 2013).

2.7 Three-dimensional structure-based proteomic approaches to detect PPIs

Nuclear magnetic resonance (NMR) is a technique which provides dynamics of structure, such as distances between atoms, the locations of atoms within a molecule, and images of protein structure. NMR gathers information on how atoms move by exploiting the magnetic properties of specific atomic nuclei and their responses to powerful magnetic fields. NMR is often paired with liquid chromatography (LC) or high-performance liquid chromatography (HPLC) (Aslam et al., 2017). One major limitation of NMR is that the limit of detection is 30 kDa, as the size of the magnet limits which proteins can be analyzed using this method.

In contrast, X-ray crystallography does not have the same size limitation as NMR. In X-ray crystallography, proteins of any size can be analyzed, and this technique can measure up to sub-atomic (<1.0 Å) levels. X-ray crystallography is considered the "gold standard" technique for the determination of three-dimensional protein structure. Diffraction patterns from the X-rays produce the repeating unit that comprises the crystal, size, and packing symmetry. X-ray crystallography can be useful for a wide range of applications, including studying protein-nucleic acid complexes and immune complexes in viral systems. The three-dimensional protein structure produced by X-ray crystallography can provide detailed information about protein-ligand interactions and enzyme mechanics, which are critical considerations for drug design (Khakurel et al., 2019). However, one major caveat to this technique is that the proteins must be highly purified and crystallized before their exposure to the X-rays.

2.8 Data repositories and databases

Several resources and databases are available from both predictive proteomic data and experimental proteomics, and encompass both known and novel protein pathways. Uniprot and Genecard are databases for everyday use. There are also databases for specific PTMs; for example, PhosphositePlus and UbiSite are both widely used for phosphorylation and ubiquitination, respectively. Databases such as the Biological General Repository for Interaction Datasets (BioGRID), the Molecular INTeraction database, and IntAct contain information regarding viral protein interactions in complexes, and are publicly available (Huttlin et al., 2015). Based on the list of provided genes and the available interactions using the IntAct database, users can visualize virus-protein interaction networks to gain a greater understanding of the biological system under investigation. Centralized data repositories allow for the data and results to be accessible to proteomics researchers and biologists alike, thus facilitating the dissemination of proteomics data.

3. Contributions of proteomics toward the discovery of KSHV and EBV virion composition

3.1 General structure of gammaherpesvirus particles

As previously mentioned, herpesviruses including KSHV and EBV are DNA viruses that replicate in the nucleus of host cells. The large double-stranded DNA genome of herpesviruses is located within a central core of the virus particle. The core containing the genomic information is encased by an icosahedral capsid, and the capsid is then surrounded by an amorphous region called the tegument (Fig. 2). The tegument contains virally-encoded proteins which are critical for early-stage infection as well as virus assembly and release from host cells (Guo et al., 2010; Lange and Damania, 2020). Finally, herpesvirus virions are encapsulated by a lipid envelope derived from host cell membranes, which contain embedded glycoproteins (Manservigi and Cassai, 1991).

Although DNA is the genetic material of gammaherpesviruses, the proteins located within the virus particle are just as important for virus propagation and survival. As stated above, many of these virion proteins are encoded by the viral genome and are located either inside (tegument and

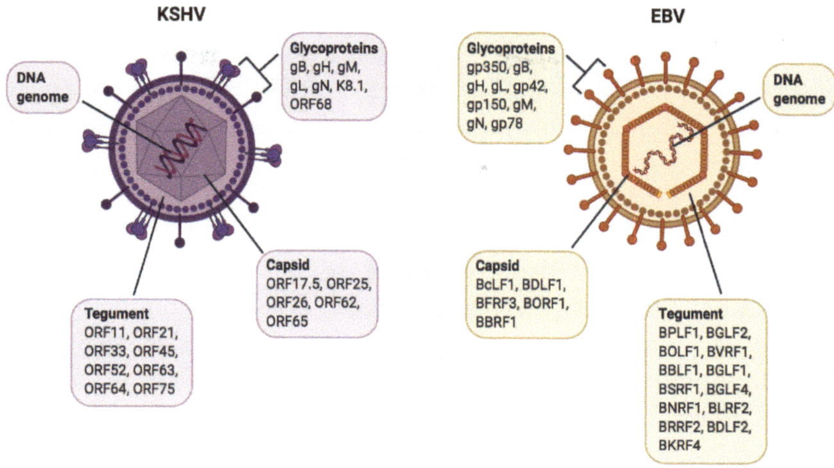

Fig. 2 Virion-associated proteins of KSHV and EBV. Viral proteins include capsid, tegument, and envelope glycoproteins. Cellular (host) proteins associated with the virus particles are omitted for clarity.

capsid proteins) or on the outside (membrane-bound glycoproteins) of the virus particle. Vital for virus attachment, entry, replication, assembly, and egress, these proteins have fascinated virologists for decades. This section will describe the contributions proteomic approaches have made in discovering and characterizing the proteins in KSHV and EBV virions.

3.2 Proteomics and KSHV virion composition

KSHV ORF45 is a viral protein encoded by an immediate early gene and was shown by Zhu and Yuan in 2003 to be a tegument protein associated with virus particles. KSHV virions were purified, lysed, resolved using a NuPAGE gel, and silver-stained to visualize protein bands. A band corresponding to the molecular weight of ORF45 was observed, and this band was then extracted and analyzed via MS to determine protein identity. The results confirmed that the ORF45 protein is contained within KSHV virus particles (Zhu and Yuan, 2003). This proteomics approach was complemented by several other methods, including immunoblotting, which demonstrated ORF45 presence alongside known KSHV structural proteins, sucrose gradient purification of virus particles, which isolated ORF45 protein within the same fraction as the known virion protein K8.1, and electron microscopy, which showed ORF45 staining of KSHV virions. Further experimentation suggested that ORF45 was a tegument protein; trypsin treatment of virus particles did not result in the degradation of ORF45 until the viral envelope had been compromised, and ORF45 was found in the tegument pellet when surface glycoproteins were liberated via detergent treatment. Thus, ORF45 was identified as a KSHV tegument protein and provides an example of how proteomics can be used to complement more targeted methods for protein identification.

Zhu and colleagues undertook a global proteomics approach to identify a larger number of KSHV virion proteins. KSHV virus particles were purified, the proteins were separated out by SDS-PAGE, and the resulting bands were subjected to HPLC-MS analysis. Excitingly, analysis of the protein data revealed 24 proteins associated with KSHV virus particles, as well as a few cellular proteins. Identified virion proteins in this study included capsid proteins (ORF17.5, ORF25, ORF26, ORF62, ORF65), envelope glycoproteins (gB, gH, gM, gL, gN, K8.1, ORF68), tegument proteins (ORF11, ORF21, ORF33, ORF45, ORF52, ORF63, ORF64, ORF75), and ORF6, ORF7, and ORF27 (Zhu et al., 2005). Interestingly, the cellular proteins β-actin and myosin were found in the tegument of KSHV virus particles, although whether packaging of these proteins was selective or non-selective was not investigated in this study. Taken together, the data obtained from these proteomic methodologies greatly advanced our understanding of the proteins encompassing the KSHV virion and set the stage for subsequent functional studies.

Around the same time, Bechtel et al. also undertook a global study of the proteins incorporated into KSHV virions. KSHV virus particles were purified and their associated proteins were separated on an SDS-PAGE gel. Subsequent bands were then analyzed using MALDI-MS. Results yielded 18 proteins of both viral and cellular origin associated with the KSHV virions. Identified proteins included capsid proteins (ORF25, ORF26, ORF62), membrane-embedded glycoproteins (gB, gH), tegument proteins (ORF21, ORF33, ORF63, ORF75), and ORF24 (Bechtel et al., 2005b). As most of these data overlap with the findings of Zhu et al., these two studies together present a comprehensive description of the proteins that are found within KSHV virions (Fig. 2). Interestingly, Bechtel et al. also discovered cellular proteins such as actin and tubulin inside the KSHV virus particles, although again, whether packaging of these proteins was intentional or random due to high abundance during virion assembly was not determined.

Similar proteomic approaches were employed to determine the virion-associated proteins of murine gammaherpesvirus 68 (MHV68), which is a closely-related virus to KSHV and EBV and is often representative of gammaherpesvirus infection and pathogenesis (Virgin et al., 1997). LC-MS/MS analysis on proteins extracted from purified MHV68 virions revealed capsid, tegument, and envelope proteins, many of which were homologues of other gammaherpesvirus virion proteins (Bortz et al., 2003). Cellular β-actin was also found inside MHV68 virions, suggesting a commonality across gammaherpesvirus assembly mechanisms. A following study also used LC-MS/MS to address the same question for MHV68, and these results greatly improved the existing dataset, more than doubling the number of proteins shown to be associated with MHV68 virions from 14 to 31 (Vidick et al., 2013).

3.3 Proteomics and EBV virion composition

A systematic study to describe the proteins contained within mature EBV virus particles was undertaken by Johanssen and colleagues using a MS-based approach. EBV virions were purified, the proteins were separated out on a gel, and the resulting bands were analyzed via LC-MS/MS. The data analysis uncovered capsid, tegument, and envelope proteins of EBV, as well as host proteins within the virion. The EBV capsid proteins identified were BcLF1, BDLF1, BFRF3, BORF1, and BBRF1; tegument proteins included BPLF1, BGLF2, BOLF1, BVRF1, BBLF1, BGLF1, BSRF1, BGLF4, BNRF1, BLRF2, BRRF2, BDLF2, and BKRF4; and glycoproteins detected in the EBV envelope were gp350, gB, gH, gL, gp42, gp150, gM, gN, and gp78 (Johannsen et al., 2004). Some of these findings confirmed previous speculation or observations made by others, while many of these proteins were novel additions to the knowledge surrounding EBV virion protein composition. Additionally, cellular proteins such as tubulin and actin were found to be associated with EBV capsids, mirroring the cellular protein composition found in KSHV and MHV68 virions. Overall, these data substantially contributed to our understanding of the protein composition of EBV virion particles (Fig. 2).

Interestingly, more recent data suggest that cytoskeletal proteins are not the only cellular components incorporated into gammaherpesvirus virions. EBV was shown to utilize the macroautophagy pathway in cells to obtain the outer membrane during virion maturation (Nowag et al., 2014). The autophagic protein LC3-II was detected in concentrated EBV virus preparations, and immunoelectron microscopy analysis revealed autophagic membranes present in virus envelopes via LC3 staining. Thus, these data indicate that EBV hijacks the autophagy pathway for lytic viral production and incorporates proteins involved in autophagy, such as LC3-II, into the virus particle during membrane acquisition.

Global proteomic approaches can be used to identify not only proteins present within virus particles, but also proteins expressed during the lytic cycle of gammaherpesviruses. Traylen et al. used proteomics to globally assess the proteins that were expressed during lytic EBV replication in cell culture. BL cells undergoing lytic viral replication were isolated, and the cellular and viral proteins present were then identified and quantified using MS. This screen resulted in 44 lytic EBV proteins, as none of these proteins had previously been linked to latency (Traylen et al., 2015). Many of these identified lytic proteins were capsid, tegument, or envelope proteins as established by others. However, the MS analysis uncovered three proteins not previously demonstrated to be detectable by this method: BBRF2 (involved in viral egress), BFLF1 (involved in viral maturation), and BSLF1 (involved in viral replication). Thus, proteomic approaches can be used to detect (i) viral proteins that compose virus particles, (ii) cellular proteins that are incorporated into virus particles, and (iii) viral proteins associated with either the latent or the lytic phase of herpesvirus lifecycles.

3.4 Proteomics and immunity

One of the hallmarks of oncogenic gammaherpesviruses is the ability to evade host immune recognition and downregulate the host immune response. Accordingly, these viruses encode multiple immunomodulatory proteins to achieve immune escape. K5, one of the immune evasion proteins encoded by KSHV, was first discovered through a genomic screen searching for viral proteins that downregulated MHC-I expression on host cells (Coscoy and Ganem, 2000). However, a more thorough understanding of proteins that were affected by K5 expression was lacking. To address this knowledge gap, comparative proteomics was used to assess differences in protein expression between cells expressing K5 versus cells lacking K5 (Bartee et al., 2006). This methodology employed SILAC, which incorporated "light" amino acids into the cells lacking K5 and "heavy" amino acids into the cells expressing K5, thus allowing identification and differentiation of sample proteins in combined MS analyses. Results of this approach identified MHC-I, activated leukocyte cell adhesion molecule (ALCAM), bone marrow stromal antigen 2 (BST-2), and syntaxin-4 as genes that were downregulated in cells expressing K5 compared to the control cells. This observed downregulation of ALCAM, BST-2, and syntaxin-4 in the presence of K5 was then independently verified using immunoblotting, immunofluorescence, and flow cytometry techniques. In sum, this study identified novel targets for the KSHV immunomodulatory protein, K5, and demonstrated for the first time the utility of quantitative proteomics for this purpose.

Proteomics can also be used to evaluate immune system recognition and antibody responses to gammaherpesvirus infection. Zheng et al. designed protein microarrays containing 92 KSHV and 82 EBV proteins embedded on a chip. These microarrays were then used to screen plasma isolated from healthy donors, EBV-negative B cell lymphoma patients, and KS/HIV-positive patients. Recognition of the viral proteins by IgG and IgA antibodies was assessed. Results showed that ORF73 and ORF38 were the most-recognized antigens by IgG in the KS/HIV-positive samples only, confirming the utility of ORF73 for KSHV serology testing and identifying ORF38 as a novel serology marker for KSHV infection (Zheng et al., 2011). In contrast to KSHV, EBV-IgG reactivity was detected in all but one individual. EBNA1, EBNA3B/3C, and the capsid protein BFRF3 were among the EBV antigens most recognized by IgG antibodies. Additionally, IgA reactivity against EBV BCRF1, EBNA1, and the early protein BRRF1 was detected in 100% of the samples. Altogether, this study demonstrates how proteomics can be used to design a specific and sensitive assay to measure humoral immunity against gammaherpesvirus infection and to determine the viral antigens predominantly recognized by the immune system.

4. Contributions of proteomics toward elucidating virus-host PPIs

There are several proteomic approaches used to explore virus-host interactions. Identifying virus-host interactions at sequential steps of the viral lifecycle may reveal virus-dependent signaling pathways that could be targeted by novel antiviral compounds to prevent or treat virus-induced diseases. In the following sections, we highlight examples of proteomic approaches used to uncover some of the gammaherpesvirus-host interactions.

4.1 Protein interactions associated with latency in gammaherpesviruses

Both EBV and KSHV attach to host cells via clathrin-mediated endocytosis. KSHV binds to cell surface heparin sulfate-like moieties (Wang et al., 2001), and enters cells via envelope glycoproteins, cellular integrins, the cystine-glutamate transporter (xCT), and CD98. Similarly, EBV binds to B cells via gp350/220, and then fuses with the host cell plasma membrane via gH, gL, and gB (Hutt-Fletcher, 2007). Using co-IP assays, Wang et al. identified neuropilin 1 (NRP1) as a novel EBV cellular entry factor that interacts with EBV gB. Interestingly, knockdown of NRP1 in NPC cells resulted in a significant decrease in EBV entry, suggesting that NRP1 is involved in EBV infection of epithelial cells in the nasopharynx (Wang et al., 2015). Subsequently, the same group identified Ephrin receptor A2 (EphA2) as an entry receptor for EBV on epithelial cells. By using co-IP, GST pull-down, and immunofluorescence assays, the authors showed that EphA2 interacted with the EBV fusion machinery (gH/gL/gB), similar to NRP1 (Zhang et al., 2018).

Once KSHV makes its way to the nucleus, ORF73, also known as the latency-associated nuclear antigen (LANA), attaches the viral episome to the host cell chromosome, where the virus remains in the latent cycle until reactivated (Aneja and Yuan, 2017). Each host cell harbors multiple copies of this KSHV episome, which is replicated alongside cellular DNA during cell division, allowing for viral persistence (Cesarman et al., 2019). During latent infection, infectious virions are not produced; however, a few genes are transcribed during KSHV latency (Dittmer et al., 1998). These latency-associated transcripts, or latent genes, include ORF73/LANA, ORF72/v-Cyclin, ORF71/v-FLIP, Kaposins, and microRNAs (miRNAs), which all function to help maintain the growth and survival of the infected cell (Mesri et al., 2010; Purushothaman et al., 2015). In EBV, the homolog to LANA is Epstein-Barr nuclear antigen-1 (EBNA-1).

As LANA regulates latency in immune cells, further elucidation of LANA-specific associations via proteomics may uncover interactions critical for the development of cancer. To determine if LANA is arginine-methylated, Campbell et al. screened a library consisting of eight of the nine mammalian protein arginine methyltransferase (PRMT) recombinant viral proteins to determine if they could function as methylation transferases of LANA in vitro by a methylation (^{3}H autoradiograph) assay. Using GST fusion protein mutants, the authors uncovered a single PRMT1 methylation site on LANA at Arg-20 and confirmed this finding by mutating LANA at this site (R20K) and comparing methylation levels to those of wild-type (WT) LANA. Results of this study showed that LANA R20K was considerably less methylated than WT LANA and, additionally, that PRMT1 associated with LANA in the cells, further suggesting that PRMT1 methylates LANA at Arg-20 (Campbell et al., 2012).

Like LANA, the EBV homolog EBNA-1 is present in both the latent and lytic lifecycles of the virus. To understand some critical interacting regulators of EBNA-1, Holowaty and colleagues utilized TAP-tagged EBNA-1. This TAP-tagged EBNA-1, and subsequent confirmation with co-IP, identified USP7/HAUSP, a cellular deubiquitinating protease, as a stable interacting protein with EBNA-1 (Holowaty et al., 2003). Other top interacting proteins identified were casein kinase II subunit α (CSNK2A2), HAUSP/USP7, NAP1, template-activating factor-Iβ/SET, CK2, and PRMT5. This study demonstrates the successful use of an alternative proteomic tagging technique for the identification of cellular proteins bound by EBNA-1. Adding to these findings, EBNA-1 was shown to also interact with the cellular ribosomal proteins RPL4 and nucleolin, facilitating stability of the EBV genome during latency (Shen et al., 2016). Finally, Wang et al. recently demonstrated that EBNA-1 interacts with and inhibits the cellular protein STUB1, which helps to suppress EBV lytic gene expression and maintain viral genomic stability in latently-infected host cells (Wang et al., 2020).

Upon EBV infection of B cells, EBNA-2 and EBNA-LP help to establish the latency program as well as maintain the growth and survival of the host cells (Kempkes et al., 1995; Szymula et al., 2018). Using two-dimensional PAGE and MALDI-MS techniques, Schlee et al. compiled a list of cellular proteins that were modulated by EBNA-2 in B cells. Results included the upregulation of protein synthesis and degradation factors as well as cell growth factors, while proteins involved in apoptosis and cytoskeleton arrangement were downregulated (Schlee et al., 2004). Additionally, co-IP and MS-based assays revealed that truncated EBNA-LP interacts with multiple subunits of the protein phosphatase PP2A in an inhibitory manner (Chelouah et al., 2018; Garibal et al., 2007). Thus, by inhibiting PP2A, EBV exerts both anti-apoptotic and pro-oncogenic effects in infected cells.

Latent membrane protein 1 (LMP-1) is an EBV transmembrane protein important for both successful EBV infection as well as primary B cell activation (Kieser and Sterz, 2015; Pratt et al., 2012). LMP-1 is a functional homolog of cellular CD40 (Graham et al., 2010), and is required for EBV-mediated B cell transformation (Soni et al., 2007). To identify novel PPIs between LMP-1 and cellular proteins, Song et al. utilized a Y2H strategy and identified interferon regulatory factor 7 (IRF7) as a top interacting protein with LMP-1 (Song et al., 2008). Co-IP confirmed IRF7 interacts with LMP-1, and the authors hypothesized that this interaction might function to evade host immune detection by silencing transcriptional activation of virus-inducible cellular genes. Using pull-down assays, Bentz et al. demonstrated that LMP-1 interacts with Ubc9, a cellular factor involved in protein modification. Furthermore, LMP-1 was shown to increase the sumoylation (a PTM) of host proteins through Ubc9, which led to an increase in cellular migration and oncogenic potential (Bentz et al., 2011). Additionally, using a global proteomics approach, DeKroon et al. examined the cellular proteins that were modulated by EBV LMP-1 and LMP-2A expression. The authors grouped the resulting data into pathways, and the cellular areas most affected by LMP-1 and LMP-2A included ubiquitination, vesicle formation/trafficking (including extracellular vesicles, which will be discussed in a later section), and cellular migration (DeKroon et al., 2018). Finally, Tworkoski and Raab-Traub performed a kinase screen to identify cellular kinases that were activated by LMP-1. Results of the screen demonstrated that LMP-1 phosphorylates and activates insulin-like growth factor 1 receptor (IGF1R), allowing for increased cell proliferation and loss of contact inhibition, both of which are characteristic of transformed cells (Tworkoski and Raab-Traub, 2015). Collectively, these examples demonstrate how proteomic approaches can be applied to uncover details in gammaherpesvirus lifecycles that might otherwise go unnoticed using an alternative approach.

Comparable to LMP-1 of EBV, KSHV encodes the transmembrane protein K1, which is necessary for transforming B cells and is similar to a B cell receptor (BCR) in its immunoreceptor tyrosine-based activation motif (ITAM) (Wen and Damania, 2010b). Proteomics has enabled insights into critical interactions which allow for K1 to regulate KSHV persistence in host cells. Lee et al. used KSHV-infected cell lysates labeled with [^{35}S] methionine-cysteine and the GST pull-down method to identify K1 binding proteins. Results of these assays revealed Lyn, Syk, Grb2, Vav, and PLCy2 as binding partners of tyrosine-phosphorylated K1 (Lee et al., 2005). Using TAP-MALDI-TOF, we found that K1 resists cellular apoptosis by interacting with heat shock protein 90 (Hsp90), and we confirmed this result through reverse co-IPs, confocal microscopy, and pharmacological inhibition of Hsp90 in both 293 cells and B cells (Wen and Damania, 2010a). Additionally, in a separate study, we performed TAP-MS to identify K1 binding partners. We identified 5′ adenosine monophosphate-activated protein kinase (AMPK) as a K1 interacting protein and found that this K1-AMPK interaction enhanced cell survival during periods of stress (Anders et al., 2016). Using pull-downs of GST fusion proteins containing K1 followed by LC-MS, Choi et al. determined that K1 interacted with cellular PYCR and showed that this K1-PYCR interaction increased cell growth in vitro and increased tumor size in mice (Choi et al., 2020).

Altogether, these studies helped to reveal intricate ways K1 interacts with cellular proteins to enable resistance to apoptosis, promote cellular growth and survival, increase angiogenesis, and facilitate tumor development.

KSHV expresses four proteins similar to the cellular interferon regulatory factors (IRFs) called viral IRFs (vIRFs), which help to antagonize the host cell antiviral response (Giffin and Damania, 2014). vIRF3, also called LANA-2, is encoded by ORF K10.5 and is the only vIRF expressed during latency (Rivas et al., 2001). Using pull-down assays, Lubyova et al. found that vIRF3 interacts with cellular IRF3 and IRF7 (Lubyova et al., 2004). The same group subsequently performed Y2H and GST pull-down assays in PEL cells and found that vIRF3 associated with the c-Myc suppressor protein MM-1alpha as well as the CDK4 promoter region (Lubyova et al., 2007).

In vitro pull-down of GST fusions of vIRF1 revealed both IRF1 and p300 as vIRF1 interacting partners, and identified a location on vIRF1 involved in overcoming the cellular interferon response and cell death (Burysek et al., 1999a). Nakamura and colleagues found vIRF1 to interact with p53 by using co-IP deletion mutants, and this interaction hindered growth arrest (Nakamura et al., 2001). Similarly, Seo et al. identified this vIRF1–p53 interaction by using GST pull-down assays, in vitro co-IP, and domain mapping. A luciferase reporter assay with vIRF1 revealed that the vIRF1–p53 interaction regulated p53 and p21 gene transcription (Seo et al., 2001). Finally, we used MS to identify vIRF1 interacting proteins and found that HERC5, an ISG15 ligase, interacted with vIRF1, and that this interaction resulted in an increase in KSHV reactivation and the number of infectious viral particles (Jacobs et al., 2015).

vIRF2 is another KSHV homolog of cellular IRFs. Burysek et al. made a GST fusion protein containing vIRF2 and performed an in vitro binding assay. Results of this assay determined that vIRF2 interacts with cellular IRF1, IRF2, ICSBP, the N-terminal domain of RelA (p65), and the C-terminal domain of p300 (Burysek et al., 1999b). A couple of years later, Burysek and Pitha used the GST pull-down method along with ^{35}S-labeling to identify vIRF2 protein interactions and found that protein kinase R (PKR) interacted with vIRF2. Studies showed that vIRF2 inhibited PKR autophosphorylation and phosphorylation of downstream PKR substrates as measured by a kinase assay (Burýšek and Pitha, 2001).

K15 is a latent KSHV protein, and several groups have utilized proteomics to understand the molecular interactions of how K15 modulates host cells to aid in virus survival and persistence. Sharp et al. found that K15 binds the cellular protein HAX-1 by using Y2H, GST pull-down assays, and co-IPs of the C-terminal domain of K15 (Sharp et al., 2002). As the receptor function of K15 is constitutively active once phosphorylated, Abere et al. examined potential phosphorylated K15 (pK15) binding partners. IP-MS of pK15 identified phosphatidylinositol 4-phosphate 3-kinase (PI3K)-C2α as a pK15 interacting protein. The authors then confirmed this pK15 interaction with PI3K-C2α using reverse co-IP in 293 cells, GST pull-down assays, and confocal microscopy. Functionally, siRNA depletion of PI3K-C2α was found to reduce KSHV lytic replication (Abere et al., 2018).

4.2 Protein interactions associated with reactivation from latency in gammaherpesviruses

Similar to other herpesviruses, KSHV can reactivate from latency under conditions that are stressful to the host or the surrounding cellular environment, such as inflammation, hypoxia, infection, or immunosuppression (Purushothaman et al., 2015). In culture, KSHV-infected cells can be chemically induced to reactive from latency by using drugs such as TPA and sodium butyrate (Aneja and Yuan, 2017; Miller et al., 1997). KSHV encodes a replication and transcription activator known as RTA or ORF50. RTA is a switch that is both necessary and sufficient to initiate the lytic cycle of KSHV, and K8/Kb-ZIP is an immediate-early gene that suppresses RTA function (Izumiya et al., 2003). By applying proteomic methods such as affinity chromatography and Y2H, several research groups demonstrated that K8/Kb-ZIP interacts with HDAC1/2 (Martínez and Tang, 2012; Tomita et al., 2004). This interaction with HDAC1/2 likely keeps the chromosome condensed and inaccessible for RTA binding to occur.

Conserved among all herpesviruses is an mRNA transcript accumulation (MTA) protein, which is a viral regulatory protein that, like RTA, allows for lytic replication to occur (Majerciak and Zheng, 2015). In KSHV, this protein is ORF57, and the EBV homolog is also called MTA or EB2. By interacting with an assortment of cellular proteins, ORF57 promotes the expression of viral genes. However, it was unclear at which stage of the viral lifecycle ORF57 was functionally active: pre- or post-transcription. Boyne et al. addressed this gap in knowledge by determining if ORF57 associates with either the pre-transcriptional human transcription/export (hTREX) mRNA processing complex or the post-transcriptional exon junction complex (EJC). Immunoaffinity assays showed that only hTREX components

(Aly, CBP80, TAP, and UAP56) interacted with MTA and not the EJC components, suggesting that the hTREX, not the EJC, is recruited to intron-less viral transcripts (Boyne et al., 2008).

In addition to MTA, EBV also encodes RTA and ZTA, which are two viral transcription factors essential for lytic replication (Chevallier-Greco et al., 1986). Through co-IP and LC-MS assays, EBV ZTA was found to interact with several cellular heat shock proteins, RNA binding proteins, and single-stranded DNA binding proteins (Wiedmer et al., 2008). One of these proteins, the mitochondrial DNA binding protein mtSSB, was shown to enhance EBV replication. Interestingly, when a mutation known to inhibit EBV replication was introduced into ZTA, mtSSB was no longer able to associate with ZTA, thus demonstrating that this mutation reduced viral replication due to lack of mtSSB binding. Along the same lines, Zhou et al. employed IP- and MS-based approaches to characterize the ZTA protein interaction network in BL cells. Both viral and cellular proteins were shown to interact with ZTA, two of which were of the nuclear factor of activated T cells family (NFATC1 and NFATC2). By binding to these cellular NFAT proteins, EBV modulates NFAT signaling, potentially as a mechanism to control ZTA expression levels in infected cells (Zhou et al., 2020).

To gain a perspective of how EBV modulates host cells on a global scale, Calderwood and colleagues utilized the Y2H system to create an "ORFeome" interaction data set consisting of both EBV-EBV and EBV-host interactions (Calderwood et al., 2007). HOMER3 was one of the 13 human proteins that interacted with several EBV proteins. HOMER3 interacted with nine EBV baits, including BFRF1 (nuclear membrane homolog of the alphaherpesvirus UL34), BILF1 (gp64; a G-protein coupled receptor), BLLF1 (gp350/220), BKRF2 (gL), BBRF3 (gM), BDLF3 (gH; formerly gp85), the replication and transcription factor BLRF1, the capsid binding protein BGLF2, and the tegument protein BFLF2.

Using AP-MS, we identified host proteins that interacted with the KSHV immunomodulatory protein vIL6. Hypoxia upregulated protein 1 (HYOU1) was shown to be a novel vIL6-interacting protein, and HYOU1 expression enhanced cell survival, migration, and signaling (Giffin et al., 2014). Similarly, but on a global scale, Davis et al. used AP-MS to identify over 500 interactions between KSHV and host cell proteins. Of note, ORF24 was shown to interact with cellular RNA polymerase II (RNA pol II). This interaction was not limited to KSHV, as the ORF24 homologues in EBV, HCMV, and MHV68 all interacted with RNA pol II as well, suggesting a conserved function among these herpesviruses. Finally, the KSHV ORF24–host RNA pol II interaction was necessary for the transcription of late genes to occur, indicating that the completion of the KSHV life cycle depends on this viral interaction with the host cell (Davis et al., 2015).

Similar global proteomic techniques have been employed for studying EBV modulation of host cell proteins during infection. Using a global and comparative proteomics approach, Ersing et al. quantified thousands of cellular and EBV proteins that were modulated in BL cells during the lytic phase of the EBV lifecycle. Noteworthy findings included viral targeting of B cell receptors for degradation by an EBV early protein, and activation of innate immune signaling pathways in the B cells (Ersing et al., 2017). The authors then extended these findings for inclusion in comparative analyses with other related herpesviruses. Results of this comparison showed that eight cellular RNA binding proteins were downregulated by both EBV and HCMV, while neuroligins 1 and 4X were downregulated by three human herpesviruses (EBV, HCMV, and KSHV), suggesting that these cell adhesion proteins might play a global role in herpesvirus biology.

5. Contributions of proteomics toward elucidating virus-virus PPIs

To date, much has been learned about the protein interactions that occur between herpesviruses and host cells. To help elucidate the interactions among herpesvirus proteins, Uetz and colleagues generated protein interaction maps for KSHV and VZV through Y2H bait and prey arrays. For KSHV, the authors examined over 12,000 viral protein interactions utilizing both fragment and full-length proteins. Results of this screen identified 123 KSHV protein-protein interactions that were non-redundant, and ~30% of these interactions were novel (Uetz et al., 2006). Of note, ORF50/RTA, the latent-to-lytic switch, was found to interact with ORF75 by several methodologies.

In an effort to better understand the protein-protein interactions that involve KSHV tegument proteins, Rozen et al. investigated global protein-protein interactions within KSHV virus particles using Y2H and co-IP assays. Results of these studies showed that ORF45 interacted with multiple tegument proteins, including ORF11, ORF21, ORF27, ORF33, ORF63, ORF64, and ORF75 (Rozen et al., 2008). Next, the authors used co-IP assays to examine the interaction network of ORF64 and ORF52 with other KSHV tegument components. ORF64 interacted with the most tegument proteins (ORFs 11, 21, 27, 33, 45, 52, 63, 64N, 64M, and 75) compared to ORF53, which interacted with only ORF45 and ORF75. Thus, these data indicate potentially essential roles for ORF45 and ORF64 in the KSHV virion, as their interaction networks were the most extensive. Additionally, Rozen et al. examined potential ORF64 and ORF52 interactions with the glycoproteins gB, gH, gM, gL, and gN by using co-IP and immunoblotting for the FLAG-tagged glycoproteins. ORF52 interacted with glycoproteins gM and gN, while ORF64 interacted with glycoproteins gB, gH, and gM, further strengthening the observations that ORF64 is promiscuous when it comes to interactions and could, therefore, be of significant importance for viral infectivity.

Finally, Fossum and colleagues systematically tested all three herpesvirus subfamilies by Y2H to determine if interactions are conserved between viral protein homologues. They added the interactomes of HSV-1, murine CMV, and EBV to their existing dataset of KSHV and VZV. The results indicated that viral homolog functionality could be more conserved than sequence (Fossum et al., 2009). Overall, studies involving proteomics have given researchers a clearer picture of the viral signaling network, allowing for additional insights into viral infectivity and persistence.

6. Determination of the protein content of exosomes secreted from gammaherpesvirus-infected cells

Exosomes are approximately 40–150 nm in diameter and deliver cargo such as proteins and miRNAs to nearby cells. Exosomes are classified as a form of extracellular vesicles (EVs) and originate from internal membrane budding of late endosomes within the cell. These exosomes can then either fuse with the lysosome for content degradation, or with the multivesicular body (MVB) for content release at the plasma membrane (Fig. 3). Additionally, exosomes carry trafficking proteins, such as tetraspanins, that can be utilized for exosome identification and experimental isolation. In the context of viral infection, it has been shown that EBV-infected cells can produce exosomes containing viral miRNAs (Pegtel et al., 2010). These viral miRNAs can then be delivered to neighboring cells, where modulation of cell signaling and oncogenic pathways can occur (Fig. 3).

Latent membrane protein 1 (LMP-1) of EBV has been detected in EVs from infected cells (Houali et al., 2007; Meckes et al., 2010). Similarly, LMP-1 was identified in EVs harvested from mice bearing NPC (Flanagan et al., 2003). Additionally, it was found that LMP-1 was packaged

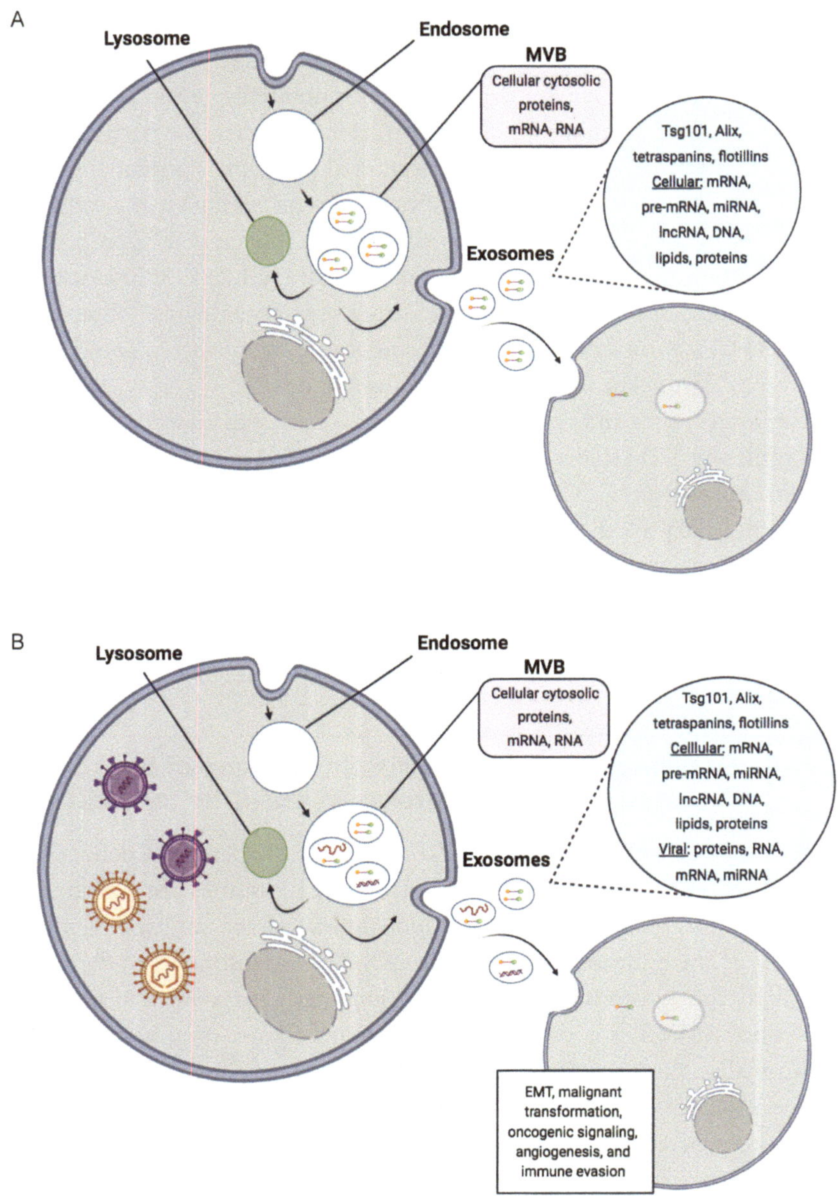

Fig. 3 Differences in exosome content secretion from uninfected and gammaherpesvirus-infected cells. (A) Exosome biogenesis and secretion from uninfected cells. Multivesicular bodies (MVBs) are formed by fusion with early endosomes. Next, the MVBs fuse either with the lysosome for degradation, or with the plasma membrane, which releases the exosomal content into the extracellular space. The exosomal content contains proteins encoded by tumor susceptibility gene 101 (Tsg101), apoptosis-linked gene-2 interacting protein

(Continued)

into EVs via a mechanism involving the cellular protein TRAF2, and that these LMP-1/TRAF2 complexes were involved in oncogenic NF-κB signaling (Verweij et al., 2015). As mentioned above, numerous studies have detected cellular and viral miRNAs in EVs, and it is thought these miRNAs are intentional to sidestep immune detection and to alter cellular functions in favor of cellular survival and, subsequently, the development of cancer (Chugh et al., 2013; Rainy et al., 2016; Schwab et al., 2015). Overall, gammaherpesviruses utilize multiple strategies to commandeer exosomes to aid in their continued propagation (Sadeghipour and Mathias, 2017).

To interrogate if exosome content was altered by infection with EBV or KSHV, Meckes et al. purified exosomes taken from (i) an uninfected B cell line, (ii) cells infected with EBV, (iii) cells infected with KSHV, and (iv) cells dually infected with EBV and KSHV, and compared exosomal components by MS data processing (Meckes et al., 2013). Additionally, the authors confirmed the LC-MS/MS data with immunoblotting and unique clustering pattern analysis. The resulting data suggested that EBV and KSHV infection influences exosome content, as over 360 proteins were identified in the virally-infected exosomes that were absent in the exosomes generated from uninfected cells.

6.1 Implications of gammaherpesvirus modulation of exosome content

Gammaherpesviruses escape the antiviral activities of the host immune response in part by interfering with cellular signaling pathways designed to alert the cell to invading pathogens (Liu et al., 2019). A recent study suggested that gammaherpesviruses might also escape immune detection by using exosomes. McNamara et al. analyzed the EV delivery system in the context of gammaherpesvirus infection by isolating EVs from both a KSHV-positive PEL cell line (BCBL1) and KSHV-positive PEL patient

Fig. 3—Cont'd X (Alix), tetraspanins, and flotillins, which can all be used for exosome identification. Additionally, multiple RNA species such as cellular mRNA, microRNA (miRNA), and long non-coding RNA (lncRNA) can be found within exosomes. Finally, exosomes can also contain DNA and lipids. (B) Exosome biogenesis and secretion from gammaherpesvirus-infected cells. In addition to the above content found in exosomes secreted from uninfected cells, viral proteins and viral nucleic acids can also be found within exosomes from virus-infected cells. The content of these exosomes often influences the surrounding cellular environment to result in epithelial-to-mesenchymal transition (EMT), malignant transformation, oncogenic signaling, angiogenesis, and immune evasion.

fluid and comparing them to the EVs present in normal plasma. KSHV-infected EVs circumvented the innate immune response and did not induce activation of antiviral cellular responses, including INF-β, TLR, RIG-I, or the cGAS/STING pathway (McNamara et al., 2019). Additionally, KSHV-infected EVs were shown to activate MEK/ERK signaling, which resulted in endothelial cell activation, proliferation, and migration. Thus, there are several functional advantages for KSHV to utilize EVs, including immune evasion and the modulation of the surrounding cellular environment in favor of viral spread.

Additionally, Meckes et al. analyzed KHSV- and EBV-modified exosome components with Ingenuity Pathway Analysis (IPA) software in order to determine if the exosome alterations exhibited oncogenic properties. While the IPA data in this study exposed differences in exosome cargo based on viral influence, the data also revealed similarities in exosomal content-mediated alteration of signaling cascades involved in cellular survival and the synthesis of proteins (Meckes et al., 2013). These data from Meckes and colleagues further support previous studies, which altogether demonstrate that modification of exosome cargo by KSHV or EBV can dysregulate cellular functions and lead to tumor development (Fig. 3).

7. Clinical implementation of proteomics for the diagnosis and treatment of KSHV- and EBV-associated cancers

7.1 Cancers associated with oncogenic human herpesviruses

7.1.1 KSHV

As mentioned previously, KSHV infection is associated with several different cancers, including Kaposi's sarcoma (KS), multicentric Castleman's disease (MCD), and primary effusion lymphoma (PEL) (see Table 1). KS is a highly vascularized endothelial cell tumor, which results in characteristic reddish-brown lesions on the skin. KS predominantly manifests on the extremities (arms/hands and legs/feet) or mucosal surfaces of infected individuals, but can sometimes be found internally on organs such as the spleen and liver (Siegel et al., 1969; Valls et al., 1991). KS is more prevalent in males and is the most common KSHV-associated cancer (Goncalves et al., 2017b). Interestingly, it has been shown that KSHV prevalence is not ubiquitous across populations, in contrast to many human herpesviruses (Gao et al., 1996). Instead, KSHV exhibits the highest prevalence in sub-Saharan

Table 1 Cancers associated with oncogenic human herpesviruses and their treatments.

Cancer	Origin/description	Treatment
KSHV		
Kaposi's sarcoma	Highly vascularized endothelial cell tumor, usually found on the extremities, often associated with HIV infection	HAART; systemic therapy in HAART-resistant cases (pegylated liposomal doxorubicin, pomalidomide)
Multicentric Castleman's disease	Plasmablastic tumor in lymph nodes, excessive inflammatory profile	Rituximab (anti-CD20 antibody); rituximab plus doxorubicin; valganciclovir plus zidovudine
Primary effusion lymphoma	B cell lymphoma, body cavities fill with malignant B cell effusions	Chemotherapeutic cocktails (CHOP, dose-adjusted EPOCH); HAART
EBV		
Post-transplant lymphoma	B cell lymphoma, occurs in immunocompromised individuals after organ transplant	Tapering or removal of immunosuppressive drugs
Burkitt's lymphoma	Germinal center B cell tumor, highly proliferative	Chemotherapy
Hodgkin's lymphoma	Only some forms associated with EBV infection, characterized by presence of Hodgkin/Reed-Sternberg cells (B cells)	Chemotherapy and radiotherapy
Nasopharyngeal carcinoma	Epithelial cell tumor of the nasopharynx, specific geographic distribution	Surgery to remove the tumor, followed by radiotherapy

Africa (Goncalves et al., 2017a), and KS is the most frequently occurring cancer in some of these countries (Dollard et al., 2010; Parkin, 2006).

While there are several different forms of MCD, the plasmablastic variant of MCD is the form associated with KSHV infection (KSHV-MCD) (Soulier et al., 1995). This lymphoproliferative cancer originates in lymph node plasmablasts and is often characterized by excessive inflammation, including robust production of the pro-inflammatory cytokine IL-6 (Nishimoto et al., 2005; Oksenhendler et al., 2000; Yoshizaki et al., 1989). Interestingly, many KSHV-MCD cases co-exist with KS in the same tissue (Naresh et al., 2008). In contrast to KS, where KSHV mainly exists in its latent (non-replicative) form in infected cells, KSHV-MCD exhibits a more pronounced lytic phenotype (Asahi-Ozaki et al., 2006). These data suggest that more virus replication is occurring in KSHV-MCD compared to KS, which might contribute to the aggressiveness of this cancer.

PEL is a highly aggressive but rare form of non-Hodgkin lymphoma (NHL) characterized by KSHV-infected B cells undergoing rapid proliferation. PEL is a body cavity malignancy where anatomical sites surrounding the heart (pericardial), lungs (pleural), and abdomen (peritoneal) become filled with these B cell effusions (Nador et al., 1996). PEL normally presents as an effusion tumor with no solid mass, but there have been reports of solid PEL tumors in the literature (Carbone et al., 2005; Chadburn et al., 2004). Each PEL cell contains 50–100 copies of the KSHV genome (Renne et al., 1996), which are passed along in the form of episomes to dividing daughter cells. Most patients with PEL are co-infected with EBV (Boulanger et al., 2005; Nador et al., 1996), but the implications of this co-infection for PEL disease progression and outcome are largely unknown.

It is worthwhile to mention that, although KSHV can infect anyone, symptomatic KSHV infection and the development of the above-mentioned cancers usually occur in the context of immunosuppression (Thakker and Verma, 2016). Examples of immunosuppression include HIV co-infection or transplant recipients receiving immunosuppressive drugs to help acceptance of the graft. One treatment avenue for these KSHV-associated cancers is the reinstatement of immunocompetence, and this will be discussed further in the next section.

7.1.2 EBV

Like KSHV, EBV is an oncogenic virus, and is associated with four cancers that will be discussed herein: post-transplant lymphoma (PTL), Burkitt's lymphoma (BL), Hodgkin's lymphoma (HL), and nasopharyngeal carcinoma (NPC) (Raab-Traub, 2007) (see Table 1). As its name suggests, PTL occurs in patients who have undergone bone marrow or solid organ transplant and have suppressed immune function, usually due to irradiation therapy and/or immunosuppressive drugs. PTL is a B cell lymphoma, arising in immunosuppressed individuals largely due to the lack of a virus-clearing cytotoxic T cell response (Green and Michaels, 2013; Meij et al., 2003; Smets et al., 2002). The likelihood of PTL developing after transplant depends on many factors, including the age of the organ recipient, whether the recipient was EBV-positive and/or CMV-positive at the time of transplant, and the type of organ received (Allen and Preiksaitis, 2009; Cockfield, 2001). Finally, children exhibit a higher occurrence of PTL compared to adults; children are more likely to be EBV-negative at the time of transplant, and primary EBV infection has been shown to be an important driver in the manifestation of PTL (Ho et al., 1988; Ho et al., 1985).

Burkitt's lymphoma (BL) was originally characterized by Denis Burkitt, who observed unusual facial tumors in the jaws of children in Uganda (Burkitt, 1958). BL is a highly proliferative B cell tumor of germinal center origin. The epidemiology of this disease suggests an association with malaria, as BL endemic regions closely overlap those of malaria (Burkitt, 1971). Unlike the causative agent of malaria (*P. falciparum*), EBV cannot be spread by mosquitoes. However, recent data suggest that infection with *P. falciparum* malaria is a risk factor for BL, due to dysregulation of activation-induced cytidine deaminase (AID) function in B cells and subsequent oncogenesis (Robbiani et al., 2015; Thorley-Lawson et al., 2016; Torgbor et al., 2014). Although first described in children, BL can develop in both children and adults. Accordingly, BL accounts for 1–5% of all non-Hodgkin lymphomas in adults and is observed more frequently in males (Boerma et al., 2004; Linch, 2012). BL can be either EBV-positive or EBV-negative, and EBV-positive BL will be further discussed in this section.

HL is characterized by the presence of distinct cells called Hodgkin/Reed-Sternberg (HRS) cells, which are B cells originating from various stages of development (Kuppers et al., 1994). There are many different forms of HL, only some of which are associated with EBV infection. Studies have shown that EBV infection of HRS cells and their precursors can contribute to the development of HL by providing anti-apoptotic signals, thus allowing the B cells to proliferate and escape programmed cell death (Bechtel et al., 2005a; Chaganti et al., 2005; Mancao et al., 2005).

NPC is an epithelial tumor of the nasopharynx that, similar to BL, displays a distinct geographic distribution (Raab-Traub, 2007). NPC is most prevalent in southern China, southeast Asia, parts of Africa, and Alaska (Mahdavifar et al., 2016). Aside from EBV infection, other risk factors for NPC include alcohol use, tobacco use, and HPV infection (Lubin et al., 2009; Mork et al., 2001). Establishing cell lines and in vivo models for NPC has proven difficult over the years, hindering the ability to develop novel treatments (Cheung et al., 1999; Tsao et al., 2017). Recently, however, several new tumor xenografts and cell lines were added to the NPC research pipeline, providing hope for further progress on elucidating NPC biology and methods to target this malignancy (Lin et al., 2018; Yip et al., 2019).

7.2 Current treatment methodologies

7.2.1 *Antiretroviral therapy and immune suppression reduction*

Many, but not all, of the cancers associated with KSHV and EBV infection arise in HIV-positive or otherwise immunocompromised individuals. Therefore, one of the main treatment options for some of these cancers, such as KS and PEL, revolves around improving immune function. For HIV-positive patients, this is achieved through the use of highly active antiretroviral therapy (HAART). By lowering HIV viral load and improving T cell count and function, the HIV-infected individual's immune system is more prepared to fight the cancer. In most KS cases, treatment with HAART is sufficient, and surgical resection of the lesions as well as radiotherapy are now rarely used (Yarchoan and Uldrick, 2018). HAART treatment alone has been shown to improve KS disease burden and lengthen survival time in patients (Bower et al., 2009; Mosam et al., 2012). Additionally, it has been suggested that HAART treatment can lead to better control of KSHV replication and a more robust antibody response toward KSHV, which could in turn improve KS disease progression (Wilkinson et al., 2002). Finally, HAART has been shown to reduce the risk of death by over 80% in HIV-positive men diagnosed with either KS or NHL (Tam et al., 2002). Unfortunately, not all HIV-positive KS patients will respond to HAART. In instances of HAART-resistant KS, systemic therapies such as pegylated liposomal doxorubicin (Northfelt et al., 1998) and pomalidomide (Polizzotto et al., 2016) can be used (Yarchoan and Uldrick, 2018).

The mainstay treatment for PTL is the lessening or removal of drugs that suppress the immune system post-transplant. This approach is more effective in pediatric patients compared to adults, and sometimes fails due to tumor heterogeneity/unresponsiveness, advanced disease stage, advanced patient age, or complications with rejection (Allen and Preiksaitis, 2009; Reshef et al., 2011). Interestingly, this immunosuppression reduction approach is most successful when the tumor is still under EBV control (Green and Michaels, 2013).

7.2.2 *Chemotherapy and radiotherapy*

Chemotherapy has been a frontline treatment for cancer for decades. As more chemotherapeutic agents are developed and brought into the clinic, the combinations in which these drugs can be used against different malignancies constantly changes. Drugs such as prednisone or methotrexate can be added to chemotherapeutic regimens to lessen inflammation and reduce the side effects from the other drugs. Chemotherapy cocktails aim to be potent enough to act against the tumor at a concentration that is not toxic to the vital tissues of the patient. The combination therapies CHOP (cyclophosphamide, doxorubicin, vincristine, and prednisone) and dose-adjusted EPOCH (cyclophosphamide, doxorubicin, vincristine, prednisone, and etoposide) are used as treatments for PEL, usually in combination with HAART (Goncalves et al., 2017b). Similar chemotherapeutic regimens are used in the treatment of BL; generally, EBV-positive and EBV-negative BL are treated alike, except in the context of immunotherapy or clinical trials (Dunleavy, 2018; Kanakry and Ambinder, 2013). For HIV-positive lymphoma patients, providers must carefully consider any potential drug interactions between HAART and the combination chemotherapy being used. The composition of the drug cocktails, as well as the timing of drug administration, can be modified as needed to help prevent adverse events from occurring (Yarchoan and Uldrick, 2018).

As a solid tumor, treatment for NPC includes surgery to remove the tumor mass if possible, followed by radiotherapy. Radiotherapy has been recognized as the frontline treatment for NPC, but has proven difficult to manage due to the proximity of the tumor to anatomical sites such as the brainstem, brain, salivary glands, and retinas (Holliday and Frank, 2016). Although efficacious, radiotherapy alone can be hindered by cytotoxic side effects, metastasis, or radio-resistance (Au et al., 2018; Yeo et al., 2018). For this reason, combining radiotherapy with chemotherapy or immunotherapy may improve patient outcomes.

7.2.3 Immunotherapy

Both KSHV and EBV are B cell tropic; thus, many of the cancers associated with these viruses are B cell lymphomas. As such, one drug that has shown marked results in the clinic is the monoclonal antibody rituximab. Rituximab targets the CD20 antigen on B cells, but this antibody can also have an effect on some CD20-negative B cell lymphomas, possibly by altering signaling pathways in the tumor microenvironment (Carbone et al., 2015). PEL cells usually do not express normal B cell markers, including CD20, and rituximab has been largely unusable against this lymphoma in the context of KSHV infection (Nador et al., 1996). In contrast, rituximab has shown extremely promising results against KSHV-MCD, with remission achieved in over 90% of patients in a phase II trial (Bower et al., 2007). The safety profile of this drug combined with its clinical success resulted in rituximab being recommended as the primary treatment for

KSHV-MCD, and the five-year progression-free survival after rituximab treatment has been estimated to be 70–90% (Goncalves et al., 2017a; Pria et al., 2017). Additionally, rituximab can be combined with doxorubicin for treatment of KSHV-MCD (Yarchoan and Uldrick, 2018). Finally, rituximab recently showed promise for treatment of EBV-PTL. Two clinical studies demonstrated a complete response in 70% or more of PTL patients, and the drug was well-tolerated (Trappe et al., 2017; Zhu et al., 2019).

Immunotherapy is still in its infancy in the oncology world, and many exciting treatment avenues remain in the context of gammaherpesvirus-associated cancers. One of these avenues is anti-PD1 therapy, which has shown incredible promise in the treatment of other cancers such as melanoma and metastatic non-small cell lung cancer (Hamid et al., 2013; Sui et al., 2018). Current efforts are underway to evaluate the safety and efficacy of anti-PD1 therapy for KS as well as EBV-associated NPC and HL (Broussard and Damania, 2019; Cao et al., 2019; Carbone et al., 2018; Makowska et al., 2019).

7.2.4 Antiviral drugs

As both KSHV and EBV are herpesviruses, drugs targeting the herpesvirus lifecycle can be used in an attempt to control the underlying cause of the cancer. These drugs include ganciclovir and its pro-drug valganciclovir, as well as zidovudine. KSHV lytic genes ORF36 and ORF21 phosphorylate ganciclovir and zidovudine, respectively, to toxic tri-phosphorylated forms, thus disabling herpesvirus replication (Cannon et al., 1999; Gustafson et al., 2000). Ganciclovir and valganciclovir have been used for treatment of KSHV-MCD, with varying rates of success (Casper et al., 2004; Kantarci et al., 2016; Speicher et al., 2014; Uldrick et al., 2011). However, using valganciclovir in combination with zidovudine resulted in 86% of KSHV-MCD patients demonstrating a positive clinical response (Uldrick et al., 2011). Ganciclovir is also included in treatment for PTL, although no clinical studies have definitively demonstrated efficacy of ganciclovir in this context (Green and Michaels, 2013). It should be noted that these drugs only target the lytic phase of the virus lifecycle, and do not affect the latent virus reservoir. As such, treatment with these antiviral drugs can control but not eradicate the virus, and recurrence can occur.

Taken together, although different treatment options exist for KSHV- and EBV-associated cancers, for many patients these are not the solution. These cancers can develop resistance to therapy or become too aggressive for these treatments to have much effect. Despite our above arsenal, the prognosis for some of these cancers remains poor. Hence, new diagnostic and treatment options are increasingly sought after, and proteomic analyses offer a powerful tool for cancer therapeutic discovery.

7.3 Clinical uses of proteomics in cancer diagnosis and treatment

7.3.1 Proteomics in cancer diagnosis and detection

Proteomic approaches encompass a rapidly expanding and exciting field for cancer diagnosis and personalized treatment. Many cancers are detected later, often after secondary spread to distal tissues. Implementation of biomarker assays using blood or saliva samples that can detect cancers early while still in the formative stages would greatly improve clinical outcomes for cancer patients and significantly ease healthcare burdens. To date, no biomarker assays have been clinically established for KSHV- or EBV-associated cancers, but potential candidates are in development. Using LC-MS and MALDI-TOF approaches, Gloghini et al. found an array of proteins that were specifically secreted by PEL cell lines and were also identified in PEL patient samples and KSHV-associated solid tumors (Gloghini et al., 2014). Examples of these specifically secreted proteins included catalase, which promotes B cell growth, and cellular nucleic acid-binding protein (CNBP), which is involved in transcription regulation. Thus, PEL "secretomes" are enriched with PEL-specific proteins that could potentially be used for diagnostic purposes. Additionally, multiple NHL studies have been undertaken to examine the serum and/or plasma of patients prior to lymphoma diagnosis in the hopes of uncovering novel protein biomarkers for this class of cancer (Vendrame and Martínez-Maza, 2011). Increased levels of cytokines such as IL-6 and TNFα, cytokine receptors such as soluble CD27 and CD30, and immunoglobulins such as circulating free light chains have all been implicated as potential NHL biomarkers (Breen et al., 2011; Landgren et al., 2010; Rabkin et al., 2011). Finally, analyses of the NPC proteome, secretome, and patient serum proteome using various MS-based approaches have demonstrated that protein profiling can be used to distinguish NPC from non-cancer samples, and identified CCL5 as a novel NPC therapeutic candidate and biomarker (Cao et al., 2010; Feng et al., 2011; Ho et al., 2006; Lin et al., 2013).

Although MS-based proteomic approaches have the capacity to yield many novel cancer biomarkers in the lab, overall, their clinical translation has been disappointing. These clinical failures can be due to study design, discrepancies in sample collection/quality, the assays used, or control issues

(You et al., 2018). While reliable biomarkers are still in development for KSHV- and EBV-associated cancers, proteomics has nevertheless been successfully used to identify diagnostic biomarkers for other cancers with clinical success. Prostate cancer (PCa) often progresses undetected until the later stages of malignancy, when treatment is often futile (Catalona, 2018). Therefore, it was of great interest to develop a biomarker assay that could detect PCa prior to the onset of symptoms. Currently, the measurement of prostate-specific antigen (PSA) levels in the blood is used for PCa screening. PSA is a protein secreted by epithelial cells in the prostate, and higher PSA levels in the serum are indicative of PCa (Catalona et al., 1991). Additionally, this screening assay can detect PCa prior to palpable disease (Stormont et al., 1993). Using radioimmunoassays, Killian et al. showed in 1986 that PSA was the most reliable biomarker for PCa progression, and the PSA assay has progressively developed since this discovery (Killian et al., 1986). Although somewhat controversial due to potential over-diagnosis and unnecessary treatment, PSA remains one of the most widely-used cancer biomarker assays and has led to a decline in PCa deaths (Catalona, 2018).

The glycoprotein alpha-fetoprotein (AFP) is the most commonly used and studied biomarker for hepatocellular carcinoma (HCC) (Carr et al., 2018). While not all HCC cases can be detected via AFP screening, it has been estimated that 25% of non-viral HCC can be identified earlier using AFP plus imaging compared to imaging techniques alone (Worland et al., 2018). However, AFP levels in the serum can be elevated for reasons other than cancer, such as benign liver disease, and for this reason the specificity of elevated AFP for HCC diagnosis remains a concern. To help improve this specificity issue, a post-translationally modified isoform of AFP called AFP-L3 can be used, which better differentiates HCC from benign liver disorders (Aoyagi et al., 2002). Recently, a high-throughput protein microarray that requires minimal sample volume was developed to detect both AFP and AFP-L3 (Wu et al., 2018). Although not widely implemented, examining AFP-L3 serum levels can potentially improve HCC diagnosis.

Like AFP, carcinoembryonic antigen (CEA) is a protein normally produced by fetal gastrointestinal cells but can be present at low levels in adults (Kim et al., 2018). Elevated CEA levels can be indicative of cancer, including colorectal cancer (CRC). CEA is not specific enough to be useful as a diagnostic marker for CRC, but it has shown promise alongside imaging as a marker for CRC recurrence post-resection (Nicholson et al., 2015). Additionally, it has been suggested that CEA levels measured at the time of surgery for CRC hepatic metastases are indicative not only of prognosis but also the choice of supplemental chemotherapeutic strategies (Chiang et al., 2019). Finally, conjugation of tagged antibodies to CEA proteins can allow for better imaging to help guide surgical procedures post-metastasis, and radiolabeled anti-CEA antibodies can be used for delivery of tumor-specific therapy (Meredith et al., 1996; Meyer et al., 2009; Tiernan et al., 2013).

Cancer antigen 125 (CA-125) is commonly used as a biomarker for ovarian cancer. Like prostate cancer, ovarian cancer often progresses undetected until later stages when achieving cure is unlikely, highlighting the need for earlier detection methods (Scholler and Urban, 2007). CA-125, a membrane-bound glycoprotein, was first established as an ovarian cancer biomarker through radioimmunoassays performed in the 1980's (Canney et al., 1984). As with most biomarkers, there are issues with specificity, as conditions other than ovarian cancer can lead to elevated CA-125 levels in the blood. Despite this, CA-125 screening is commonly used as a biomarker (i) alongside imaging for diagnosis of suspected ovarian cancer cases, (ii) for monitoring response to treatment post-diagnosis, and (iii) for monitoring disease recurrence post-surgery. There have been recent attempts to improve the sensitivity and specificity of CA-125 screening for ovarian cancer. One such attempt used quantitative proteomics methods to identify candidate autoantibodies highly expressed in ovarian cancer but not healthy controls (Karabudak et al., 2013). These autoantibody markers were shown to be specific across a range of ovarian cancer disease stages as well as CA-125 levels. Thus, combining CA-125 screening with alternative screening methods such as autoantibody markers could improve testing efficiency.

Although not an exhaustive list of cancer biomarkers currently used in the clinic, these data demonstrate how analysis of proteins that differ between healthy vs. cancer patients can be developed into assays that enable earlier detection of malignancies and improved survival outcomes for patients.

7.3.2 Proteomics in cancer treatment and drug discovery

There have been numerous studies to date that have used proteomic approaches to identify novel targets for treatment of KSHV- and EBV-associated cancers. We recently utilized multiplexed inhibitor bead-mass spectrometry (MIB/MS) to profile the kinome of PEL cells against healthy B cell controls. This method runs cell lysates through a column containing kinase inhibitors attached to beads. The beads bind functionally expressed kinases, which can then be eluted off, identified, and quantified using MS. This kinome screen resulted in identification of several kinases that were significantly upregulated in PEL cells but not in healthy B cells. We went on to demonstrate that inhibition of one these kinases, Tyro3, resulted in decreased PEL cell survival in vitro and decreased tumor burden in vivo (Wong et al., 2019). Thus, inhibition of Tyro3 represents a novel therapeutic strategy for PEL.

Morphoproteomics is a novel approach to large-scale protein analysis; it combines histology, molecular biology, and chemistry to pinpoint specific pathways which drive a specific cancer's phenotype, thus enabling identification of druggable targets in a personalized manner (Tan, 2008). Using this morphoproteomic approach in a published case study, Salihi et al. identified cyclo-oxygenase 2 (COX2) and Enhancer of Zeste homolog 2 (EZH2) as drivers of the pathogenic phenotype exhibited by a patient with KS (Salihi et al., 2016). They were then able to recommend personalized therapy based on these findings, which included treatment with metformin and/or celecoxib, both of which can inhibit COX2/EZH2 expression and KSHV replication.

While proteomic analyses can identify novel therapeutic targets in cancer, these techniques can also be used to help discern novel treatment options in the context of an existing druggable target. For example, heat shock protein 90 (Hsp90) is known to play a role in tumorigenesis, including KSHV- and EBV-associated lymphomas (Calderwood et al., 2006). Given this knowledge, Nayar et al. inhibited Hsp90 in PEL cells and examined the resulting Hsp90 "interactome," looking for proteins that interacted with Hsp90, which could then be targeted. Interacting proteins were identified using unbiased affinity capture, which utilized beads conjugated to the Hsp90 inhibitor, followed by MS analysis of the captured proteins. The screen revealed multiple anti-apoptotic proteins interacting with Hsp90, leading to the hypothesis that inhibition of apoptosis was driving the PEL phenotype. Cell viability assays showed that PEL cell death was induced by inhibiting members of the BCL family of anti-apoptotic proteins, and that combinatorial therapy against both Hsp90 and BCL proteins resulted in a synergistic effect on PEL cell death (Nayar et al., 2013). Hsp90 inhibition was also shown to be efficacious against KS (Chen et al., 2012). Thus, probing protein-protein interactions can uncover targetable oncogenic pathways or novel drug treatment combinations for virus-associated malignancies.

To date, there are a multitude of reports using proteomics to uncover novel treatment avenues for EBV-associated cancers, especially for NPC. Here, we will highlight some of these reports for NPC, as well as EBV-BL. Using two-dimensional gel electrophoresis (2DGE) and subsequent MS, Tang et al. identified amyloid-β precursor protein (APP) as a secreted protein from an NPC cell line. They then went on to show that antibody blocking of secretory APP resulted in decreased tumor growth and migration, suggesting that APP helps to drive NPC progression and thus warrants further investigation into APP as a novel therapeutic target (Tang et al., 2010). Other groups have performed comparative analyses using a nasopharyngeal (NP) epithelial cell line alongside an NPC cell line to identify differentially expressed proteins between the healthy and cancerous cells. 2DGE/MS analysis showed that the NPC cells upregulated glycolysis compared to the control NP cell line, as expression of the glycolytic enzymes PKM2 and TPI1 was increased in NPC cells as well as glucose uptake and the secretion of lactic acid (Huang et al., 2015). Thus, targeting the glycolytic pathway in NPC is a possible therapeutic target, since NPC cells rely more heavily on glycolysis compared to normal NP epithelial cells. A similar comparative approach was employed by Li and colleagues, who analyzed differential protein expression in the stroma of NPC vs. normal NP mucosa (NNM) isolated from patients. Laser capture microdissection was used to separate stroma cells from NPC or NNM patient tissue, and then the proteins were separated using 2DGE and identified using MALDI-TOF-MS. Periostin, a protein involved in cell communication and signaling, was significantly upregulated in the stroma of NPC compared to NNM (Li et al., 2012). Overexpression of periostin in an NPC cell line led to increased invasion, and analysis of NPC patient samples showed that increased expression of periostin was correlated with poorer survival outcomes. Taken together, these data suggest that increased periostin levels are involved in NPC metastasis and that expression of periostin can help predict NPC survival. These findings were later confirmed by an independent cohort study, highlighting the potential of periostin as a novel target for NPC (Wei et al., 2018).

Comparative proteomic analyses have also been used in EBV-BL to identify virally-regulated proteins that could potentially be targeted. 2DGE/MS analysis in an EBV-transformed BL cell line showed down-regulation of members of the COMM domain (COMMD) family of proteins, which negatively regulate NF-κB signaling pathway components (Gkiafi and Panayotou, 2011). The authors hypothesized that downregulation of COMMD proteins and the subsequent upregulation of NF-κB signaling was one mechanism by which EBV infection results in oncogenesis, although this hypothesis was not formally tested in this study. While intriguing, more work is needed to elucidate the role of COMMD proteins in EBV tumor formation and the potential of these proteins as therapeutic targets.

Phosphoproteomic analyses can reveal signaling pathways important in cancer and help identify key regulatory proteins in these pathways. Rituximab and obinutuzumab are both anti-CD20 antibodies; however, they recognize distinct epitopes. Obinutuzumab has been shown to reduce tumor burden in rituximab-resistant EBV-positive BL in vivo and result in enhanced survival compared to rituximab treatment in rituximab-susceptible mice (Awasthi et al., 2015). To help unravel the mechanism behind these different treatment outcomes, Awasthi et al. used a global phosphoproteomics approach. BL cell lines, including EBV-positive BL, were treated with either rituximab or obinutuzumab and subsequent phosphorylated peptides were enriched using metal oxide affinity chromatography and analyzed via MS. Results revealed many differentially phosphorylated proteins, including proteins involved in B cell receptor signaling and NK cell-mediated cytotoxicity. One of these B cell signaling proteins, PLGC-2, was upregulated after treatment with obinutuzumab vs. rituximab, and knockdown of PLGC-2 resulted in increased BL cell growth and cell viability after obinutuzumab treatment compared to rituximab treatment (Awasthi et al., 2017). These data suggest that the differences in treatment outcomes between these two anti-CD20 antibodies could be explained in part by differential phosphorylation signatures on B cell receptor signaling molecules and invite future investigations of these signaling molecules as therapeutic targets. For NPC, a comparative phosphoproteomic analysis showed that EphA2 phosphorylation was significantly increased in metastatic NPC compared to non-metastatic NPC, and a phosphokinase antibody array revealed that AKT/Stat3 signaling was responsible for phospho-EphA2-mediated NPC invasiveness (Li et al., 2019). Therefore, inhibition of phospho-EphA2 and potentially other molecules in the AKT/Stat3 pathway are possible targets for treatment of NPC.

Finally, proteomics can be used to uncover novel mechanisms of resistance in cancer cells and to identify targets that could reinstate susceptibility to radiotherapy or chemotherapy. An example of this approach is a 2018 study by Li et al., who identified mitogen-activated protein kinase

15 (MAPK15) as a novel mediator of radiotherapy resistance in NPC. In an effort to screen for possible modulators of radioresistance, peptides derived from radiosensitive and radioresistant NPC cell lysates were fractionated via reverse-phase HPLC and analyzed via MS. Differential protein expression between the susceptible and resistant cell lines was then compiled, and MAPK15 was the most abundant protein kinase in the radioresistant cells (Li et al., 2018). Subsequent knockdown of MAPK15 in the resistant cell line rendered the cells more susceptible to therapy, and overexpression of MAPK15 in the sensitive cell line enhanced resistance to therapy. Lastly, the mechanism driving this radioresistance was examined, and the data suggested that MAPK15 controls radioresistance by limiting the accumulation of reactive oxygen species and improving repair of damaged DNA. In sum, these elegant experiments indicate that inhibition of MAPK15 sensitizes treatment-resistant NPC to radiotherapy, providing an exciting therapeutic addition, which hopefully translates to clinical application.

Although beyond the scope of this chapter, MS-based proteomic approaches that screen not only the overall proteome but also the phosphoproteome, glycoproteome, kinome, interactome, and secretome of cancer are widely employed to identify novel therapeutic targets and drugs for many other cancer types aside from those associated with human gammaherpesvirus infection. We refer the reader to Dias et al. for a comprehensive review on this topic (Dias et al., 2016). Overall, this is an exciting time for cancer research. As we move forward toward personalized cancer medicine, studying the proteome of cancer on an individual basis will hopefully help lead to more targeted therapies with fewer off-target effects and improved patient outcomes.

8. Conclusions and future perspectives

Proteomic techniques, many of which are based in mass spectrometry or microarray technology, have substantially advanced our understanding of gammaherpesvirus biology and their associated malignancies. Proteomics has significantly aided in the identification of the proteins contained within KSHV and EBV virus particles and the exosomes secreted from infected cells, as well as the interaction network that exists between gammaherpesvirus proteins and host cells. Proteomic approaches have also led to a better understanding of how gammaherpesvirus proteins contribute to immune evasion, thus allowing successful viral persistence within infected cells, and how targeting both viral and host proteins can be beneficial from a therapeutic perspective.

As such, proteomic approaches represent a powerful methodology for pushing the fields of cancer diagnostics and treatment forward. Immunotherapy and personalized medicine are bright futures of oncology. Immune responses are influenced by different factors, including genetics and the surrounding environment. Hence, every individual's immune system is different, and every individual's cancer is different. A blanket approach to cancer treatment, therefore, has many shortcomings. Precision medicine treats each individual's cancer as a stand-alone malignancy and enables targeted therapy with a lower chance of resistance. Ideally, genomics and proteomics will be combined to inform personalized cancer treatment. Using genomics, one can screen individual cancers and determine (i) the genetic landscape of the cancer and how it might have formed, (ii) if genetic mutations are present that would indicate or contraindicate certain treatment choices, and (iii) the extent of genetic variation and the impact this variation could have on protein structure and function. Proteomic screening would complement this genetic information, allowing for (i) identification of upregulated or dysregulated pathways in the cancer, which can inform treatment choices, (ii) identification of proteins and/or post-translational modifications that are driving the cancer, leading to novel therapeutic targets, and (iii) identification of resistant signatures, which could indicate selection of an alternative therapy. Comparative proteomic analyses can also be used to identify potential biomarkers for different cancers. The earlier cancer is detected, the better the prognosis, and since many cancers do not elicit symptoms until later stages of malignancy, having a sensitive and specific screening assay for each cancer would save many lives. By comparing cancerous tissue to normal tissue, proteomics can reveal differentially expressed proteins that might be present in blood or saliva prior to symptoms, which could indicate the beginnings of cancer formation. Although most biomarkers identified in the lab fail to translate to clinical application, the increasing improvement of the selectivity and sensitivity of proteomic approaches will enable identification of more robust candidates with better clinical success.

In conclusion, proteomics has greatly advanced our understanding of oncogenic human gammaherpesviruses and their associated cancers, leading to new discoveries in their biology, virus-host interactions, and treatment methodologies. We look forward to the future and the innovations that these powerful techniques will bring to the field of oncogenic viruses.

Acknowledgments

This work was supported by National Institutes of Health (NIH) grants CA019014, DE028211, CA096500, CA239583, and CA163217 to BD. DLC is supported by the Caroline H. and Thomas S. Royster Fellowship from the Graduate School at the University of North Carolina at Chapel Hill and a NCI diversity supplement. MCW is supported by NIH grant 5T32CA009156-44. BD is a Leukemia and Lymphoma Society Scholar and a Burroughs Wellcome Fund Investigator in Infectious Disease. We apologize to all the authors whose work we could not discuss due to length restrictions. All figures were made using BioRender.

Conflict of interest

The authors declare no conflict of interest.

References

Abere, B., Samarina, N., Gramolelli, S., Ruckert, J., Gerold, G., Pich, A., Schulz, T.F., 2018. Kaposi's sarcoma-associated herpesvirus nonstructural membrane protein pK15 recruits the class II phosphatidylinositol 3-kinase PI3K-C2alpha to activate productive viral replication. J. Virol. 92, e00544-18.

Aebersold, R., Mann, M., 2003. Mass spectrometry-based proteomics. Nature 422, 198–207.

Aebersold, R., Mann, M., 2016. Mass-spectrometric exploration of proteome structure and function. Nature 537, 347–355.

Allen, U., Preiksaitis, J., 2009. Epstein-barr virus and posttransplant lymphoproliferative disorder in solid organ transplant recipients. Am. J. Transplant. 9 (Suppl. 4), S87–S96.

Anders, P.M., Zhang, Z., Bhende, P.M., Giffin, L., Damania, B., 2016. The KSHV K1 protein modulates AMPK function to enhance cell survival. PLoS Pathog. 12, e1005985.

Aneja, K.K., Yuan, Y., 2017. Reactivation and lytic replication of Kaposi's sarcoma-associated herpesvirus: an update. Front. Microbiol. 8, 613.

Aoyagi, Y., Mita, Y., Suda, T., Kawai, K., Kuroiwa, T., Igarashi, M., Kobayashi, M., Waguri, N., Asakura, H., 2002. The fucosylation index of serum alpha-fetoprotein as useful prognostic factor in patients with hepatocellular carcinoma in special reference to chronological changes. Hepatol. Res. 23, 287.

Asahi-Ozaki, Y., Sato, Y., Kanno, T., Sata, T., Katano, H., 2006. Quantitative analysis of Kaposi sarcoma-associated herpesvirus (KSHV) in KSHV-associated diseases. J Infect Dis 193, 773–782.

Aslam, B., Basit, M., Nisar, M.A., Khurshid, M., Rasool, M.H., 2017. Proteomics: technologies and their applications. J. Chromatogr. Sci. 55, 182–196.

Au, K.H., Ngan, R.K.C., Ng, A.W.Y., Poon, D.M.C., Ng, W.T., Yuen, K.T., Lee, V.H.F., Tung, S.Y., Chan, A.T.C., Sze, H.C.K., et al., 2018. Treatment outcomes of nasopharyngeal carcinoma in modern era after intensity modulated radiotherapy (IMRT) in Hong Kong: a report of 3328 patients (HKNPCSG 1301 study). Oral Oncol. 77, 16–21.

Awasthi, A., Ayello, J., Van de Ven, C., Elmacken, M., Sabulski, A., Barth, M.J., Czuczman, M.S., Islam, H., Klein, C., Cairo, M.S., 2015. Obinutuzumab (GA101) compared to rituximab significantly enhances cell death and antibody-dependent cytotoxicity and improves overall survival against CD20(+) rituximab-sensitive/−resistant Burkitt lymphoma (BL) and precursor B-acute lymphoblastic leukaemia (pre-B-ALL): potential targeted therapy in patients with poor risk CD20(+) BL and pre-B-ALL. Br. J. Haematol. 171, 763–775.

Awasthi, A., Rolland, D.C.M., Ayello, J., van de Ven, C., Basrur, V., Conlon, K., Fermin, D., Barth, M.J., Klein, C., Elenitoba-Johnson, K.S.J., et al., 2017. A comparative global phosphoproteomics analysis of obinutuzumab (GA101) versus rituximab (RTX) against RTX sensitive and resistant Burkitt lymphoma (BL) demonstrates differential phosphorylation of signaling pathway proteins after treatment. Oncotarget 8, 113895–113909.

Bartee, E., McCormack, A., Fruh, K., 2006. Quantitative membrane proteomics reveals new cellular targets of viral immune modulators. PLoS Pathog. 2, e107.

Bechtel, D., Kurth, J., Unkel, C., Kuppers, R., 2005a. Transformation of BCR-deficient germinal-center B cells by EBV supports a major role of the virus in the pathogenesis of Hodgkin and posttransplantation lymphomas. Blood 106, 4345–4350.

Bechtel, J.T., Winant, R.C., Ganem, D., 2005b. Host and viral proteins in the virion of Kaposi's sarcoma-associated herpesvirus. J. Virol. 79, 4952–4964.

Bentz, G.L., Whitehurst, C.B., Pagano, J.S., 2011. Epstein-Barr virus latent membrane protein 1 (LMP1) C-terminal-activating region 3 contributes to LMP1-mediated cellular migration via its interaction with Ubc9. J. Virol. 85, 10144–10153.

Boerma, E.G., van Imhoff, G.W., Appel, I.M., Veeger, N.J., Kluin, P.M., Kluin-Nelemans, J.C., 2004. Gender and age-related differences in Burkitt lymphoma—epidemiological and clinical data from The Netherlands. Eur. J. Cancer 40, 2781–2787.

Bortz, E., Whitelegge, J.P., Jia, Q., Zhou, Z.H., Stewart, J.P., Wu, T.T., Sun, R., 2003. Identification of proteins associated with murine gammaherpesvirus 68 virions. J. Virol. 77, 13425–13432.

Boulanger, E., Gerard, L., Gabarre, J., Molina, J.M., Rapp, C., Abino, J.F., Cadranel, J., Chevret, S., Oksenhendler, E., 2005. Prognostic factors and outcome of human herpesvirus 8-associated primary effusion lymphoma in patients with AIDS. J. Clin. Oncol. 23, 4372–4380.

Bower, M., Powles, T., Williams, S., Davis, T.N., Atkins, M., Montoto, S., Orkin, C., Webb, A., Fisher, M., Nelson, M., et al., 2007. Brief communication: rituximab in HIV-associated multicentric Castleman disease. Ann. Intern. Med. 147, 836–839.

Bower, M., Weir, J., Francis, N., Newsom-Davis, T., Powles, S., Crook, T., Boffito, M., Gazzard, B., Nelson, M., 2009. The effect of HAART in 254 consecutive patients with AIDS-related Kaposi's sarcoma. AIDS 23, 1701–1706.

Boyne, J.R., Colgan, K.J., Whitehouse, A., 2008. Recruitment of the complete hTREX complex is required for Kaposi's sarcoma–associated herpesvirus intronless mRNA nuclear export and virus replication. PLoS Pathog. 4, e1000194.

Breen, E.C., Hussain, S.K., Magpantay, L., Jacobson, L.P., Detels, R., Rabkin, C.S., Kaslow, R.A., Variakojis, D., Bream, J.H., Rinaldo, C.R., et al., 2011. B-cell stimulatory cytokines and markers of immune activation are elevated several years prior to the diagnosis of systemic AIDS-associated non-Hodgkin B-cell lymphoma. Cancer Epidemiol. Biomarkers Prev. 20, 1303–1314.

Broussard, G., Damania, B., 2019. KSHV: immune modulation and immunotherapy. Front. Immunol. 10, 3084.

Brückner, A., Polge, C., Lentze, N., Auerbach, D., Schlattner, U., 2009. Yeast two-hybrid, a powerful tool for systems biology. Int. J. Mol. Sci. 10, 2763–2788.

Burkitt, D., 1958. A sarcoma involving the jaws in African children. Br. J. Surg. 46, 218–223.

Burkitt, D.P., 1971. Epidemiology of Burkitt's lymphoma. Proc. R. Soc. Med. 64, 909–910.

Burýšek, L., Pitha, P.M., 2001. Latently expressed human herpesvirus 8-encoded interferon regulatory factor 2 inhibits double-stranded RNA-activated protein kinase. J. Virol. 75, 2345–2352.

Burysek, L., Yeow, W.S., Lubyova, B., Kellum, M., Schafer, S.L., Huang, Y.Q., Pitha, P.M., 1999a. Functional analysis of human herpesvirus 8-encoded viral interferon regulatory factor 1 and its association with cellular interferon regulatory factors and p300. J. Virol. 73, 7334–7342.

Burysek, L., Yeow, W.S., Pitha, P.M., 1999b. Unique properties of a second human herpesvirus 8-encoded interferon regulatory factor (vIRF-2). J. Hum. Virol. 2, 19–32.

Calderwood, S.K., Khaleque, M.A., Sawyer, D.B., Ciocca, D.R., 2006. Heat shock proteins in cancer: chaperones of tumorigenesis. Trends Biochem. Sci. 31, 164–172.

Calderwood, M.A., Venkatesan, K., Xing, L., Chase, M.R., Vazquez, A., Holthaus, A.M., Ewence, A.E., Li, N., Hirozane-Kishikawa, T., Hill, D.E., et al., 2007. Epstein-Barr virus and virus human protein interaction maps. Proc. Natl. Acad. Sci. U. S. A. 104, 7606–7611.

Campbell, M., Chang, P.-C., Huerta, S., Izumiya, C., Davis, R., Tepper, C.G., Kim, K.Y., Shevchenko, B., Wang, D.-H., Jung, J.U., et al., 2012. Protein arginine methyltransferase 1-directed methylation of Kaposi sarcoma-associated herpesvirus latency-associated nuclear antigen. J. Biol. Chem. 287, 5806–5818.

Canney, P.A., Moore, M., Wilkinson, P.M., James, R.D., 1984. Ovarian cancer antigen CA125: a prospective clinical assessment of its role as a tumour marker. Br. J. Cancer 50, 765–769.

Cannon, J.S., Hamzeh, F., Moore, S., Nicholas, J., Ambinder, R.F., 1999. Human herpesvirus 8-encoded thymidine kinase and phosphotransferase homologues confer sensitivity to ganciclovir. J. Virol. 73, 4786–4793.

Cao, S.M., Yu, J.K., Chen, Q.Y., Li, N.W., Xiang, Y.Q., Qian, C.N., Hu, X., Zhang, C.Q., Xie, D., Guo, X., 2010. Detection of nasopharyngeal carcinoma using surface-enhanced laser desorption and ionization mass spectrometry profiles of the serum proteome. Chin. J. Cancer 29, 721–728.

Cao, C., Wei, Q., Tang, X., Jia, Y., Sun, X., Li, W., Hu, Q., Chen, X., 2019. PD-1 and PD-L1 in locoregionally advanced nasopharyngeal carcinoma: substudy of a randomized phase III trial. Head Neck 41, 1427–1433.

Carbone, A., Gloghini, A., Vaccher, E., Cerri, M., Gaidano, G., Dalla-Favera, R., Tirelli, U., 2005. Kaposi's sarcoma-associated herpesvirus/human herpesvirus type 8-positive solid lymphomas: a tissue-based variant of primary effusion lymphoma. J. Mol. Diagn. 7, 17–27.

Carbone, A., De Paoli, P., Gloghini, A., Vaccher, E., 2015. KSHV-associated multicentric Castleman disease: a tangle of different entities requiring multitarget treatment strategies. Int. J. Cancer 137, 251–261.

Carbone, A., Gloghini, A., Carlo-Stella, C., 2018. Are EBV-related and EBV-unrelated Hodgkin lymphomas different with regard to susceptibility to checkpoint blockade? Blood 132, 17–22.

Carr, B.I., Akkiz, H., Uskudar, O., Yalcin, K., Guerra, V., Kuran, S., Karaogullarindan, U., Altintas, E., Ozakyol, A., Tokmak, S., et al., 2018. HCC with low- and normal-serum alpha-fetoprotein levels. Clin. Pract. (Lond.) 15, 453–464.

Casper, C., Nichols, W.G., Huang, M.L., Corey, L., Wald, A., 2004. Remission of HHV-8 and HIV-associated multicentric Castleman disease with ganciclovir treatment. Blood 103, 1632–1634.

Catalona, W.J., 2018. Prostate cancer screening. Med. Clin. North Am. 102, 199–214.

Catalona, W.J., Smith, D.S., Ratliff, T.L., Dodds, K.M., Coplen, D.E., Yuan, J.J., Petros, J.A., Andriole, G.L., 1991. Measurement of prostate-specific antigen in serum as a screening test for prostate cancer. N. Engl. J. Med. 324, 1156–1161.

Cesarman, E., Damania, B., Krown, S.E., Martin, J., Bower, M., Whitby, D., 2019. Kaposi sarcoma. Nat. Rev. Dis. Primers. 5, 9.

Chadburn, A., Hyjek, E., Mathew, S., Cesarman, E., Said, J., Knowles, D.M., 2004. KSHV-positive solid lymphomas represent an extra-cavitary variant of primary effusion lymphoma. Am. J. Surg. Pathol. 28, 1401–1416.

Chaganti, S., Bell, A.I., Pastor, N.B., Milner, A.E., Drayson, M., Gordon, J., Rickinson, A.B., 2005. Epstein-Barr virus infection in vitro can rescue germinal center B cells with inactivated immunoglobulin genes. Blood 106, 4249–4252.

Chahrour, O., Cobice, D., Malone, J., 2015. Stable isotope labelling methods in mass spectrometry-based quantitative proteomics. J. Pharm. Biomed. Anal. 113, 2–20.
Chelouah, S., Cochet, E., Couvé, S., Balkaran, S., Robert, A., May, E., Ogryzko, V., Wiels, J., 2018. New interactors of the truncated EBNA-LP protein identified by mass spectrometry in P3HR1 Burkitt's lymphoma cells. Cancers (Basel) 10, 12.
Chen, W., Sin, S.H., Wen, K.W., Damania, B., Dittmer, D.P., 2012. Hsp90 inhibitors are efficacious against Kaposi Sarcoma by enhancing the degradation of the essential viral gene LANA, of the viral co-receptor EphA2 as well as other client proteins. PLoS Pathog. 8, e1003048.
Cheung, S.T., Huang, D.P., Hui, A.B., Lo, K.W., Ko, C.W., Tsang, Y.S., Wong, N., Whitney, B.M., Lee, J.C., 1999. Nasopharyngeal carcinoma cell line (C666-1) consistently harbouring Epstein-Barr virus. Int. J. Cancer 83, 121–126.
Cheung, M.C., Pantanowitz, L., Dezube, B.J., 2005. AIDS-related malignancies: emerging challenges in the era of highly active antiretroviral therapy. Oncologist 10, 412–426.
Chevallier-Greco, A., Manet, E., Chavrier, P., Mosnier, C., Daillie, J., Sergeant, A., 1986. Both Epstein-Barr virus (EBV)-encoded trans-acting factors, EB1 and EB2, are required to activate transcription from an EBV early promoter. EMBO J. 5, 3243–3249.
Chiang, J.M., Hung, H.Y., You, J.F., Chiang, S.F., Lee, C.F., Chou, H.S., Lee, W.C., Chan, K.M., 2019. Applicability of postoperative carcinoembryonic antigen levels in determining post-liver-resection adjuvant chemotherapy regimens for colorectal cancer hepatic metastasis. Medicine (Baltimore) 98, e17696.
Choi, U.Y., Lee, J.J., Park, A., Zhu, W., Lee, H.R., Choi, Y.J., Yoo, J.S., Yu, C., Feng, P., Gao, S.J., et al., 2020. Oncogenic human herpesvirus hijacks proline metabolism for tumorigenesis. Proc. Natl. Acad. Sci. U. S. A. 117, 8083–8093.
Chugh, P.E., Sin, S.H., Ozgur, S., Henry, D.H., Menezes, P., Griffith, J., Eron, J.J., Damania, B., Dittmer, D.P., 2013. Systemically circulating viral and tumor-derived microRNAs in KSHV-associated malignancies. PLoS Pathog. 9, e1003484.
Cockfield, S.M., 2001. Identifying the patient at risk for post-transplant lymphoproliferative disorder. Transpl. Infect. Dis. 3, 70–78.
Coscoy, L., Ganem, D., 2000. Kaposi's sarcoma-associated herpesvirus encodes two proteins that block cell surface display of MHC class I chains by enhancing their endocytosis. Proc. Natl. Acad. Sci. U. S. A. 97, 8051–8056.
Davis, Z.H., Verschueren, E., Jang, G.M., Kleffman, K., Johnson, J.R., Park, J., Von Dollen, J., Maher, M.C., Johnson, T., Newton, W., et al., 2015. Global mapping of herpesvirus-host protein complexes reveals a transcription strategy for late genes. Mol. Cell 57, 349–360.
De Clercq, E., Holy, A., 2005. Acyclic nucleoside phosphonates: a key class of antiviral drugs. Nat. Rev. Drug Discov. 4, 928–940.
DeKroon, R.M., Gunawardena, H.P., Edwards, R., Raab-Traub, N., 2018. Global proteomic changes induced by the Epstein-Barr virus oncoproteins latent membrane protein 1 and 2A. MBio 9. e00959-18.
Dias, M.H., Kitano, E.S., Zelanis, A., Iwai, L.K., 2016. Proteomics and drug discovery in cancer. Drug Discov. Today 21, 264–277.
Dittmer, D., Lagunoff, M., Renne, R., Staskus, K., Haase, A., Ganem, D., 1998. A cluster of latently expressed genes in Kaposi's sarcoma-associated herpesvirus. J. Virol. 72, 8309–8315.
Dollard, S.C., Butler, L.M., Jones, A.M., Mermin, J.H., Chidzonga, M., Chipato, T., Shiboski, C.H., Brander, C., Mosam, A., Kiepiela, P., et al., 2010. Substantial regional differences in human herpesvirus 8 seroprevalence in sub-Saharan Africa: insights on the origin of the "Kaposi's sarcoma belt". Int. J. Cancer 127, 2395–2401.
Dunham, W.H., Mullin, M., Gingras, A.C., 2012. Affinity-purification coupled to mass spectrometry: basic principles and strategies. Proteomics 12, 1576–1590.

Dunleavy, K., 2018. Approach to the diagnosis and treatment of adult Burkitt's lymphoma. J. Oncol. Pract. 14, 665–671.
Ersing, I., Nobre, L., Wang, L.W., Soday, L., Ma, Y., Paulo, J.A., Narita, Y., Ashbaugh, C.W., Jiang, C., Grayson, N.E., et al., 2017. A temporal proteomic map of Epstein-Barr virus lytic replication in B cells. Cell Rep. 19, 1479–1493.
Feng, X., Zhang, J., Chen, W.N., Ching, C.B., 2011. Proteome profiling of Epstein-Barr virus infected nasopharyngeal carcinoma cell line: identification of potential biomarkers by comparative iTRAQ-coupled 2D LC/MS-MS analysis. J. Proteomics 74, 567–576.
Flanagan, J., Middeldorp, J., Sculley, T., 2003. Localization of the Epstein-Barr virus protein LMP 1 to exosomes. J. Gen. Virol. 84, 1871–1879.
Fossum, E., Friedel, C.C., Rajagopala, S.V., Titz, B., Baiker, A., Schmidt, T., Kraus, T., Stellberger, T., Rutenberg, C., Suthram, S., et al., 2009. Evolutionarily conserved herpesviral protein interaction networks. PLoS Pathog. 5, e1000570.
Gao, S.J., Kingsley, L., Li, M., Zheng, W., Parravicini, C., Ziegler, J., Newton, R., Rinaldo, C.R., Saah, A., Phair, J., et al., 1996. KSHV antibodies among Americans, Italians and Ugandans with and without Kaposi's sarcoma. Nat. Med. 2, 925–928.
Garibal, J., Hollville, E., Bell, A.I., Kelly, G.L., Renouf, B., Kawaguchi, Y., Rickinson, A.B., Wiels, J., 2007. Truncated form of the Epstein-Barr virus protein EBNA-LP protects against caspase-dependent apoptosis by inhibiting protein phosphatase 2A. J. Virol. 81, 7598–7607.
Giffin, L., Damania, B., 2014. KSHV: pathways to tumorigenesis and persistent infection. Adv. Virus Res. 88, 111–159.
Giffin, L., Yan, F., Ben Major, M., Damania, B., 2014. Modulation of Kaposi's sarcoma-associated herpesvirus interleukin-6 function by hypoxia-upregulated protein 1. J. Virol. 88, 9429–9441.
Gingras, A.C., Gstaiger, M., Raught, B., Aebersold, R., 2007. Analysis of protein complexes using mass spectrometry. Nat. Rev. Mol. Cell Biol. 8, 645–654.
Gkiafi, Z., Panayotou, G., 2011. Comparative proteomic analysis implicates COMMD proteins as Epstein-Barr virus targets in the BL41 Burkitt's lymphoma cell line. J. Proteome Res. 10, 2959–2968.
Gloghini, A., Volpi, C.C., Caccia, D., Gualeni, A.V., Cilia, A.M., Carbone, A., Bongarzone, I., 2014. Primary effusion lymphoma: secretome analysis reveals novel candidate biomarkers with potential pathogenetic significance. Am. J. Pathol. 184, 618–630.
Goncalves, P.H., Uldrick, T.S., Yarchoan, R., 2017a. HIV-associated Kaposi sarcoma and related diseases. AIDS 31, 1903–1916.
Goncalves, P.H., Ziegelbauer, J., Uldrick, T.S., Yarchoan, R., 2017b. Kaposi sarcoma herpesvirus-associated cancers and related diseases. Curr. Opin. HIV AIDS 12, 47–56.
Graham, J.P., Arcipowski, K.M., Bishop, G.A., 2010. Differential B-lymphocyte regulation by CD40 and its viral mimic, latent membrane protein 1. Immunol. Rev. 237, 226–248.
Green, M., Michaels, M.G., 2013. Epstein-Barr virus infection and posttransplant lymphoproliferative disorder. Am. J. Transplant. 13 (Suppl. 3), 41–54 (quiz 54).
Guo, H., Shen, S., Wang, L., Deng, H., 2010. Role of tegument proteins in herpesvirus assembly and egress. Protein Cell 1, 987–998.
Gustafson, E.A., Schinazi, R.F., Fingeroth, J.D., 2000. Human herpesvirus 8 open reading frame 21 is a thymidine and thymidylate kinase of narrow substrate specificity that efficiently phosphorylates zidovudine but not ganciclovir. J. Virol. 74, 684–692.
Gygi, S.P., Rist, B., Gerber, S.A., Turecek, F., Gelb, M.H., Aebersold, R., 1999. Quantitative analysis of complex protein mixtures using isotope-coded affinity tags. Nat. Biotechnol. 17, 994–999.
Hamid, O., Robert, C., Daud, A., Hodi, F.S., Hwu, W.J., Kefford, R., Wolchok, J.D., Hersey, P., Joseph, R.W., Weber, J.S., et al., 2013. Safety and tumor responses with lambrolizumab (anti-PD-1) in melanoma. N. Engl. J. Med. 369, 134–144.

Harper, S., Speicher, D.W., 2011. Purification of proteins fused to glutathione S-transferase. Methods Mol. Biol. 681, 259–280.
Ho, M., Miller, G., Atchison, R.W., Breinig, M.K., Dummer, J.S., Andiman, W., Starzl, T.E., Eastman, R., Griffith, B.P., Hardesty, R.L., et al., 1985. Epstein-Barr virus infections and DNA hybridization studies in posttransplantation lymphoma and lymphoproliferative lesions: the role of primary infection. J Infect Dis 152, 876–886.
Ho, M., Jaffe, R., Miller, G., Breinig, M.K., Dummer, J.S., Makowka, L., Atchison, R.W., Karrer, F., Nalesnik, M.A., Starzl, T.E., 1988. The frequency of Epstein-Barr virus infection and associated lymphoproliferative syndrome after transplantation and its manifestations in children. Transplantation 45, 719–727.
Ho, D.W., Yang, Z.F., Wong, B.Y., Kwong, D.L., Sham, J.S., Wei, W.I., Yuen, A.P., 2006. Surface-enhanced laser desorption/ionization time-of-flight mass spectrometry serum protein profiling to identify nasopharyngeal carcinoma. Cancer 107, 99–107.
Holliday, E.B., Frank, S.J., 2016. Proton therapy for nasopharyngeal carcinoma. Chin. Clin. Oncol. 5, 25.
Holowaty, M.N., Zeghouf, M., Wu, H., Tellam, J., Athanasopoulos, V., Greenblatt, J., Frappier, L., 2003. Protein profiling with Epstein-Barr nuclear antigen-1 reveals an interaction with the herpesvirus-associated ubiquitin-specific protease HAUSP/USP7. J. Biol. Chem. 278, 29987–29994.
Houali, K., Wang, X., Shimizu, Y., Djennaoui, D., Nicholls, J., Fiorini, S., Bouguermouh, A., Ooka, T., 2007. A new diagnostic marker for secreted Epstein-Barr virus encoded LMP1 and BARF1 oncoproteins in the serum and saliva of patients with nasopharyngeal carcinoma. Clin. Cancer Res. 13, 4993–5000.
Huang, P.Y., Zeng, T.T., Li, M.Q., Ban, X., Zhu, Y.H., Zhang, B.Z., Mai, H.Q., Zhang, L., Guan, X.Y., Li, Y., 2015. Proteomic analysis of a nasopharyngeal carcinoma cell line and a nasopharyngeal epithelial cell line. Tumori 101, 676–683.
Hutt-Fletcher, L.M., 2007. Epstein-Barr virus entry. J. Virol. 81, 7825–7832.
Huttlin, E.L., Ting, L., Bruckner, R.J., Gebreab, F., Gygi, M.P., Szpyt, J., Tam, S., Zarraga, G., Colby, G., Baltier, K., et al., 2015. The BioPlex network: a systematic exploration of the human interactome. Cell 162, 425–440.
Izumiya, Y., Lin, S.F., Ellison, T., Chen, L.Y., Izumiya, C., Luciw, P., Kung, H.J., 2003. Kaposi's sarcoma-associated herpesvirus K-bZIP is a coregulator of K-Rta: physical association and promoter-dependent transcriptional repression. J. Virol. 77, 1441–1451.
Jacobs, S.R., Stopford, C.M., West, J.A., Bennett, C.L., Giffin, L., Damania, B., 2015. Kaposi's sarcoma-associated herpesvirus viral interferon regulatory factor 1 interacts with a member of the interferon-stimulated gene 15 pathway. J. Virol. 89, 11572–11583.
Jha, H.C., Banerjee, S., Robertson, E.S., 2016. The role of gammaherpesviruses in cancer pathogenesis. Pathogens 5, 18.
Johannsen, E., Luftig, M., Chase, M.R., Weicksel, S., Cahir-McFarland, E., Illanes, D., Sarracino, D., Kieff, E., 2004. Proteins of purified Epstein-Barr virus. Proc. Natl. Acad. Sci. U. S. A. 101, 16286–16291.
Kanakry, J.A., Ambinder, R.F., 2013. EBV-related lymphomas: new approaches to treatment. Curr. Treat. Options Oncol. 14, 224–236.
Kantarci, F.E., Eren, R., Gundogan, C., Huq, G.E., Dogu, M.H., Suyani, E., 2016. A HHV-8 positive, HIV negative multicentric Castleman disease treated with R-CEOP chemotherapy and valganciclovir combination. J. Infect. Chemother. 22, 483–485.
Karabudak, A.A., Hafner, J., Shetty, V., Chen, S., Secord, A.A., Morse, M.A., Philip, R., 2013. Autoantibody biomarkers identified by proteomics methods distinguish ovarian cancer from non-ovarian cancer with various CA-125 levels. J. Cancer Res. Clin. Oncol. 139, 1757–1770.

Kempkes, B., Spitkovsky, D., Jansen-Dürr, P., Ellwart, J.W., Kremmer, E., Delecluse, H.J., Rottenberger, C., Bornkamm, G.W., Hammerschmidt, W., 1995. B-cell proliferation and induction of early G1-regulating proteins by Epstein-Barr virus mutants conditional for EBNA2. EMBO J. 14, 88–96.

Khakurel, K.P., Angelov, B., Andreasson, J., 2019. Macromolecular nanocrystal structural analysis with electron and X-rays: a comparative review. Molecules 24, 3490.

Kieser, A., Sterz, K.R., 2015. The latent membrane protein 1 (LMP1). Curr. Top. Microbiol. Immunol. 391, 119–149.

Killian, C.S., Emrich, L.J., Vargas, F.P., Yang, N., Wang, M.C., Priore, R.L., Murphy, G.P., Chu, T.M., 1986. Relative reliability of five serially measured markers for prognosis of progression in prostate cancer. J. Natl. Cancer Inst. 76, 179–185.

Kim, S.S., Donahue, T.R., Girgis, M.D., 2018. Carcinoembryonic antigen for diagnosis of colorectal cancer recurrence. JAMA 320, 298–299.

Kuppers, R., Rajewsky, K., Zhao, M., Simons, G., Laumann, R., Fischer, R., Hansmann, M.L., 1994. Hodgkin disease: Hodgkin and Reed-Sternberg cells picked from histological sections show clonal immunoglobulin gene rearrangements and appear to be derived from B cells at various stages of development. Proc. Natl. Acad. Sci. U. S. A. 91, 10962–10966.

Landgren, O., Goedert, J.J., Rabkin, C.S., Wilson, W.H., Dunleavy, K., Kyle, R.A., Katzmann, J.A., Rajkumar, S.V., Engels, E.A., 2010. Circulating serum free light chains as predictive markers of AIDS-related lymphoma. J. Clin. Oncol. 28, 773–779.

Lange, P., Damania, B., 2020. Kaposi sarcoma-associated herpesvirus (KSHV). Trends Microbiol. 28, 236–237.

Larance, M., Lamond, A.I., 2015. Multidimensional proteomics for cell biology. Nat. Rev. Mol. Cell Biol. 16, 269–280.

Lee, B.S., Lee, S.H., Feng, P., Chang, H., Cho, N.H., Jung, J.U., 2005. Characterization of the Kaposi's sarcoma-associated herpesvirus K1 signalosome. J. Virol. 79, 12173–12184.

Li, M., Li, C., Li, D., Xie, Y., Shi, J., Li, G., Guan, Y., Li, M., Zhang, P., Peng, F., et al., 2012. Periostin, a stroma-associated protein, correlates with tumor invasiveness and progression in nasopharyngeal carcinoma. Clin. Exp. Metastasis 29, 865–877.

Li, Z., Li, N., Shen, L., Fu, J., 2018. Quantitative proteomic analysis identifies MAPK15 as a potential regulator of radioresistance in nasopharyngeal carcinoma cells. Front. Oncol. 8, 548.

Li, J.Y., Xiao, T., Yi, H.M., Yi, H., Feng, J., Zhu, J.F., Huang, W., Lu, S.S., Zhou, Y.H., Li, X.H., et al., 2019. S897 phosphorylation of EphA2 is indispensable for EphA2-dependent nasopharyngeal carcinoma cell invasion, metastasis and stem properties. Cancer Lett. 444, 162–174.

Lin, S.J., Chang, K.P., Hsu, C.W., Chi, L.M., Chien, K.Y., Liang, Y., Tsai, M.H., Lin, Y.T., Yu, J.S., 2013. Low-molecular-mass secretome profiling identifies C-C motif chemokine 5 as a potential plasma biomarker and therapeutic target for nasopharyngeal carcinoma. J. Proteomics 94, 186–201.

Lin, W., Yip, Y.L., Jia, L., Deng, W., Zheng, H., Dai, W., Ko, J.M.Y., Lo, K.W., Chung, G.T.Y., Yip, K.Y., et al., 2018. Establishment and characterization of new tumor xenografts and cancer cell lines from EBV-positive nasopharyngeal carcinoma. Nat. Commun. 9, 4663.

Linch, D.C., 2012. Burkitt lymphoma in adults. Br. J. Haematol. 156, 693–703.

Liu, Q., Rao, Y., Tian, M., Zhang, S., Feng, P., 2019. Modulation of innate immune signaling pathways by herpesviruses. Viruses 11, 572.

Lubin, J.H., Purdue, M., Kelsey, K., Zhang, Z.F., Winn, D., Wei, Q., Talamini, R., Szeszenia-Dabrowska, N., Sturgis, E.M., Smith, E., et al., 2009. Total exposure and exposure rate effects for alcohol and smoking and risk of head and neck cancer: a pooled analysis of case-control studies. Am. J. Epidemiol. 170, 937–947.

Lubyova, B., Kellum, M.J., Frisancho, A.J., Pitha, P.M., 2004. Kaposi's sarcoma-associated herpesvirus-encoded vIRF-3 stimulates the transcriptional activity of cellular IRF-3 and IRF-7. J. Biol. Chem. 279, 7643–7654.

Lubyova, B., Kellum, M.J., Frisancho, J.A., Pitha, P.M., 2007. Stimulation of c-Myc transcriptional activity by vIRF-3 of Kaposi sarcoma-associated herpesvirus. J. Biol. Chem. 282, 31944–31953.

Mahdavifar, N., Towhidi, F., Makhsosi, B.R., Pakzad, R., Moini, A., Ahmadi, A., Lotfi, S., Salehiniya, H., 2016. Incidence and mortality of nasopharynx cancer and its relationship with human development index in the world in 2012. World J. Oncol. 7, 109–118.

Majerciak, V., Zheng, Z.M., 2015. KSHV ORF57, a protein of many faces. Viruses 7, 604–633.

Makowska, A., Braunschweig, T., Denecke, B., Shen, L., Baloche, V., Busson, P., Kontny, U., 2019. Interferon beta and Anti-PD-1/PD-L1 checkpoint blockade cooperate in NK cell-mediated killing of nasopharyngeal carcinoma cells. Transl. Oncol. 12, 1237–1256.

Mancao, C., Altmann, M., Jungnickel, B., Hammerschmidt, W., 2005. Rescue of "crippled" germinal center B cells from apoptosis by Epstein-Barr virus. Blood 106, 4339–4344.

Manners, O., Murphy, J.C., Coleman, A., Hughes, D.J., Whitehouse, A., 2018. Contribution of the KSHV and EBV lytic cycles to tumourigenesis. Curr. Opin. Virol. 32, 60–70.

Manservigi, R., Cassai, E., 1991. The glycoproteins of the human herpesviruses. Comp. Immunol. Microbiol. Infect. Dis. 14, 81–95.

Martínez, F.P., Tang, Q., 2012. Leucine zipper domain is required for Kaposi sarcoma-associated herpesvirus (KSHV) K-bZIP protein to interact with histone deacetylase and is important for KSHV replication. J. Biol. Chem. 287, 15622–15634.

McNamara, R.P., Chugh, P.E., Bailey, A., Costantini, L.M., Ma, Z., Bigi, R., Cheves, A., Eason, A.B., Landis, J.T., Host, K.M., et al., 2019. Extracellular vesicles from Kaposi sarcoma-associated herpesvirus lymphoma induce long-term endothelial cell reprogramming. PLoS Pathog. 15, e1007536.

Meckes Jr., D.G., Shair, K.H., Marquitz, A.R., Kung, C.P., Edwards, R.H., Raab-Traub, N., 2010. Human tumor virus utilizes exosomes for intercellular communication. Proc. Natl. Acad. Sci. U. S. A. 107, 20370–20375.

Meckes Jr., D.G., Gunawardena, H.P., Dekroon, R.M., Heaton, P.R., Edwards, R.H., Ozgur, S., Griffith, J.D., Damania, B., Raab-Traub, N., 2013. Modulation of B-cell exosome proteins by gamma herpesvirus infection. Proc. Natl. Acad. Sci. U. S. A. 110, E2925–E2933.

Meij, P., van Esser, J.W., Niesters, H.G., van Baarle, D., Miedema, F., Blake, N., Rickinson, A.B., Leiner, I., Pamer, E., Lowenberg, B., et al., 2003. Impaired recovery of Epstein-Barr virus (EBV)—specific CD8+ T lymphocytes after partially T-depleted allogeneic stem cell transplantation may identify patients at very high risk for progressive EBV reactivation and lymphoproliferative disease. Blood 101, 4290–4297.

Meredith, R.F., Khazaeli, M.B., Plott, W.E., Grizzle, W.E., Liu, T., Schlom, J., Russell, C.D., Wheeler, R.H., LoBuglio, A.F., 1996. Phase II study of dual 131I-labeled monoclonal antibody therapy with interferon in patients with metastatic colorectal cancer. Clin. Cancer Res. 2, 1811–1818.

Mesri, E.A., Cesarman, E., Boshoff, C., 2010. Kaposi's sarcoma and its associated herpesvirus. Nat. Rev. Cancer 10, 707–719.

Meyer, T., Gaya, A.M., Dancey, G., Stratford, M.R., Othman, S., Sharma, S.K., Wellsted, D., Taylor, N.J., Stirling, J.J., Poupard, L., et al., 2009. A phase I trial of radioimmunotherapy with 131I-A5B7 anti-CEA antibody in combination with combretastatin-A4-phosphate in advanced gastrointestinal carcinomas. Clin. Cancer Res. 15, 4484–4492.

Miller, G., Heston, L., Grogan, E., Gradoville, L., Rigsby, M., Sun, R., Shedd, D., Kushnaryov, V.M., Grossberg, S., Chang, Y., 1997. Selective switch between latency and lytic replication of Kaposi's sarcoma herpesvirus and Epstein-Barr virus in dually infected body cavity lymphoma cells. J. Virol. 71, 314–324.

Mork, J., Lie, A.K., Glattre, E., Hallmans, G., Jellum, E., Koskela, P., Moller, B., Pukkala, E., Schiller, J.T., Youngman, L., et al., 2001. Human papillomavirus infection as a risk factor for squamous-cell carcinoma of the head and neck. N. Engl. J. Med. 344, 1125–1131.

Mosam, A., Shaik, F., Uldrick, T.S., Esterhuizen, T., Friedland, G.H., Scadden, D.T., Aboobaker, J., Coovadia, H.M., 2012. A randomized controlled trial of highly active antiretroviral therapy versus highly active antiretroviral therapy and chemotherapy in therapy-naive patients with HIV-associated Kaposi sarcoma in South Africa. J. Acquir. Immune Defic. Syndr. 60, 150–157.

Nador, R.G., Cesarman, E., Chadburn, A., Dawson, D.B., Ansari, M.Q., Sald, J., Knowles, D.M., 1996. Primary effusion lymphoma: a distinct clinicopathologic entity associated with the Kaposi's sarcoma-associated herpes virus. Blood 88, 645–656.

Nakamura, H., Li, M., Zarycki, J., Jung, J.U., 2001. Inhibition of p53 tumor suppressor by viral interferon regulatory factor. J. Virol. 75, 7572–7582.

Naresh, K.N., Rice, A.J., Bower, M., 2008. Lymph nodes involved by multicentric Castleman disease among HIV-positive individuals are often involved by Kaposi sarcoma. Am. J. Surg. Pathol. 32, 1006–1012.

Nayar, U., Lu, P., Goldstein, R.L., Vider, J., Ballon, G., Rodina, A., Taldone, T., Erdjument-Bromage, H., Chomet, M., Blasberg, R., et al., 2013. Targeting the Hsp90-associated viral oncoproteome in gammaherpesvirus-associated malignancies. Blood 122, 2837–2847.

Nicholson, B.D., Shinkins, B., Pathiraja, I., Roberts, N.W., James, T.J., Mallett, S., Perera, R., Primrose, J.N., Mant, D., 2015. Blood CEA levels for detecting recurrent colorectal cancer. Cochrane Database Syst. Rev. 2015, CD011134.

Nishimoto, N., Kanakura, Y., Aozasa, K., Johkoh, T., Nakamura, M., Nakano, S., Nakano, N., Ikeda, Y., Sasaki, T., Nishioka, K., et al., 2005. Humanized anti-interleukin-6 receptor antibody treatment of multicentric Castleman disease. Blood 106, 2627–2632.

Northfelt, D.W., Dezube, B.J., Thommes, J.A., Miller, B.J., Fischl, M.A., Friedman-Kien, A., Kaplan, L.D., Du Mond, C., Mamelok, R.D., Henry, D.H., 1998. Pegylated-liposomal doxorubicin versus doxorubicin, bleomycin, and vincristine in the treatment of AIDS-related Kaposi's sarcoma: results of a randomized phase III clinical trial. J. Clin. Oncol. 16, 2445–2451.

Nowag, H., Guhl, B., Thriene, K., Romao, S., Ziegler, U., Dengjel, J., Munz, C., 2014. Macroautophagy proteins assist Epstein Barr virus production and get incorporated into the virus particles. EBioMedicine 1, 116–125.

Oksenhendler, E., Carcelain, G., Aoki, Y., Boulanger, E., Maillard, A., Clauvel, J.P., Agbalika, F., 2000. High levels of human herpesvirus 8 viral load, human interleukin-6, interleukin-10, and C reactive protein correlate with exacerbation of multicentric castleman disease in HIV-infected patients. Blood 96, 2069–2073.

Pagano, J.S., Whitehurst, C.B., Andrei, G., 2018. Antiviral drugs for EBV. Cancer 10, 197.

Parkin, D.M., 2006. The global health burden of infection-associated cancers in the year 2002. Int. J. Cancer 118, 3030–3044.

Pegtel, D.M., Cosmopoulos, K., Thorley-Lawson, D.A., van Eijndhoven, M.A., Hopmans, E.S., Lindenberg, J.L., de Gruijl, T.D., Würdinger, T., Middeldorp, J.M., 2010. Functional delivery of viral miRNAs via exosomes. Proc. Natl. Acad. Sci. U. S. A. 107, 6328–6333.

Polizzotto, M.N., Uldrick, T.S., Wyvill, K.M., Aleman, K., Peer, C.J., Bevans, M., Sereti, I., Maldarelli, F., Whitby, D., Marshall, V., et al., 2016. Pomalidomide for symptomatic Kaposi's sarcoma in people with and without HIV infection: a phase I/II study. J. Clin. Oncol. 34, 4125–4131.

Pratt, Z.L., Zhang, J., Sugden, B., 2012. The latent membrane protein 1 (LMP1) oncogene of Epstein-Barr virus can simultaneously induce and inhibit apoptosis in B cells. J. Virol. 86, 4380–4393.

Pria, A.D., Pinato, D., Roe, J., Naresh, K., Nelson, M., Bower, M., 2017. Relapse of HHV8-positive multicentric Castleman disease following rituximab-based therapy in HIV-positive patients. Blood 129, 2143–2147.

Purushothaman, P., Uppal, T., Verma, S.C., 2015. Molecular biology of KSHV lytic reactivation. Viruses 7, 116–153.

Raab-Traub, N., 2007. EBV-induced oncogenesis. In: Campadelli-Fiume, G., Arvin, A., Mocarski, E., et al. (Eds.), Human Herpesviruses: Biology, Therapy, and Immunoprophylaxis. Cambridge University Press, Cambridge.

Rabkin, C.S., Engels, E.A., Landgren, O., Schuurman, R., Camargo, M.C., Ruth, P., Goedert, J.J., 2011. Circulating cytokine levels, Epstein-Barr viremia, and risk of acquired immunodeficiency syndrome-related non-Hodgkin lymphoma. Am. J. Hematol. 86, 875–878.

Rainy, N., Zayoud, M., Kloog, Y., Rechavi, O., Goldstein, I., 2016. Viral oncomiR spreading between B and T cells is employed by Kaposi's sarcoma herpesvirus to induce non-cell-autonomous target gene regulation. Oncotarget 7, 41870–41884.

Renne, R., Lagunoff, M., Zhong, W., Ganem, D., 1996. The size and conformation of Kaposi's sarcoma-associated herpesvirus (human herpesvirus 8) DNA in infected cells and virions. J. Virol. 70, 8151–8154.

Reshef, R., Vardhanabhuti, S., Luskin, M.R., Heitjan, D.F., Hadjiliadis, D., Goral, S., Krok, K.L., Goldberg, L.R., Porter, D.L., Stadtmauer, E.A., et al., 2011. Reduction of immunosuppression as initial therapy for posttransplantation lymphoproliferative disorder(bigstar). Am. J. Transplant. 11, 336–347.

Rivas, C., Thlick, A.E., Parravicini, C., Moore, P.S., Chang, Y., 2001. Kaposi's sarcoma-associated herpesvirus LANA2 is a B-cell-specific latent viral protein that inhibits p53. J. Virol. 75, 429–438.

Robbiani, D.F., Deroubaix, S., Feldhahn, N., Oliveira, T.Y., Callen, E., Wang, Q., Jankovic, M., Silva, I.T., Rommel, P.C., Bosque, D., et al., 2015. Plasmodium infection promotes genomic instability and AID-dependent B cell lymphoma. Cell 162, 727–737.

Rozen, R., Sathish, N., Li, Y., Yuan, Y., 2008. Virion-wide protein interactions of Kaposi's sarcoma-associated herpesvirus. J. Virol. 82, 4742–4750.

Sadeghipour, S., Mathias, R.A., 2017. Herpesviruses hijack host exosomes for viral pathogenesis. Semin. Cell Dev. Biol. 67, 91–100.

Salihi, S.A., Buryanek, J., Rios, A.A., Brown, R.E., 2016. Morphoproteomics identifies etiopathogenetic correlates of HHV-8-associated Kaposi's sarcoma and provides pathways with therapeutic options: a case study. Ann. Clin. Lab. Sci. 46, 537–543.

Schlee, M., Krug, T., Gires, O., Zeidler, R., Hammerschmidt, W., Mailhammer, R., Laux, G., Sauer, G., Lovric, J., Bornkamm, G.W., 2004. Identification of Epstein-Barr virus (EBV) nuclear antigen 2 (EBNA2) target proteins by proteome analysis: activation of EBNA2 in conditionally immortalized B cells reflects early events after infection of primary B cells by EBV. J. Virol. 78, 3941–3952.

Scholler, N., Urban, N., 2007. CA125 in ovarian cancer. Biomark. Med. 1, 513–523.

Schwab, A., Meyering, S.S., Lepene, B., Iordanskiy, S., van Hoek, M.L., Hakami, R.M., Kashanchi, F., 2015. Extracellular vesicles from infected cells: potential for direct pathogenesis. Front. Microbiol. 6, 1132.

Seo, T., Park, J., Lee, D., Hwang, S.G., Choe, J., 2001. Viral interferon regulatory factor 1 of Kaposi's sarcoma-associated herpesvirus binds to p53 and represses p53-dependent transcription and apoptosis. J. Virol. 75, 6193–6198.

Sharp, T.V., Wang, H.W., Koumi, A., Hollyman, D., Endo, Y., Ye, H., Du, M.Q., Boshoff, C., 2002. K15 protein of Kaposi's sarcoma-associated herpesvirus is latently expressed and binds to HAX-1, a protein with antiapoptotic function. J. Virol. 76, 802–816.

Shen, C.L., Liu, C.D., You, R.I., Ching, Y.H., Liang, J., Ke, L., Chen, Y.L., Chen, H.C., Hsu, H.J., Liou, J.W., et al., 2016. Ribosome protein L4 is essential for Epstein-Barr virus nuclear antigen 1 function. Proc. Natl. Acad. Sci. U. S. A. 113, 2229–2234.

Siegel, J.H., Janis, R., Alper, J.C., Schutte, H., Robbins, L., Blaufox, M.D., 1969. Disseminated visceral Kaposi's sarcoma. Appearance after human renal homograft operation. JAMA 207, 1493–1496.

Smets, F., Latinne, D., Bazin, H., Reding, R., Otte, J.B., Buts, J.P., Sokal, E.M., 2002. Ratio between Epstein-Barr viral load and anti-Epstein-Barr virus specific T-cell response as a predictive marker of posttransplant lymphoproliferative disease. Transplantation 73, 1603–1610.

Song, Y.-J., Izumi, K.M., Shinners, N.P., Gewurz, B.E., Kieff, E., 2008. IRF7 activation by Epstein–Barr virus latent membrane protein 1 requires localization at activation sites and TRAF6, but not TRAF2 or TRAF3. Proc. Natl. Acad. Sci. U. S. A. 105, 18448–18453.

Soni, V., Cahir-McFarland, E., Kieff, E., 2007. LMP1 TRAFficking activates growth and survival pathways. Adv. Exp. Med. Biol. 597, 173–187.

Soulier, J., Grollet, L., Oksenhendler, E., Cacoub, P., Cazals-Hatem, D., Babinet, P., d'Agay, M.F., Clauvel, J.P., Raphael, M., Degos, L., et al., 1995. Kaposi's sarcoma-associated herpesvirus-like DNA sequences in multicentric Castleman's disease. Blood 86, 1276–1280.

Speicher, D.J., Sehu, M.M., Mollee, P., Shen, L., Johnson, N.W., Faoagali, J.L., 2014. Successful treatment of iatrogenic multicentric Castleman's disease arising due to recrudescence of HHV-8 in a liver transplant patient. Am. J. Transplant. 14, 1207–1213.

Stormont, T.J., Farrow, G.M., Myers, R.P., Blute, M.L., Zincke, H., Wilson, T.M., Oesterling, J.E., 1993. Clinical stage B0 or T1c prostate cancer: nonpalpable disease identified by elevated serum prostate-specific antigen concentration. Urology 41, 3–8.

Sui, H., Ma, N., Wang, Y., Li, H., Liu, X., Su, Y., Yang, J., 2018. Anti-PD-1/PD-L1 therapy for non-small-cell lung cancer: toward personalized medicine and combination strategies. J. Immunol. Res. 2018, 6984948.

Sutandy, F.X., Qian, J., Chen, C.S., Zhu, H., 2013. Overview of protein microarrays. Curr. Protoc. Protein Sci. 72, 27.1.1–27.1.16. Chapter 27, Unit 27 1.

Szymula, A., Palermo, R.D., Bayoumy, A., Groves, I.J., Ba Abdullah, M., Holder, B., White, R.E., 2018. Epstein-Barr virus nuclear antigen EBNA-LP is essential for transforming naïve B cells, and facilitates recruitment of transcription factors to the viral genome. PLoS Pathog. 14, e1006890.

Tam, H.K., Zhang, Z.F., Jacobson, L.P., Margolick, J.B., Chmiel, J.S., Rinaldo, C., Detels, R., 2002. Effect of highly active antiretroviral therapy on survival among HIV-infected men with Kaposi sarcoma or non-Hodgkin lymphoma. Int. J. Cancer 98, 916–922.

Tan, D., 2008. Morphoproteomics: a novel approach to identify potential therapeutic targets in cancer patients. Int. J. Clin. Exp. Pathol. 1, 331–332.

Tang, C.E., Guan, Y.J., Yi, B., Li, X.H., Liang, K., Zou, H.Y., Yi, H., Li, M.Y., Zhang, P.F., Li, C., et al., 2010. Identification of the amyloid beta-protein precursor and cystatin C as novel epidermal growth factor receptor regulated secretory proteins in nasopharyngeal carcinoma by proteomics. J. Proteome Res. 9, 6101–6111.

Tao, W.A., Aebersold, R., 2003. Advances in quantitative proteomics via stable isotope tagging and mass spectrometry. Curr. Opin. Biotechnol. 14, 110–118.

Thakker, S., Verma, S.C., 2016. Co-infections and pathogenesis of KSHV-associated malignancies. Front. Microbiol. 7, 151.

Thorley-Lawson, D., Deitsch, K.W., Duca, K.A., Torgbor, C., 2016. The link between *Plasmodium falciparum* malaria and endemic Burkitt's lymphoma-new insight into a 50-year-old enigma. PLoS Pathog. 12, e1005331.

Tiernan, J.P., Perry, S.L., Verghese, E.T., West, N.P., Yeluri, S., Jayne, D.G., Hughes, T.A., 2013. Carcinoembryonic antigen is the preferred biomarker for in vivo colorectal cancer targeting. Br. J. Cancer 108, 662–667.
Tomita, M., Choe, J., Tsukazaki, T., Mori, N., 2004. The Kaposi's sarcoma-associated herpesvirus K-bZIP protein represses transforming growth factor β signaling through interaction with CREB-binding protein. Oncogene 23, 8272–8281.
Torgbor, C., Awuah, P., Deitsch, K., Kalantari, P., Duca, K.A., Thorley-Lawson, D.A., 2014. A multifactorial role for *P. falciparum* malaria in endemic Burkitt's lymphoma pathogenesis. PLoS Pathog. 10, e1004170.
Trappe, R.U., Dierickx, D., Zimmermann, H., Morschhauser, F., Mollee, P., Zaucha, J.M., Dreyling, M.H., Duhrsen, U., Reinke, P., Verhoef, G., et al., 2017. Response to rituximab induction is a predictive marker in B-cell post-transplant lymphoproliferative disorder and allows successful stratification into rituximab or R-CHOP consolidation in an international, prospective, multicenter phase II trial. J. Clin. Oncol. 35, 536–543.
Traylen, C., Ramasubramanyan, S., Zuo, J., Rowe, M., Almohammad, R., Heesom, K., Sweet, S.M., Matthews, D.A., Sinclair, A.J., 2015. Identification of Epstein-Barr virus replication proteins in Burkitt's lymphoma cells. Pathogens 4, 739–751.
Tsao, S.W., Tsang, C.M., Lo, K.W., 2017. Epstein-Barr virus infection and nasopharyngeal carcinoma. Philos. Trans. R. Soc. Lond. B Biol. Sci. 372, 20160270.
Tworkoski, K., Raab-Traub, N., 2015. LMP1 promotes expression of insulin-like growth factor 1 (IGF1) to selectively activate IGF1 receptor and drive cell proliferation. J. Virol. 89, 2590–2602.
Uetz, P., Dong, Y.-A., Zeretzke, C., Atzler, C., Baiker, A., Berger, B., Rajagopala, S.V., Roupelieva, M., Rose, D., Fossum, E., et al., 2006. Herpesviral protein networks and their interaction with the human proteome. Science 311, 239–242.
Uldrick, T.S., Polizzotto, M.N., Aleman, K., O'Mahony, D., Wyvill, K.M., Wang, V., Marshall, V., Pittaluga, S., Steinberg, S.M., Tosato, G., et al., 2011. High-dose zidovudine plus valganciclovir for Kaposi sarcoma herpesvirus-associated multicentric Castleman disease: a pilot study of virus-activated cytotoxic therapy. Blood 117, 6977–6986.
Valls, C., Canas, C., Turell, L.G., Pruna, X., 1991. Hepatosplenic AIDS-related Kaposi's sarcoma. Gastrointest. Radiol. 16, 342–344.
Vendrame, E., Martínez-Maza, O., 2011. Assessment of pre-diagnosis biomarkers of immune activation and inflammation: insights on the etiology of lymphoma. J. Proteome Res. 10, 113–119.
Verweij, F.J., de Heus, C., Kroeze, S., Cai, H., Kieff, E., Piersma, S.R., Jimenez, C.R., Middeldorp, J.M., Pegtel, D.M., 2015. Exosomal sorting of the viral oncoprotein LMP1 is restrained by TRAF2 association at signalling endosomes. J. Extracell. Vesicles 4, 26334.
Vidick, S., Leroy, B., Palmeira, L., Machiels, B., Mast, J., Francois, S., Wattiez, R., Vanderplasschen, A., Gillet, L., 2013. Proteomic characterization of murid herpesvirus 4 extracellular virions. PLoS One 8, e83842.
Virgin 4th, H.W., Latreille, P., Wamsley, P., Hallsworth, K., Weck, K.E., Dal Canto, A.J., Speck, S.H., 1997. Complete sequence and genomic analysis of murine gammaherpesvirus 68. J. Virol. 71, 5894–5904.
Wang, F.Z., Akula, S.M., Pramod, N.P., Zeng, L., Chandran, B., 2001. Human herpesvirus 8 envelope glycoprotein K8.1A interaction with the target cells involves heparan sulfate. J. Virol. 75, 7517–7527.
Wang, H.B., Zhang, H., Zhang, J.P., Li, Y., Zhao, B., Feng, G.K., Du, Y., Xiong, D., Zhong, Q., Liu, W.L., et al., 2015. Neuropilin 1 is an entry factor that promotes EBV infection of nasopharyngeal epithelial cells. Nat. Commun. 6, 6240.
Wang, Y., Du, S., Zhu, C., Wang, C., Yu, N., Lin, Z., Gan, J., Guo, Y., Huang, X., He, Y., et al., 2020. STUB1 is targeted by the SUMO-interacting motif of EBNA1 to maintain Epstein-Barr virus latency. PLoS Pathog. 16, e1008447.

Wei, Y.C., Yang, S.F., Chang, S.L., Chen, T.J., Lee, S.W., Chen, H.S., Lin, L.C., Li, C.F., 2018. Periostin overexpression is associated with worse prognosis in nasopharyngeal carcinoma from endemic area: a cohort study. Onco. Targets. Ther. 11, 3205–3213.

Wen, K.W., Damania, B., 2010a. Hsp90 and Hsp40/Erdj3 are required for the expression and anti-apoptotic function of KSHV K1. Oncogene 29, 3532–3544.

Wen, K.W., Damania, B., 2010b. Kaposi sarcoma-associated herpesvirus (KSHV): molecular biology and oncogenesis. Cancer Lett. 289, 140–150.

Wiedmer, A., Wang, P., Zhou, J., Rennekamp, A.J., Tiranti, V., Zeviani, M., Lieberman, P.M., 2008. Epstein-Barr virus immediate-early protein Zta co-opts mitochondrial single-stranded DNA binding protein to promote viral and inhibit mitochondrial DNA replication. J. Virol. 82, 4647–4655.

Wilkinson, J., Cope, A., Gill, J., Bourboulia, D., Hayes, P., Imami, N., Kubo, T., Marcelin, A., Calvez, V., Weiss, R., et al., 2002. Identification of Kaposi's sarcoma-associated herpesvirus (KSHV)-specific cytotoxic T-lymphocyte epitopes and evaluation of reconstitution of KSHV-specific responses in human immunodeficiency virus type 1-infected patients receiving highly active antiretroviral therapy. J. Virol. 76, 2634–2640.

Wong, J.P., Stuhlmiller, T.J., Giffin, L.C., Lin, C., Bigi, R., Zhao, J., Zhang, W., Bravo Cruz, A.G., Park, S.I., Earp, H.S., et al., 2019. Kinome profiling of non-Hodgkin lymphoma identifies Tyro3 as a therapeutic target in primary effusion lymphoma. Proc. Natl. Acad. Sci. U. S. A. 116, 16541–16550.

Worland, T., Harrison, B., Delmenico, L., Dowling, D., 2018. Hepatocellular carcinoma screening utilising serum alpha-fetoprotein measurement and abdominal ultrasound is more effective than ultrasound alone in patients with non-viral cirrhosis. J. Gastrointest. Cancer 49, 476–480.

Wu, M., Liu, H., Liu, Z., Liu, C., Zhang, A., Li, N., 2018. Analysis of serum alpha-fetoprotein (AFP) and AFP-L3 levels by protein microarray. J. Int. Med. Res. 46, 4297–4305.

Yarchoan, R., Uldrick, T.S., 2018. HIV-associated cancers and related diseases. N. Engl. J. Med. 378, 1029–1041.

Yeo, E.L.L., Li, Y.Q., Soo, K.C., Wee, J.T.S., Chua, M.L.K., 2018. Combinatorial strategies of radiotherapy and immunotherapy in nasopharyngeal carcinoma. Chin. Clin. Oncol. 7, 15.

Yip, Y.L., Lin, W.T., Deng, W., Tsang, C.M., Tsao, S.W., 2019. Chapter 5—establishment of nasopharyngeal carcinoma cell lines, patient-derived xenografts, and immortalized nasopharyngeal epithelial cell lines for nasopharyngeal carcinoma and Epstein–Barr virus infection studies. In: Lee, A.W.M., Lung, M.L., Ng, W.T. (Eds.), Nasopharyngeal Carcinoma. Academic Press, pp. 85–107.

Yoshizaki, K., Matsuda, T., Nishimoto, N., Kuritani, T., Taeho, L., Aozasa, K., Nakahata, T., Kawai, H., Tagoh, H., Komori, T., et al., 1989. Pathogenic significance of interleukin-6 (IL-6/BSF-2) in Castleman's disease. Blood 74, 1360–1367.

You, J., Kao, A., Dillon, R., Croner, L.J., Benz, R., Blume, J.E., Wilcox, B., 2018. A large-scale and robust dynamic MRM study of colorectal cancer biomarkers. J. Proteomics 187, 80–92.

Zhang, H., Li, Y., Wang, H.B., Zhang, A., Chen, M.L., Fang, Z.X., Dong, X.D., Li, S.B., Du, Y., Xiong, D., et al., 2018. Ephrin receptor A2 is an epithelial cell receptor for Epstein-Barr virus entry. Nat. Microbiol. 3, 1–8.

Zheng, D., Wan, J., Cho, Y.G., Wang, L., Chiou, C.J., Pai, S., Woodard, C., Zhu, J., Liao, G., Martinez-Maza, O., et al., 2011. Comparison of humoral immune responses to Epstein-Barr virus and Kaposi's sarcoma-associated herpesvirus using a viral proteome microarray. J Infect Dis 204, 1683–1691.

Zhou, Y., Heesom, K., Osborn, K., AlMohammed, R., Sweet, S.M., Sinclair, A.J., 2020. Identifying the cellular interactome of Epstein-Barr virus lytic regulator Zta reveals cellular targets contributing to viral replication. J. Virol. 94. e00927-19.

Zhu, F.X., Yuan, Y., 2003. The ORF45 protein of Kaposi's sarcoma-associated herpesvirus is associated with purified virions. J. Virol. 77, 4221–4230.
Zhu, F.X., Chong, J.M., Wu, L., Yuan, Y., 2005. Virion proteins of Kaposi's sarcoma-associated herpesvirus. J. Virol. 79, 800–811.
Zhu, C.Y., Zhao, S.S., Wang, X.K., Wang, L., Wang, F.Y., Fang, S., Liu, Z.X., Guan, L.X., Liu, Y.C., Ding, Y., et al., 2019. Outcome of rituximab-based treatment for post-transplant lymphoproliferative disorder after allogeneic hematopoietic stem cell transplantation: a single-center experience. Ann. Transplant. 24, 175–184.

FSC
www.fsc.org
MIX
Papier | Fördert
gute Waldnutzung
FSC® C083411

Kolibri 360 GmbH
Ferdinand-Jühlke-Straße 7
99095 Erfurt, Deutschland
produktsicherheit@kolibri360.de